Serengeti

Dynamics of an Ecosystem

Contributors

B. C. R. Bertram—The Research
Centre, Kings College, Cambridge,
CB2 1ST, England.

J. D. Bygott—Sub-Department of
Animal Behavior, Madingley,
Cambridge, England.

J. J. R. Grimsdell—African Wildlife
Leadership Foundation, P.O. Box
48177, Nairobi, Kenya.

J. P. Hanby—Sub-Department of
Animal Behavior, Madingley,
Cambridge, England.

R. Hilborn—Institute of Animal
Resource Ecology, University of
British Columbia, Vancouver, B.C.,
V6T 1W5, Canada.

D. C. Houston—Department of
Zoology, Glasgow University,
Glasgow, G12 8QQ, Scotland.

P. J. Jarman and M. V. Jarman—
School of Natural Resources,
University of New England,
Armidale, New South Wales,
2351, Australia.

L. Maddock—Marine Biological
Association, The Laboratory,
Citadel Hill, Plymouth, England.

S. J. McNaughton—Department of
Biology, Syracuse University, 130
College Place, Syracuse, New York
13210, USA.

M. Norton-Griffiths—Ecosystems
Ltd., P.O. Box 30239, Nairobi,
Kenya.

C. J. Pennycuick—Department of
Zoology, Bristol University,
Woodland Road, Bristol BS8 1VG,
England.

A. R. E. Sinclair—Institute of Animal
Resource Ecology, Department of
Zoology, University of British
Columbia, Vancouver, B.C., V6T
1W5, Canada.

Serengeti

Dynamics of an Ecosystem

Edited by A. R. E. Sinclair
and M. Norton-Griffiths

The University of Chicago Press
Chicago and London

A. R. E. Sinclair is assistant professor at the
Institute of Animal Resource Ecology, Depart-
ment of Zoology, University of British
Columbia. He is the author of *The African
Buffalo,* also published by the University of
Chicago Press.
M. Norton-Griffiths was staff ecologist at
the Serengeti Research Institute from 1969 to
1973. He heads Ecosystems Ltd., an ecological
consulting firm in Kenya.

The University of Chicago Press, Chicago 60637
The University of Chicago Press, Ltd., London
© 1979 by The University of Chicago
All rights reserved. Published 1979
Printed in the United States of America
83 82 81 80 79   8 7 6 5 4 3 2 1

Library of Congress Cataloging in Publication Data
Main entry under title:
Serengeti, dynamics of an ecosystem.
Includes bibliographies and index.
1.   Animal ecology—Tanzania—Serengeti Plain.
2. Serengeti National Park.   I.   Sinclair, Anthony
Ronald Entrican.   II.   Norton-Griffiths, M.
QL337.T3S43   574.5'264   79–10146
ISBN 0–226–76028–6

Contents

Preface

The Serengeti is by good fortune one of those few areas that has not suffered the ravages of modern man. It contains the largest herds of grazing mammals in the world, and, as a wildlife spectacle, is second to none. It is an area of plains and open woodland in northern Tanzania and southern Kenya, East Africa. Olduvai Gorge, famous for the fossil remains of early man, cuts through it. Overlooking the plains in the east are the volcanoes which created Ngorongoro Crater, and to the west lies Lake Victoria. Part of the area is protected as the Serengeti National Park in Tanzania, which is adjoined to the smaller Masai Mara Park in Kenya. East of the Serengeti Park, the eastern plains and Ngorongoro Crater are administered by the Ngorongoro Conservation Unit, which caters for both the wildlife and Masai cattle herders.

This area, which we call the Serengeti-Mara ecosystem, became the center of attention for a group of scientists. Starting in the 1950s and expanding in the 1960s with the formation of the Serengeti Research Institute, the group studied a wide range of topics involving the biology, soils, geology, and climate. Their aim was to obtain sufficient understanding of how the ecosystem worked, so that they could advise on the proper conservation of the area. This book reports on those studies. It has three main purposes.

Firstly, it is an attempt to bring together and synthesize some aspects of the processes and patterns in an ecosystem that have been studied over the past twenty years. Since Bernhard and Michael Grzimek made their first audacious attempts to study this enormous ecosystem in 1957, scientific monitoring has expanded in scope and become considerably more complex. We realized that only by piecing together the various facets of this long-term research could we hope to develop an idea of how a particular community functions.

There have been other studies of ecosystems, notably in the International Biological Program, in which each component is described in great detail; but these studies generally present a static description, a mean value over the period of study. Where temporal variations have been noted, it has usually not been possible to analyze causation because the fluctuations were not experimental in nature. The greater time period of the Serengeti work, combined with the fact that the initial fluctuations were induced, presented us with the unique situation of analyzing a large-scale ecosystem experiment.

This book is a first attempt to see where we stand. The chapters are mostly independent contributions, with the authors following their own interests and concentrating on what they think most important. Most chapters were read and commented on by biologists unfamiliar with Africa, in order to give an outside perspective. Three chapters (1, 12, 13) synthesize, with different degrees of detail, what is presented in this volume and elsewhere. Chapter 2 provides a general description of the area. Chapter 3 is concerned with the way grasslands respond to the impact of the grazers. Chapters 4–8 are concerned with the dynamics of the grazing populations, their numbers, movements, feeding behavior, and social organization, and their interaction with the vegetation. Chapters 9 and 10 are concerned with similar aspects of the predators, in particular their social and feeding behavior, and their response to changes in herbivore populations. Chapter 11 discusses the role of the avian scavengers and how they manage to compete with mammal predators. Of course, there has been much other work not directly related to the present theme, and we document this in the bibliography. For historical and conservation reasons, research has concentrated on certain aspects, notably large mammals, to the exclusion of others. There are, therefore, large gaps in our knowledge, but we hope that this book will draw attention to those areas, highlighting areas that require high-priority investigation.

The second purpose of the book is to provide the information and rationale for a sound, long-term management plan for the Serengeti ecosystem. Until now, management has been based upon either intuition or short-term studies conducted in response to local ecological crises, such as elephants' "damaging" mature trees. We hope that by looking at the various components of the ecosystem together, we will be able to provide sounder recommendations for management. In some cases the solutions to possible problems are quite different from those that might have been suggested based on the results of short-term studies too narrow to have provided a proper perspective. Our aims are, first, to provide suggestions designed to prevent problems; and secondly, in the event of a problem's arising, to suggest how to approach it through the "principle of experimental management." This approach of experimental management is equally applicable to parks and reserves in North America and elsewhere —areas for which "Master Plans for Management" are normally drawn up by administrative personnel unfamiliar with ecosystem processes, rather than by ecologists.

The third purpose of the book is to demonstrate that there is at least some tangible return for the money spent on research, in the form of a synthesis of results which can be used as a basis for management proposals. Considerable financial support has been provided by funding agencies, but so far most published results are scattered through the literature and are fragmentary in nature. At the same time, we want to extend a plea for further support; without it, the value of the previous monitoring will, to some extent, be lost. We have established that large-scale changes are taking place and that the ecosystem is entering a critical stage which requires close monitoring. Yet, at the present time (1978), all monitoring of the major aspects has ceased because of a lack of funding. The Tanzania National Parks have had to assume financial responsibility for the Serengeti Research Institute, but they have not the resources to keep it going adequately. Developing countries cannot be expected to divert urgently needed money from social projects to the comparative luxury of ecological monitoring in national parks. Yet these parks (and especially the Serengeti) are a world asset, and it is the responsibility of the developed countries to provide for their care and upkeep. Funding was available in the sixties during the economic boom and when social and political harmony prevailed in East Africa. Now conditions are considerably more difficult, but these conditions make external aid more necessary than ever. Without it, the Serengeti as a natural ecosystem may not even exist in a few years' time.

Acknowledgments

The Tanzania National Park board of trustees, the directors John Owen, Soloman ole Saibull, and Derek Bryceson, and wardens A. Field, M. Turner, S. Stevenson, and David Babu provided the initiative for setting up and encouraging research over the years. They have set an example rarely seen elsewhere that research is needed before management. We would like to thank them for their help throughout.

We acknowledge the contribution of all the scientists to our understanding of the system; we regret there are too many to mention individually. In addition to the contributors, several have helped us with this book, particularly George Schaller, Hans Kruuk, Patrick Duncan, Dirk Kreulen, and Hugo De Wit.

We asked a number of referees, not associated with work in East Africa, to review the chapters; G. Caughley, W. C. Clark, M. I. Dyer, N. Gilbert, N. R. Liley, W. E. Neill, R. M. Peterman, and J. N. M. Smith provided valuable comment. Many of our ideas developed in discussion with colleagues, and we would like to acknowledge their contribution, particularly that of W. E. Neill, W. C. Clark, H. Croze, D. Herlocker, and R. Pellew.

The Serengeti Research Institute was built with a major grant from the Fritz Thyssen Stiftung, and additional buildings were provided by the Caesar Kleburg Foundation. Operating costs were provided by the Ford Foundation, Canadian International Development Agency (CIDA), and Tanzania National Parks. We would also like to mention the Royal Society; the African Wildlife Leadership Foundation, Washington, D.C.; the New York Zoological Society; the Caesar Kleburg Foundation of Texas A & M University; the Max-Planck Institute, West Germany; the Netherlands Foundation of Tropical Research (WOTRO); and the East African Wildlife Society, all of which contributed to the work of the institute. The National Research Council of Canada and the New York Zoological Society have also supported one of us (A. R. E. Sinclair) in recent Serengeti work. Prof. B. Grzimek and Dr. John Owen both generated considerable financial support—their combined interest initiated the research in the Serengeti. Professors Niko Tinbergen, G. P. Baerends, J. W. S. Pringle, A. Msangi, and W. Wickler provided considerable advice and support, together with the other members of the Scientific Council. Dr. H. F. Lamprey, as first director of the institute (1966–72), deserves much of the credit for setting up and coordinating the research, and T. Mcharo continued it until 1976.

Anne Sinclair and Ann Norton-Griffiths helped us in numerous ways. We thank them for their long-suffering understanding.

A. R. E. Sinclair

One Dynamics of the Serengeti
Ecosystem

Process and Pattern

Rinderpest struck East Africa in 1890, and in two years 95 percent of
the buffalo and wildebeest there had died. So began a series of events of
such profound ecological importance that the repercussions are still being
felt today. These events have provided us with a rare opportunity to
understand an ecosystem.

A fundamental issue arising with ecosystem investigations concerns the
kind and degree of linkage among components. This is important for two
reasons: firstly, much ecological theory emphasizes single processes as the
dominant mediators of observed pattern. Thus, there are predator-prey
stability theories (Nicholson 1933; Huffaker 1970; Murdoch and Oaten
1975), niche-partitioning, competition theories (Schoener 1974; Connell
1975), density-independent fluctuation theories (Andrewartha and Birch
1954; Reddingius and den Boer 1970), and phenotypic polymorphism
and dispersal theories (Chitty 1967; Myers and Krebs 1971). While the
effects of these processes undoubtedly predominate in some instances, the
extent to which each affects overall ecosystem patterns is known for few
natural situations. Furthermore, it is now clear that theories involving
the effects of several processes simultaneously can be used to predict the
existence of systems with complex behavior and multiple equilibria similar to those observed in nature (Ricker 1954; Holling 1973; Austin and
Cook 1974; Noy-Meir 1975; Sutherland 1974; Southwood and Comins
1976; May 1977; Peterman, Clark, and Holling 1978). Again, there
are too few cases where actual changes in linkage types have been identified, rather than hypothesized about, as the cause of changes in ecosystem
equilibria.

Secondly, the extent to which organisms are ecologically linked or un-
linked is important for the conservation and management of natural regions. Obviously, if a species can fluctuate in number without affecting

many other species in the system, this presents far less of a management problem than if the repercussions affect most of the community. Since man is creating disturbances in most environments now, it is imperative that we learn how to predict their impact, and the likelihood that they will shift the system's equilibrium. (Holling et al. 1978).

To do this, it is necessary to make use of experiments (whether induced by nature or by humans) on an ecosystem-wide scale (Walters and Hilborn 1976; Larkin 1972). Simple descriptions of relatively undisturbed ecosystems give no indication of the processes involved, nor the extent of their action. But large-scale experiments in the form of perturbations are rarely studied in the necessary detail, so our knowledge remains limited. The Serengeti region, however, experienced a major perturbation in 1962, at a time when a multidisciplinary research team began monitoring the ecosystem. Because the ultimate objectives were conservation of large-mammal fauna and its habitats, there are some conspicuous gaps in our knowledge of the system. Nevertheless, sufficient information has accumulated over the past twenty years, as illustrated in the succeeding chapters, to allow an initial attempt at answering three main questions: To what extent are different components linked? What are the dominant processes forming the linkage? and How are these processes affected by the environment?

The various processes such as predation, interference, or dispersal are really different aspects of one major process, that of resource acquisition. Individual organisms have requirements of food, shelter from extremes of climate, and mates; and they also need to avoid becoming food for someone else. In pursuing these ends, individuals affect others by removing food or by displacing others from an area where food, shelter, or mates might be found, or by eating them: all these effects are negative for other individuals, and all are interrelated. For convenience, these effects have been labeled as exploitation competition, interference competition, or predation, and there has been a tendency to regard them as alternatives. In fact, they are part of the single process of promoting one's genes in the next generation at the expense of other, unrelated individuals (Dawkins 1976).

In this chapter, I shall describe the main events that have been observed in the Serengeti and the conclusions derived from them. Following chapters will review in greater detail different aspects of the ecosystem. The Serengeti is a savannah area close to the equator in eastern Africa. It is dominated by large ungulates, particularly wildebeest, which migrate annually in response to seasonal changes in rainfall. To save

space, I refer the reader to chapter 2 for a full description of the Serengeti environment.

The Perturbations

The Great Rinderpest

To understand present-day events, it is necessary to go back ninety years
to the first great rinderpest epizootic. The following account is condensed
mainly from Ford (1971) with additional information from Branagan
and Hammond (1965) and W. Plowright (pers. comm.). Rinderpest is
a virus whose natural host is cattle. It was enzootic in the steppes of Asia,
and, throughout the past thousand years, periodic eruptions spread into
Europe as a result of wars and human invasions. This accounts for its
introduction to Africa, south of the Sahara. It is commonly believed that
it was brought in by cattle in the Italian invasion of Ethiopia in 1889,
but an alternative view suggests it came in with Russian cattle from the
Black Sea region as early as 1884 during the relief of General Gordon
in Khartoum.

The disease spread swiftly west and south, with the first noticeable
affects appearing in East Africa in 1890. It reached the Cape Province of
South Africa in 1896. The cattle population of East Africa succumbed
rapidly, and, by 1892, 95 percent had died. The pastoral and nomadic
peoples were those most affected by this, because they depended on cattle
for their livelihood. Famine developed, and under these conditions,
endemic diseases, such as smallpox, became epidemic. Tribes in the
Serengeti area include the Masai in the eastern woodlands, plains, and
Ngorongoro Crater; and the pastoralists inhabiting Maswa and Sukumaland (fig. 1.1) to the southwest of the park. Both areas were severely
affected by the rinderpest. The Masai, in particular, were devastated, for
they depended entirely on their cattle. After the death of their cattle, they
first took to raiding cattle from the Sukuma peoples, for rustling was a
way of life for them. But this merely helped to spread the disease, and
soon there were no cattle left to raid. The Masai then declined to abject
poverty and starvation. Thus Ford (1971, p. 191) quotes Langheld:

> At the first onset colossal numbers of the Masai, at least two-thirds
> of the whole tribe, were destroyed. The warriors were at first able to
> carry themselves through by hunting and petty thieving, but the
> women, children and old men were completely abandoned to misery.
> Reduced to skeletons they tottered through the steppe, feeding

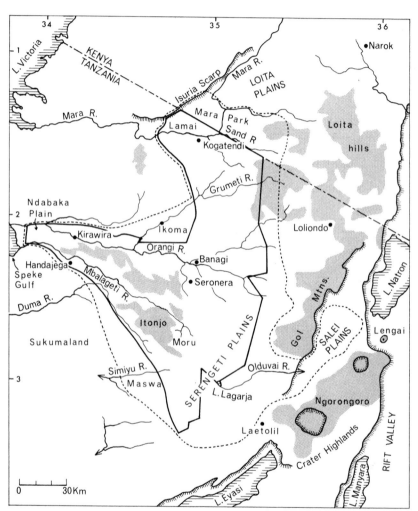

Figure 1.1

The Serengeti-Mara ecosystem is the area used by the wildebeest (*even broken line*). The Serengeti Park is shown by the heavy line. Hills are shaded.

on honey of the wil<   All warlike
undertakings went av   ly thrown back
and often did not ret   /ay. Only in a few
areas kraals remaine<   and as a result of
hunting, wide stretcl   and the Masai lived
as beggars among the

The Sukuma tribes a   they suffered less from
rinderpest, it was onl   also suffered from the in-
cursions of the Mas?   , the Sukuma abandoned their
easternmost villag<   mpt to retreat from the disease.
It was a futile   st of their cattle were dead.
   Biological   there, for, apart from the previously
mentioned   c of villages and survivors' disabilities
preven'   on and harvesting of crops. Consequently, the
fan   vitable. People were reduced to chewing the bark
                                          ation had now been so decimated that people were
                                          n a subsistence level, and famines hit them repeatedly

ect consequences of rinderpest mediated through the wild
population may have been even worse than these disasters. Before
, buffalo and other species were abundant in Sukumaland and, presumably, in the adjacent Serengeti woodlands, for Stanley and Emin
found that tracks of zebra and buffalo were common. By 1890, the buffalo
were dying from rinderpest, as were wildebeest and giraffe, and in a few
years a large proportion of the wildlife had disappeared in the woodlands (Sinclair 1977). The first effect of this was the immediate disappearance of tsetse flies (*Glossina swynnertoni*), blood-sucking flies whose
normal hosts are wild ungulates. These flies transmit blood parasites
(*Trypanosoma* spp.) that cause "sleeping sickness" in humans and cattle.
They require trees and bushes as their habitat.

The second effect resulting from the die-off of both wild and domestic
herbivores was the appearance of man-eating lions: lack of their usual
prey forced the predators to prey on humans in many areas of East
Africa. Contrary to the commonly held belief that man-eaters are old,
toothless individuals, these were not; and they were the scourge of the
country for at least a quarter of a century. The famous man-eaters of
Tsavo, Kenya, appeared in 1898, and in Uganda an outbreak occurred in
the 1920s. One lion reportedly killed eighty-four people!

Nothing encourages a population to abandon an area faster than an attack of man-eating lions. And as a result of the disappearance of these tribes of cultivators, as well as the disappearance of the grazers and browsers, grass and trees grew up, forming a dense woodland. White (1915) reports extensive thickets in the northern Serengeti, where now there is open grassland. In the abandoned cultivated areas, old rootstocks regenerated rapidly to develop thickets. Many of the wild ungulates developed immunity to rinderpest and became numerous by 1910, and these were the hosts for tsetses, whose range expanded into the new thickets. Because tsetse flies transmitted tryanosomiasis, the human population was forced to retreat still further. In Sukumaland, Swynnerton (in Ford 1971, p. 193) comments:

> in 1909, four man-eating lions became so troublesome, killing in all about thirty persons, that a great number of people deserted. Bushes grew up on their gardens and grazing land, and the tsetses, arriving later and pushing into these, made the ground to right and left of the gaps untenable by the cattle owners living there, and these also fell back. The first very general retreat in the face of the tsetse occurred ten years ago, [i.e. 1915] and since then the progress has been continuous. The tsetses unaided have proved amply sufficient to keep the people in steady retreat . . .

It was not until the 1930s that the tsetse advance was halted by mechanical bush-clearing methods. From that time, the trend was reversed, and both human and cattle populations increased (aided by vaccination against rinderpest) and expanded eastward into the Serengeti woodlands again. This advance was particularly noticeable in the 1960s and early 1970s along the western boundary of the park, both in the north and in Maswa (see chap. 13) and may have been the cause of the increased numbers of elephants seen in the park in these later years—the elephants were probably retreating from the advancing cultivation.

Removal of Rinderpest

Meanwhile, in the Serengeti and eastward in the Masai country, the originator of these ecological disturbances, rinderpest, continued to make itself felt through periodic epizootics in the wild ruminants. These occurred in 1917–18, 1923, and 1938–41; and in each case the disease spread from northern Tanzania southward. It rapidly became enzootic in the extensive cattle populations of Masailand surrounding the wild ungulate populations of the Serengeti and the rift valley. As early as 1925, veterinarians noticed mild strains of the virus appearing in cattle and wild ruminants, causing reduced mortality. Adults acquired immunity in their first year, so mortality was confined to yearlings. Only highly gregarious species, such as cattle, buffalo, and wildebeest, continued to get the disease. At least since 1933, when it was first identified, rinderpest mortality in yearling wildebeest was an annual event: it usually occurred toward the end of the dry season and early rains (October through December) after calves had lost the immunity acquired from their mothers.

After 1952, veterinarians changed their tactics from ad hoc treatment following an outbreak to generalized immunization of cattle in vulnerable areas. Although rinderpest recurred annually in subsequent years in wildebeest and buffalo, the ring of immunized cattle continued to close around the Serengeti. For two years, 1955–56, no rinderpest was detected in wild species, but it reappeared with heavy mortality in 1957–59, possibly because of the higher number of susceptible animals.

Nineteen sixty-two was the last year that rinderpest was detected in wildebeest of the Serengeti, Mara, and Ngorongoro, and 1963 was the last year for buffalo. No further infection has been detected in these or other wild ruminants because of consolidation of the immunized ring preventing reinfection of the wild species from infected cattle beyond. It is apparent that this exotic virus could not maintain itself in indigenous ruminants because a nonlethal virus/carrier-host relationship had not yet evolved: the hosts either died or developed immunity and became non-carriers.

The immediate result in wildebeest was a doubling of yearling survival from 25 percent to 50 percent, or even higher (chap. 4), which allowed the population to increase from about 250,000 in 1961 to 500,000 in 1967. The buffalo population increased at the same rate, from about 30,000 in 1961 to 50,000 in 1967. Most significantly, zebra, being nonruminants and, hence, unaffected by rinderpest, showed no change in population during this period. Removal of rinderpest, therefore, acted as a comparatively sudden perturbation affecting the dynamics of one of the major herbivores. This was the first of the perturbations that altered the ecosystem. The second followed a decade later, brought about by a change in rainfall patterns.

The Climate

The Serengeti-Mara ecosystem lies 1°–3° south of the equator in *Acacia* savannah woodland (chap. 2.) There is usually enough moisture from November through June to produce grass growth. From July to October, there is less rain, the soil dries out, and little grass growth occurs. Trees, with their deeper roots, can tap groundwater, so their phenology of growth is less affected by immediate fluctuations in climate.

A change in this seasonal pattern appears to have occurred in 1971: prior to this, average dry-season rainfall (July–October) was about 150 mm in the northern woodlands, but in 1971–76 it was consistently higher, averaging 250 mm. This produced considerably more grass growth in areas used by wildebeest at the height of the dry season, which, in turn, allowed a further population increase of both wildebeest and buffalo, the former reaching 1.3 million animals in 1977. This rainfall change also produced grass growth on the plains during the dry season, when previously there had been none.

In summary, the Serengeti area has felt the impact of three major changes in the past twenty years. The most important of these was the disappearance of rinderpest in 1962–63, allowing an eruption of the dominant grazers, wildebeest. A second change took place through the climate nearly a decade later, and this had a more widespread effect on the grazers. Finally, the gradual encroachment of cultivation, causing the shrinkage in area of natural habitats, especially for elephant, became noticeable in the 1960s.

Effects of the Perturbations

The Wildebeest Increase

The change in wildebeest numbers due to rinderpest removal has had extensive effects in the ecosystem (fig. 1.2).

At the plant trophic level, there has been a direct effect through grazing on the grasses, but the effect differs according to place and timing. On the plains, grazing occurs when grasses are growing in the wet season, so that herbage removal prevents flowering until after the grazers have left and probably lowers the competitive ability of palatable grasses. McNaughton (chap. 3) has documented changes in grass-species composition due to changes in grazing pressure, and there has almost certainly been an increase in herbs as well (pl. 18).

In the woodlands, wildebeest grazing takes place after flowering, when the grasses are dormant. Consequently, grazing has little impact on the viability of the grasses, and there has been no noticeable change in species composition. But the herbivores trample or graze the tall, dry grass so that there has been progressively less combustible material for dry-

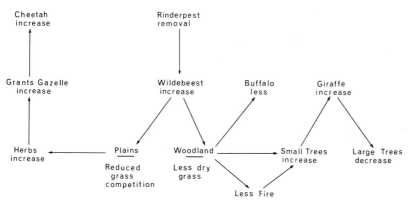

Figure 1.2   The response of various compo-
                                        nents to removal of rinderpest,
                                        showing the linkages between
                                        them.

season fires (chap. 13). Fewer fires have had the effect of allowing higher survival of regenerating trees, particularly those below 2 m. There were already many old rootstocks well established, but, in early years, the frequent and extensive fires killed the above-ground shoots from the previous wet season. Hence, increased grazing eventually allowed trees to reach a height where they could withstand any subsequent fires. This appears to repeat the vegetation changes described in the early years of this century following the first rinderpest panzootic.

The effects of wildebeest on the vegetation extend beyond their immediate food supply, by altering the competitive balance among the plants and thereby indirectly affecting both herbs and trees. This change, in turn, has affected the immediate competitors of wildebeest as well as noncompetitive herbivores. Of the competing grazers, the strongest evidence for change comes from buffalo. Wildebeest eat some of the buffalo dry-season food and trample a lot more. As a result, the buffalo population, which originally increased after rinderpest elimination, leveled off during 1970–73, while the wildebeest population continued to increase.

Effects on other competitors are less evident. Zebra numbers may have declined slightly in the 1960s, but the figures are not significant (chap. 4). This period was comparatively dry, and zebra, with their ability to digest low-quality forage, may not have felt the impact of competition. However, zebra have not increased in the 1970s, though conditions have been good, which suggests that wildebeest, by their more efficient grazing behavior, were keeping zebra numbers down through food competition.

Indirect effects on other herbivores, although rather more unquantifiable, are nonetheless noticeable. Grant's gazelle eat herbs almost exclusively (pl. 18), and are tolerant of dry conditions, so they can remain on the plains for a large part of the year. Their numbers in the 1960s (c. 30,000—Norton-Griffiths, pers. comm.) were low compared to the Thomson's gazelle (c. 725,000—Bradley 1977). The 1978 census suggests that Grant's gazelle are considerably more abundant than previously (c. 52,000), and this could be the result of the increase in herbs on the plains. Thus, wildebeest may have had a positive effect on this species. In the central woodlands, there has been an increase in the giraffe population (appendix A), which has been attributed to the increase in regenerating *Acacia* trees. Female giraffe, in particular, prefer to eat small *Acacia* bushes (pl. 40) rather than tall trees (R. Pellew, pers. comm.). And it was into the central woodlands that the wildebeest expanded their range (chap. 5), so wildebeest probably had a positive effect on the browsers. Indeed, simulation models suggest that this is likely (chap. 13).

The wildebeest is the main food species in the diet of the dominant predators, hyena and lion. With a tripling of wildebeest numbers in the 1960s, one might have expected an increase in predator populations, so it is interesting to note that both Kruuk (1972) and Schaller (1972) found predator numbers constant at that time. The reason for this lack of numerical response lies in the migratory behavior of the herbivores: in any one area, wildebeest are present for only part of the year, moving to other areas as the season changes. Both hyena and lion are confined to fixed territories, so when the migrants leave, the predators have to switch to less abundant resident prey (chaps. 9, 10). Consequently, these predator populations are not regulated by their migrant prey, nor do the predators regulate their main prey, that is, the migrant herbivores are uncoupled from their main predators.

While wildebeest have little effect on hyena and lion populations, they may have indirect effects on other predators. The cheetah (pl. 33) population on the plains is thought to be increasing (chap. 10), possibly due to increasing numbers of gazelles, particularly Grant's gazelle. Vultures are likely to have benefited directly, particularly the Ruppell's griffon and white-backed vultures (pl. 39), which scavenge carcasses of the migratory ungulates (chap. 11).

Essentially, the wildebeest population increase shows there is significant competition between ecologically similar herbivores, and facilitation of noncompeting herbivores through changes in the vegetation. Thus, there are both negative and positive effects among herbivores. Conversely, there is only a weak interaction between wildebeest and its own predators.

The Rainfall Increase

The change in climate occurred nearly a decade after the removal of rinderpest, so it is possible to distinguish between the effects of the two in some cases.

The distribution of rain through the year changed: more fell in the dry season, and less in the wet season; total rainfall did not increase. The first effect of this (fig. 1.3) was to triple the amount of dry-season green grass production in the northern woodlands (chap. 4). Changes of dry-season production on the plains were perhaps even more significant; whereas it was negligible during the 1960s, sufficient rain fell to moisten the soil and allow some growth during the 1970s.

Increase in grass production had two major effects on the grazing herbivores. Firstly, among wildebeest, dry-season mortality due to undernutri-

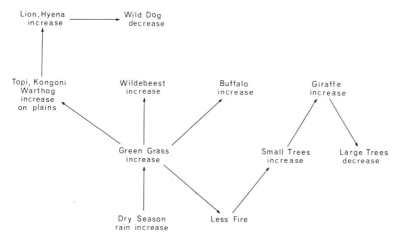

Figure 1.3

The response of various components to the increase of dry-season rainfall, showing how they are linked.

tion virtually ceased, and this promoted a further increase in population. In fact, food production was so high that in 1977, with 1.3 million wildebeest, the amount of food per individual was the same as in 1961, when there were only a fifth as many wildebeest. After a two-year lag, the buffalo population also increased, probably due to a relaxation of competition with wildebeest (chap. 4).

The other major effect on herbivores was seen on the plains (chap. 10). Whereas a few Grant's gazelle, oryx, and ostrich subsisted there in the 1960s, there was a significant increase in numbers of topi, kongoni, warthog, and both gazelle species using the plains in the dry season during the 1970s. Oryx have expanded their range westward, buffalo have colonized the woodland fringe along Olduvai Gorge, previously too dry for them, and I even saw waterbuck (normally a riverine species) in May 1977 on the short-grass plains near Gol kopjes.

Resident grazers of the central woodlands, such as impala, topi, and kongoni, have not changed in numbers in the 1970s possibly because wildebeest now graze the central woodlands more than they did in the 1960s (chap. 5), and so remove any increase in food there.

Events in the western corridor of the park support this possibility. In this region, the northern park boundary was moved north in 1968, thereby increasing the area, and human settlements were withdrawn from the boundary in 1974. Both these factors provided more space and grazing and greater protection from human poaching, particularly for resident species. And in this area, topi and possibly impala have shown a marked increase, and buffalo have increased faster than elsewhere (appendix A). Since wildebeest move from the west to the north, they would not have absorbed the extra food becoming available in the dry season, so their competitive effects would be less. Topi, at least, were increasing before the reduction in human predation, so their increase was not entirely due to release from predation.

The increase in dry-season grass production had an indirect effect on the browsers. Because the grass was less dry and combustible, fires became less frequent (J. Grimsdell, pers. comm.) and thus further promoted the regeneration of *Acacia* trees initiated by wildebeest grazing. As mentioned earlier, this has allowed an increase in giraffe population. This, in turn, has increased the browsing impact of giraffe on young trees, many of which are continually pruned and sculptured into strange shapes (pls. 41, 42).

At this stage, we do not know the long-term impact of giraffe browsing, but simulation studies indicate that giraffe may have an important impact

on the number of subadult trees that escape browsing and become mature
trees (chap. 13). Mature trees are pushed over and browsed by elephant.
In the past twenty years, the number of mature trees has declined, as a
result of several factors. Firstly, the encroaching human population in the
early 1960s could have compressed the elephants into the park so that tree
mortality increased. At the same time, extensive burning prevented regeneration of trees which, with a lag of ten to twenty years, would lower
recruitment to the mature-tree class. Moreover, increased browsing by
giraffe could further inhibit recruitment. These combined effects could
have resulted in the observed tree mortality exceeding recruitment (chap.
13), an effect likely to continue until the present young *Acacia* trees reach
maturity.

In the 1970s, predators on the plains have responded numerically to
the increase in resident herbivores (chap. 10). Between 1969 and 1976,
the lion population nearly doubled, and the hyena population increased
by 50 percent. In lions, on which there is more available information, the
increase was due both to more prides because of extra space, and to more
animals per pride because of more prey.

This response of the dominant predators may have been predictable.
But their effect on other predator species was more surprising. Wild dogs
have declined on the plains from over one hundred in the 1960s to only
thirty in 1977 (chap. 10), and there is a real possibility that they will
become extinct on the plains. Frame has noticed that they suffer both from
predation of young by hyenas when left with the mother while the pack
is hunting, and from competition with hyenas and lions, which can displace them from kills. If the pack is large enough, wild dogs can chase
away hyenas (pl. 38), but otherwise they lose (pl. 37). Combined with
this, they suffer from periodic outbreaks of disease (distemper, probably).
Distemper is not new, for Schaller (1972) and others observed it in the
1960s; hence, increased predation and competition from other predators
probably tipped the balance against the dogs.

Cheetahs are less vulnerable to competition, for, unlike wild dogs, they
hunt at times when other species are inactive, and they dispatch and eat
their prey quickly, thereby drawing little attention to themselves. Leopards
(pl. 34) avoid competition by carrying their prey into trees, where they
store it for several days. There the prey is inaccessible to lions and hyenas.
Neither cheetah nor leopard have shown a decline during the 1970s
(chaps. 9, 10).

In summary, the increased dry-season rainfall improved the productivity of the grasslands at a time when it is normally low or, as on the plains,

negligible. This resulted in a widespread expansion of range and population increase of both migrant and resident herbivore species. The latter was accompanied by an increase in lion and hyena populations, which, in turn, decreased the smaller population of wild dogs.

Dominant Processes and
Life History Strategies

From these two disturbances to the system, one can detect not only the extent of the linkages discussed above, but also the process involved.

Competition

The release of buffalo and wildebeest populations from rinderpest showed that initially they were both well below a level where they were limited by resources, particularly food supplies. Monitoring of the buffalo population showed that intraspecific competition for dry-season food supplies became increasingly more severe, so that numbers leveled off by the early 1970s (Sinclair 1977). Data for wildebeest, although less satisfactory, indicated a similar process in the late 1960s, when the ratio of numbers relative to available food was highest (chap. 4).

From these observations emerged the hypothesis that herbivores were regulated by their available food supplies. If this was the case, then an increase in food availability during the season when regulation was taking place—the dry season—should produce an increase in population numbers. This hypothesis was confirmed after the increase in dry-season rainfall (and therefore food) that took place in 1971 and in subsequent years. Both wildebeest and buffalo responded with a numerical increase, the latter after a two-year lag. Supporting evidence comes from the numerical increase in several other herbivore species on the plains.

Of equal importance is the increase in lion and hyena populations, as their own food supplies increased (chap. 10), pointing to intraspecific competition in the predator populations as well. Previous information confirming this comes from Schaller (1972), who observed lion cub mortality in the 1960s, when resident prey was scarce, and from Bertram (1973), who suggested expulsion of subadult females as a proximate mechanism for limiting pride size when food was scarce.

Interspecific competition between wildebeest and buffalo was particularly evident in the 1960s (Sinclair 1977), and this is likely to have contributed to the delay in the buffalo population's response to increased rainfall. Due to the great numerical disparity between these two species, buffalo have little effect upon wildebeest, but the reverse is pronounced. Interspecific competition between wildebeest and other grazers may also be important. Duncan (1975) recorded in the 1971 dry season that wildebeest almost completely removed the food in the home range of a topi group in the central woodlands, and the topi moved to another area for the remainder of that season. The wildebeest population has expanded its range into the northern and central woodlands during the dry season, thereby increasing their competitive effect on resident grazing species such as impala, topi, and kongoni. The numerical dominance of wildebeest allowed them to remove most of the increased food supply (chap. 3) that resulted from the change in rainfall. This left little food for other species, whose populations did not increase. These species suffer from competition to a greater extent than buffalo, which finds ecological refuge in forest not used by other resident grazers.

Competition between predator species occurs as they displace each other from kills. With the changing conditions on the plains, the balance appears to have tipped against wild dogs, and plains habitat may no longer be suitable for them. Wild dogs also live in the woodlands, although their numbers are unknown, and in this area hyena are relatively scarce, so dogs can probably persist.

Facilitation

Facilitation is the process whereby one species has a beneficial effect upon another. Vesey-Fitzgerald (1960) proposed that, among the herbivores, larger species altered vegetation structure in such a way that smaller herbivores could use the area. In the Serengeti, Bell (1970, 1971) showed that zebra were the first species to move into long-grass areas (pl. 19) during the dry season and suggested this resulted from their ability to digest poor-quality, highly lignified food better than ruminants like wildebeest. Bell proposed that zebra had a beneficial effect, for by removing the coarse top stems, they made the more nutritious leaves more available for the wildebeest, which followed them, and which, in turn, were followed by Thomson's gazelle. Wildebeest benefited the gazelle by reducing the grass to a short sward and making available the scattered herbs preferred by gazelle.

Analysis of these species' distribution and movements (chap. 5) indicates that the sequence of movements took place in less than a month. In the dry season, Maddock found zebra and wildebeest were closely associated with each other but that wildebeest did not necessarily follow the same migration route as zebra. Furthermore, the wildebeest population increased during the 1960s, when that of the zebra did not. Both of these observations show that the migrant wildebeest are not dependent on zebra for their existence. Hence, the facilitation hypothesis is an unlikely explanation for the close association of these two migrating populations.

Coexistence can occur because each species has its own ecological refuge: zebra use the drier northeast of the Serengeti and Mara, while wildebeest prefer the green northwest, probably excluding zebra there. Thomson's gazelle are less associated with the other two species, preferring the drier, short-grass areas in the central woodlands.

This is not to say that facilitation is absent: McNaughton (1976) has found that Thomson's gazelle prefer areas already grazed by wildebeest, and this supports the facilitation hypothesis. In general, herbivore species probably benefit from the activity of others, the degree of benefit varying between species, but it is unlikely that any herbivore is entirely dependent upon another for coexistence.

Similar remarks may be made about predators. Kruuk (1972) has dispelled the old notion that hyenas are strictly scavengers by showing that they kill a large proportion of their own prey. They may obtain some small benefit from scavenging buffalo or giraffe carcasses killed by lions, prey which they are incapable of killing for themselves. Such beneficial effects are probably more pronounced for the three jackal species, but again it is unlikely that they are dependent on other predators for their food. As for vultures, Houston (chap. 11) shows that only a small proportion of their food comes from animals killed by predators, most coming from animals that died from other causes.

Predation

Heavy grazing on growing plants of the plains has altered the species composition of the grassland community (chap. 3), probably by depressing palatable grasses and giving a competitive advantage to herbs.

Predation on the migratory herbivores appears to have had only minor effects. While wildebeest numbers increased in the 1960s, those of lion and hyena did not, though they preferred to eat wildebeest when available. The failure of predator numbers to increase showed that these species were not limited by wildebeest numbers, but, rather, by the time that wildebeest were available. When wildebeest and zebra left an area, only nonterritorial lions could follow them (pl. 28). Territory is necessary for reproduction for both lions and hyenas (pl. 29), for nomads start rearing cubs only after they have settled in an area (chap. 10). The majority of predators, then, cannot follow wildebeest and zebra, limiting them to smaller prey like gazelle, which they find difficult to catch (Schaller 1972; Elliot, Cowan, and Holling 1977), or to less numerous prey, such as topi, impala, and warthog. Predator numbers, so restricted, have little impact on the migratory herbivores. Also, predation has little impact on buffalo, giraffe, and other large species because they are difficult to kill.

Having said this, I should mention that simulation studies (chap. 12) suggest predators, in combination with rinderpest, could have kept the wildebeest population below about 400,000 animals. This accords with the censuses in 1961, when rinderpest was present (chap. 4). Since rinderpest had previously caused mortality almost every year, the predator-disease combination may have been responsible for keeping the wildebeest population in check from 1890 on. Nevertheless, this would have been an unnatural state, for rinderpest was an exotic to which the wildebeest were not adapted.

Environmental Heterogeneity and Life History Strategy

From the above discussion, it appears that most animal populations are limited by their food supplies, so the predominant linkage is through some form of competition. Only at the plant level does it appear that predation has a regulating effect.

The main reason for the predominance of competition lies in the way animals have adapted to the Serengeti environment. The peculiarities of the area are, firstly, the spatial heterogeneity of short-grass plains in one area, and long-grass woodlands in another; and secondly, the temporal heterogeneity, with short green grass appearing only on the plains during December to May, and only in the woodlands during June to November. These two features of the environment have led the dominant species in the ecosystem, wildebeest, to adopt a life history that involves large-scale movements. Indeed, it is likely this species is dominant because it uses this strategy of following the ever-changing pattern of short green grass patches.

This single strategy determines how the rest of the large mammal ecosystem is interlinked. Other herbivores can go along with the wildebeest—

Thomson's gazelle benefit by following behind, and I suspect that zebra are also obtaining some, as yet unknown, benefit. Other herbivores are forced to find other ecological refuges. The life history strategies of predators require them to be sedentary, making it possible for migrant herbivores to escape most of their predators. Because of their migration pattern, herbivores can exert a heavy grazing impact on the growing plants of the plains. They are not limited by dry-season grass biomass on the plains, so they can graze the vegetation down to a low biomass. Hence, predation predominates at this level.

This hypothesis that it is the environmental fluctuations (both spatial and temporal) which determine the types of interactions in an ecosystem predicts that a more homogeneous environment would allow predators to have a stronger impact. If wildebeest and zebra were to stay in the same area all year, not only would there be fewer of them, so reducing their impact on the plants, but they would suffer higher predation and might even be regulated by it.

This prediction can be examined by reference to the Ngorongoro ecosystem immediately to the east of the Serengeti plains. Wildebeest, zebra, and Thomson's gazelle remain in the same area all year. The impact of predation on wildebeest is almost ten times greater here compared to the Serengeti—about 10 percent of the population is killed by predators in Ngorongoro, as against 1 percent in Serengeti (Kruuk 1972; Elliot and Cowan 1978). Whether predators actually regulate the herbivore populations remains uncertain, but there is some circumstantial evidence to support such a theory. Rinderpest disappeared from Ngorongoro wildebeest, as it did from those of the Serengeti, in 1962. Censuses of the Ngorongoro population since the late 1950s have shown no consistent increase. The population has remained between 10,000 and 15,000 animals over the period from 1958 to 1977 (Turner and Watson 1964; Kruuk 1972; Estes, pers. comm.). This suggests that predators were able to absorb the extra wildebeest calves surviving once rinderpest was eliminated. There should have been an increase in predator numbers if this interpretation is correct, but data on predator numbers are lacking.

Finally, the effects of environmental heterogeneity determining species interactions can be seen in a totally separate component of the Serengeti ecosystem: the influx of considerable numbers of migrating palearctic birds is possible only because they are nomadic, following local thunderstorms in the dry season, where they find local concentrations of migrating insects feeding on the new green grass (Sinclair 1978). This almost exactly duplicates the strategy of the wildebeest. If palearctic migrants were not nomadic, they would exist in much lower numbers and perhaps not at all in this area because of competition with local resident species.

In conclusion, the Serengeti evidence shows that it is the characteristics of environmental heterogeneity and the life history strategies that organisms have evolved to accommodate this heterogeneity, which determine (1) the type of interaction between organisms, (2) whether the interactions are strong or weak, and (3) how much of the ecosystem is interlinked.

Patterns in the Community

The Vegetation

The results of the major biological processes of competition, facilitation, and predation can be seen in the pattern of species association at all trophic levels. The effects of predation are most pronounced on the grass layer. McNaughton (chap. 3) observes that both the grass family, Graminae, and herbivore family, Bovidae, show contemporaneous adaptive radiation during the Pleistocene, because of their interaction. McNaughton describes a short-term and long-term response by grasses to grazing. In the short term, grass species that can tolerate moderate grazing can actually increase their growth when grazed, through a variety of physiological adaptations. Presumably, this is beneficial to the herbivores, and may encourage them to continue grazing in the area longer. Since competition between plants is affected by grazing pressure, species less tolerant to grazing would be excluded under this system. Grasses have an advantage over palatable herbs because in the former, growth occurs from the base of the plant, which is protected from grazing damage, whereas in the latter, growth is at the apex, which is destroyed by grazing. Within the grasses, some species have greater root reserves to support regrowth than others. Those with higher reserves thrive under grazing.

Long-term changes resulting from grazing include the development of prostrate, low-growing forms with shorter internodes, smaller leaves, and more prostrate culms. McNaughton found that these differences within a species were genetically determined. Under grazing, the grass community is dominated by species with this form. When grazing was prevented, they were replaced by taller species. McNaughton (1978) describes another effect of grazing: palatable grasses can escape grazing by growing among unpalatable grasses,thereby producing "plant defense guilds." Relatively unselective grazers such as buffalo avoid swards with a high proportion of unpalatable species. Grazing, therefore, has a strong influence on the pattern of plant communities.

The Herbivores

The dominant process affecting ungulates is competition, and a number of patterns have emerged in response to it. Pennycuick (chap. 7) suggests that because dry-season food is in short supply and unpredictable in location, certain herbivores have evolved a large "foraging radius," that is, they can move long distances in search of food. To achieve this, animals need a large body size to maintain a fast gait for long periods without going into oxygen debt. Larger species have a larger radius of movement, and this allows them to win out over smaller species in competition for food. Indeed, Pennycuick suggests that running in ungulates is not only an adaptation to avoid predators, but also an adaptation that produces a large foraging radius.

Smaller species, however, can still adopt a large foraging radius, provided they can exist on the "crumbs" left by the larger species. Furthermore, grass growth stimulated by large grazers (chap. 3) should facilitate the survival of small species following in the wake of larger ones. For example, McNaughton (1976) has shown that Thomson's gazelle benefit from the previous grazing by wildebeest. Jarman and Sinclair (chap. 6) note that this strategy of smaller species following larger ones is quite common among ungulates. Smaller species need less food and are more selective than larger species. In the dry season the latter move into an ungrazed area first and take the bulk of the forage. Smaller species benefit because much of the coarse top grass is removed, leaving the more nutritious food at the base. Thus, resource partitioning is achieved by a pattern of different sizes of herbivore and different times for movements.

Jarman and Jarman (chap. 8) extend this pattern to the social organization of ungulates. Larger species cannot hide from predators, so they form large groups or herds as an antipredator behavior. Small species living on scattered food plants cannot form herds, so must develop cryptic coloration and hide. Group living imposes other behavioral requirements, such as group alertness, communication, and, in the extreme, group defense. These constraints, in turn, determine the type of strategy males adopt for obtaining mates. Males obtain dispersed females by staying with one,

forming a monogamous pair, and setting up territory. With larger herding females, males set up territories in the habitat favored by females, and
other males are excluded. This leads to intense competition between males
and the evolution of large males with horns or other characteristics for
display: it results in sexual dimorphism. Where females live in a few large
herds, as in buffalo, males stay with them and develop a dominance
hierarchy.

These are the long-term consequences of competion primarily, but
modified by predation and mating demands. Jarman and Jarman also
show that short-term events modify the social organization. For example,
seasonal changes producing greater food dispersion for impala result first
in a change in diet, then in habitat choice, and finally in a reduction in
group size. In very dry years, females become more dispersed, herds even
smaller, and territorial males find it increasingly difficult to control herds
and defend them from other males. Under these extreme conditions, most
males cease territorial behavior and the sexes mingle freely in herds. But
herds still form, indicating that groups are not solely the result of male
herding activity. Essentially, these changes reflect natural disturbances
in the balance of processes determining social organization: when competition becomes severe, other processes become relatively less important.
Also, the dynamic nature of this balance shows that these processes,
particularly competition for food and mates, act as both proximate and
ultimate determinants of social pattern.

The fact that predators take a relatively small proportion of the prey
does not mean that they are unimportant as selecting agents over evolutionary time. Both Jarman and Jarman, and Bertram (chap. 9), emphasize
the effect of predation in shaping ungulate adaptations, such as the formation of herds, avoidance behavior, and cryptic or disruptive coat colors.
Bertram also makes the point that because different predator species adopt
different hunting strategies for a common prey species, this prey cannot
evolve antipredator behavior suitable for all predators. For example, gazelles cannot be both good sprinters to avoid cheetahs, and good long-
distance runners to avoid wild dogs or hyenas. Thus, predator species may
benefit each other in the long term.

The Predators

Bertram (chap. 9) describes the pattern of feeding and social behavior
among the predators. From the responses of predator populations to

changes in resident prey numbers, I have suggested competition among predators is important, and Bertram considers this as a primary factor in shaping hunting strategy. It is instructive to compare the pattern among predators with that among ungulates. Firstly, in both groups, small or scattered food resources are associated with a solitary existence. Thus, dik-dik live on scattered herbs; leopards are solitary and hunt by stalking tightly grouped impala, or scattered reedbuck and dik-dik.

Secondly, where food is abundant, ungulates are large and form herds to avoid predation. Predators also form groups, but for a different reason —they do so in order to catch prey, not to avoid predators. Ungulates, unlike predators, do not face the problem of their food being larger than themselves and running away from them.

Sexual dimorphism in ungulates occurs when males compete for females in herds. Bertram considers the same process in lions: females remain small because this helps in hunting. Conversely, in hyena and wild dogs, large size helps in hunting as well as fighting, so both sexes are large, and there is little dimorphism.

Bertram points out that social organization evolved in different ways in lion, hyena, and wild dogs, which has led to different degrees of genetic relationship between members of a group, with that relationship greatest in wild dogs and least in hyena. This, in turn, determines other aspects of social behavior, such as the degree of dominance, who obtains mates and food, and whether young from other parents are protected and fed.

Among the vultures, Houston (chap. 11) notes that small, scattered food items are obtained by small vultures living as territorial pairs. Abundant food, in the form of large ungulate carcasses, results in gregarious behavior by large vultures. In this case, grouping is not to avoid predators, nor to cooperate in hunting; it is probably a result of the nature of the food distribution. Vultures appear to watch each other when searching, and they follow a bird that is descending to a carcass. This behavior indicates that grouping may help birds obtain information on where food is to be found.

Unlike mammal predators, vultures can follow the migrating ungulates, because gliding flight is more efficient than walking, and because vultures have good vision. This allows vultures to obtain food unavailable to mammal predators. And the long flights from their nests to migrating herds have necessitated a number of special breeding adaptations. Houston suggests that many of the characteristics associated with movements have evolved during the time when herbivore migrations were more common than in today's geographically restricted fauna.

Conclusions

Dominant Processes

As a result of the changes taking place in the Serengeti ecosystem, one can identify certain important biotic processes. Firstly, it is clear that there are strong interactions between a few components in the system, while the other components have relatively small impact. Thus, the wildebeest is clearly a dominant herbivore, and it has a strong interaction with perennial grasses that can withstand grazing. Likewise, the interactions between giraffe or elephant and trees are the major processes among the browsers. Conversely, there appears to be only a weak interaction between carnivores and migrating herbivores.

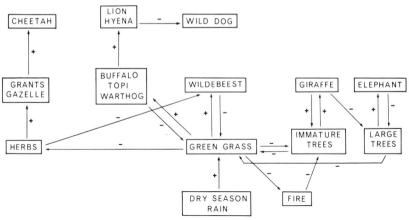

Figure 1.4   Summary of the linkages and their
                                    effects detected in the Serengeti
                                    ecosystem. Minus signs indicate
                                    that higher values in the preceding
                                    box result in lower values in the
                                    succeeding box, while plus signs
                                    result in higher values.

Secondly, these few major components have a far-reaching effect on the other components (fig. 1.4). Thus, the wildebeest has a direct predatory effect determining the composition of the grass-herb layer, and a direct competitive effect on other grazers. There are also indirect effects spreading out to herb and tree feeders, and indirectly to all the carnivores.

Giraffe and elephant both have important effects on trees, and affect grass composition and, hence, grazers indirectly. (It is not yet clear which of these browsers is more important). Essentially, if one of these components, such as wildebeest, were removed from the system, major changes in all trophic levels would result. In contrast, if buffalo or impala, for example, were to disappear, probably little change in the system would occur.

Thirdly, of the processes observed, competition is by far the most important, for it plays a part in the species composition of the herb layer, and it determines the relative abundance of both grazers and carnivores. Predation by herbivores on grasses is an important influence (chap. 3), but it is minor at other trophic levels: that is to say, predators take a considerable part of grass production, but little of the herbivore production.

Fourthly, the relative importance of biotic processes is dependent upon the degree of environmental fluctuation and unpredictability, leading to spatial and temporal heterogeneity. This is most important in the context of stability and resilience of ecosystems. Peterman, Clark, and Holling (1978) point out that ecosystems appear to have multiple equilibria, each with a domain of attraction. Moreover, they suggest the boundaries of these domains shift in response to changing variables and parameters. In the present case, resources, in the form of food supply, determine the upper equilibrium. A lower equilibrium determined by predation is not, at present, possible in the Serengeti because the exogenous variable, climate, dictates a migration strategy in wildebeest. If the climate became more suitable for a sedentary existence by wildebeest, as in Ngorongoro, then a lower equilibrium imposed by predators could appear. Furthermore, a change in a parameter, such as wildebeest yearling survival (through rinderpest effects), also appeared to determine whether a lower equilibrium could exist.

Fifthly, there are certain important feedback processes that involve the major components of the system. Some components are interactive, while others are not (see Caughley 1976). For example, a noninteractive system is seen in the regulation of wildebeest and buffalo numbers by limited food supply in the dry season. Here the plants are effectively dormant, and the grazers do not kill off or change the morphology of the plants. On the other hand, wildebeest grazing on the plains grasses during the wet season is interactive (chap. 3). In this case, grazing is on growing plants, which not only influences the morphology of the plants, but changes their species composition. Similarly, elephants kill many of the trees they browse (pl. 43), and this, in turn, affects the elephants by changing food abundance and, hence, in the long term, elephant reproduction (Laws 1969; Laws, Parker, and Johnson 1975).

Finally, competition as the major process is reflected in the pattern that is common to all consumer groups: the spatial and temporal dispersion of food seems to determine the pattern of resource utilization and social organization in herbivores, predators, and scavengers. Superimposed on this basic plan are the effects of other processes, such as competition for females and predation.

There are now a number of comparative studies of ecology and social organization of members within a taxonomic group. From the Serengeti work, I have compared three such groups, but there is a need to extend the comparison to other groups, such as primates, birds, and fish.

Interpretations for Management

Systems such as the Serengeti should be maintained in as natural a state as possible to fulfill their functions as National Parks and as ecological baseline areas (Sinclair 1977). Because ecosystems must be considered dynamic (Connell 1978), one must resist the temptation to manage them with a view toward maintaining some arbitrary status quo. The present evidence shows that there are natural negative feedback mechanisms operating between some components in the system. Indeed, research must aim at establishing whether feedbacks occur, their nature, and strength of effect. Provided these negative feedbacks are strong enough, the system can absorb disturbances without management.

For example, the increase in herbs that may be occurring on the plains, far from being undesirable, is, in fact, a necessary negative feedback on wildebeest, for herbs act to inhibit grazing and thereby relax pressure on the grasses (chap. 3). The wildebeest get less food, and this is ultimately reflected in lower calf survival and a stable population. Interference by culling the wildebeest would promote potential instability by increasing their recruitment rate and allowing a further eruption when culling ceases. Culling, therefore, is undesirable.

Secondly, there are density-dependent negative feedbacks operating on wildebeest and buffalo in the dry season, and it is likely that this applies to most of the ungulates (chap. 4, and Sinclair 1977). This feedback operates through a lack of dry-season food. Because grasses in this season are not killed by heavy grazing, it is unlikely that these grasslands (in the west and north of the park) will change much in the future. In other words this system of grass and grazers is noninteractive, making equilibrium between them possible. However, ungulate populations will fluctuate because of periodic climatic fluctuations, such as droughts. These could produce rapid declines in wildebeest and other herbivores, but these species will build up again when the food supply increases after the drought is over. Such declines are desirable and should be expected.

The complex of fire, giraffe browsing, and elephant browsing on tree survival has resulted in a mosaic of tree stands, some patches with mature trees and declining numbers, others with immature trees and increasing numbers (chap. 13). Giraffe concentrate on the immature patches and inhibit escapement of trees into the mature class; elephant concentrate on mature trees and, by pushing them over, lower their number. Fire has the greatest effect, however, by preventing regeneration of seedlings and inhibiting small tree regrowth. This system is interactive in that there is a long-term response by the trees to the effects of browsing (unlike the grazing system), so that there will be cycles of high and low tree numbers and elephant numbers. We do not yet know whether these cycles, once initiated by the perturbation when elephants were compressed into the park, are self-dampening or not. If the cycles are dampening, then there is no need to worry, except in certain local areas where management requires the preservation of mature trees. In these areas, management will have the greatest effect by controlling fires, perhaps through early-burning practices.

Predators like lions and hyenas depend on the nonmigratory herbivores for their food when migrants are not present. If the resident prey should decline in numbers, then these predator populations will be seriously affected. Poaching in the northwest of the park has had a serious effect on buffalo and other resident species (Sinclair 1977), and this is likely to be the cause of low predator numbers recorded in that area by Schaller (1972) and Kruuk (1972). This point is particularly important in the context of culling of resident prey for commercial marketing of meat and skins. Culling could seriously interfere with predators if the prey are nonmigratory: this applies to culling of topi and other resident prey in the western Serengeti, and of wildebeest and zebra in Ngorongoro. Management, therefore, should concentrate on protecting resident prey.

Finally, predators have little effect on the migratory ungulates at present. However, if these migrants were forced to curtail their movements, by remaining within the park boundaries, then the effect of predators could be more dramatic. Due to the smaller area, the prey would be lower in number, and also would remain within reach of the predators for a longer period. A decline in prey numbers due to a drought might result in a change in equilibrium: prey would become dominated by predators, and therefore would remain at a low level. This situation remains conjecture, but it emphasizes that the park itself is not a self-contained unit and that the migratory system depends on surrounding areas, whose management must be coordinated with the park's.

Future Research and Experimental Management

Understanding of how the ecosystem works is far from complete, but the present synthesis does allow us to see major areas requiring future investigation.

Firstly, there are whole elements of the system about which little is known. The impact of invertebrates on the vegetation, and of predators (birds and small carnivores) on the invertebrates, remains almost unknown. Crude measurements suggest that invertebrates have as much impact on the grasslands as do the large mammals (Sinclair 1975), and if this is correct, then fluctuations in phytophagous insects could affect the mammals.

Secondly, the present synthesis has produced a number of new hypotheses. For example (a) the wildebeest population may be depressing the populations of resident herbivores through competition; or (b) predators may be maintained at a high enough density by the migrant population that they depress populations of resident prey. Yet the functional and numerical responses of the woodland populations of hyenas and lions remain entirely unknown. Similarly, (c) it now appears the close association between zebra and wildebeest migrants is not explained by the facilitation hypothesis and that alternatives to this hypothesis must be considered. Perhaps zebra are benefiting from their association with wildebeest? The movements of these two species must now be studied more precisely. Monitoring of ungulate and predator populations, including their fertility, recruitment, and mortality over the next decade is, therefore, an essential element of future research, together with monitoring of climate and vegetation changes.

Thirdly, management should be designed in such a way that results from it can be used as part of the research program (Larkin 1972). Since most management is essentially experimental, it should be designed so that its effects can be compared with controls. For example, fire management in the *Acacia* woodlands should involve different fire treatments in different areas. Information gathered from such treatments could then be used to test current hypotheses on vegetation response to giraffe and elephant. If we are to preserve the evergreen thickets of the north (see chap. 2), we should treat different patches in different ways, in case our ideas about why they are disappearing are wrong. Thus, some should be protected from fire; others from animals; while still others should be exposed to different fire treatments, including abnormally severe burning; and others should be left untreated.

Indeed, the principal of experimental management should be a basic concern in our understanding of ecosystem processes. Yet in most national parks and reserves, it is rare that management practices are designed so that biologists can compare the effects with control areas and follow up long-term consequences. As a result, most wildlife management in North America, for example, still rests on untested myths.

References

Andrewartha, H. G., and Birch, L. C. 1954. *The distribution and abundance of animals.* Chicago: Univ. of Chicago Press.

Austin, M. P., and Cook, B. G. 1974. Ecosystem stability: a result from an abstract simulation. *J. Theoret. Biol.* 45:435–58.

Bell, R. H. V. 1970. The use of the herb layer by grazing ungulates in the Serengeti. In *Animal Populations in Relation to their Food Resources,* ed. A. Watson, pp. 111–23. Oxford: Blackwell.

Bell, R. H. V. 1971. A grazing ecosystem in the Serengeti. *Sci. Am.* 224: 86–93.

Bertram, B. C. R. 1973. Lion population regulation. *E. Afr. Wildl. J.* 11: 215–25.

Bradley, R. M. 1977. Aspects of the ecology of the Thomson's gazelle in the Serengeti National Park, Tanzania. Ph.D. dissertation, Texas A & M University.

Branagan, D., and Hammond, J. A. 1965. Rinderpest in Tanganyika: a review. *Bull. Epizoot. Dis. Afr.* 13:225–46.

Caughley, G. 1976. Plant-herbivore systems. In *Theoretical ecology,* ed. R. M. May. pp. 94–113. Philadelphia: W. B. Saunders.

Chitty, D. 1967. The natural selection of self-regulating behaviour in animal populations. *Proc. Ecol. Soc. Aust.* 2:51–78.

Connell, J. H. 1975. Some mechanisms producing structure in natural communities: a model and evidence from field experiments. In *Ecology and evolution of communities,* ed. M. L. Cody and J. M. Diamond, pp. 460–90. Cambridge, Mass.: Belknap Press.

———. 1978. Diversity in tropical rain forests and coral reefs. *Science* 199:1302–10.

Dawkins, R. 1976. *The selfish gene.* Oxford: Oxford University Press.

Duncan, P. 1975. Topi and their food supply. Ph.D dissertation, Univ. of Nairobi.

Elliott, J. P., and Cowan, I. McT. 1978. Territoriality, density, and prey of the lion in Ngorongoro Crater, Tanzania. *Can. J. Zool.,* 56:1726–34.

Elliott, J. P., Cowan, I. McT., and Holling, C. S. 1977. Prey capture of the African lion. *Can. J. Zool.* 55:1811–28.

Ford, J. 1971. *The role of the trypanosomiases in African ecology.* Oxford: Clarendon Press.

Holling, C. S. 1973. Resilience and stability of ecological systems. *Ann. Rev. Ecol. Syst.* 4:1–23.

Holling, C. S., ed. 1978. *Adaptive environmental assessment and management.* Chichester: John Wiley & Sons.

Huffaker, C. B. 1970. The phenomenon of predation and its roles in nature. In *Dynamics of populations,* ed. P. J. den Boer and G. R. Gradwell, pp. 327–43. Proc. Adv. Study Inst., Oosterbeek. Wageningen: Center for Agricultural Publishing.

Kruuk, H. 1972. *The spotted hyena.* Chicago: Univ. of Chicago Press.

Larkin, P. A. 1972. A confidential memorandum on fisheries science. In *World fisheries policy: multidisciplinary views,* ed. B. J. Rothschild, pp. 189–97. Seattle: Univ. of Washington Press.

Laws, R. M. 1969. Aspects of reproduction in the African elephant, *Loxodonta africana. J. Reprod. Fert.,* supp. 6, pp. 193–217.

Laws, R. M.; Parker, I. S. C.; and Johnstone, R. C. B. 1975. *Elephants and their habitats.* Oxford: Oxford University Press.

McNaughton, S. J. 1976. Serengeti migratory wildebeest: facilitation of energy flow by grazing. *Science* 191:92–94.

———. 1978. Serengeti ungulates: feeding selectivity influences the effectiveness of plant defense guilds. *Science* 199:806–7.

May, R. M. 1977. Thresholds and breakpoints in ecosystems with a multiplicity of stable states. *Nature* (Lond.) 269:471–77.

Murdoch, W. W., and Oaten, A. 1975. Predation and population stability. *Adv. Ecol. Res.* 9:1–131.

Myers, J. H., and Krebs, C. J. 1971. Genetic, behavioral, and reproductive attributes of dispersing field voles *Microtus pennsylvanicus* and *Microtus ochrogaster. Ecol. Monogr.* 41:53–78.

Nicholson, A. J. 1933. The balance of animal populations. *J. Anim. Ecol.* 2:132–78.

Noy-Meir, I. 1975. Stability of grazing systems: an application of predator-prey graphs. *J. Ecol.* 63:459–81.

Peterman, R. M.; Clark, W. C.; and Holling, C. S. 1978. The dynamics of resilience: shifting stability domains in fish and insect systems. In *Population dynamics,* ed. R. M. Anderson and B. D. Turner. Proc. Symp. Brit. Ecol. Soc. Oxford: Blackwell, forthcoming.

Reddingius, J., and den Boer, P. J. 1970. Simulation experiments illustrating stabilization of animal numbers by spreading of risk. *Oecologia* (Berlin) 5:250–84.

Ricker, W. E. 1954. Stock and recruitment. *J. Fish. Res. Bd.* (Canada) 11:559–623.

Schaller, G. B. 1972. *The Serengeti lion.* Chicago: Univ. of Chicago Press.

Schoener, T. W. 1974. Resource partitioning in ecological communities. *Science* 185:27–39.

Sinclair, A. R. E. 1975. The resource limitation of trophic levels in tropical grassland ecosystems. *J. Anim. Ecol.* 44:497–520.

———. 1977. *The African buffalo.* Chicago: Univ. of Chicago Press.

———. 1978. Factors affecting the food supply and breeding season of resident birds and movements of palaearctic migrants in a tropical African savannah. *Ibis,* 120:480–97.

Southwood, T. R. E., and Comins, H. N. 1976. A synoptic population model. *J. Anim. Ecol.* 45:949–65.

Sutherland, J. P. 1974. Multiple stable points in natural communities. *Amer. Nat.* 108:859–73.

Turner, M. I. M., and Watson, R. M. 1964. A census of game in Ngorongoro Crater. *E. Afr. Wildl. J.* 2:165–68.

Walters, C. J., and Hilborn, R. 1976. Adaptive control of fishing systems. *J. Fish. Res. Bd.* (Canada) 33:145–59.

White, S. E. 1915. *The rediscovered country.* New York: Doubleday, Page.

A. R. E. Sinclair

Two   The Serengeti Environment

This chapter forms the background against which the ecological events described later will be set. Although this is largely a description of the average situation, I emphasize two types of variability. Firstly, the Serengeti is spatially heterogeneous in climate, geology, soils, and vegetation. Secondly, there is variation on an evolutionary time scale which underlies the short-term events forming the present focus. The evolutionary history of the ecosystem dictates the way in which it may respond to present-day disturbances, so with this in mind, I review briefly information on the paleoecology of the region. General descriptions of the Serengeti can be found in Schaller (1972), Kruuk (1972), and Sinclair (1977).

Topography

The Serengeti-Mara ecosystem (34° to 36°E, 1° 15′ to 3° 30′ S) has traditionally been accepted as that area influenced by the migratory wildebeest population (fig. 1.1). This system, covering 25,000 km², has natural boundaries that effectively prevent emigration or immigration of large mammals. In the north, the ecosystem is delimited by the Loita plains of Kenya, which are dry and inhospitable at the time when migrants move north. The eastern boundary is formed by the forested Loita hills and their southern extension, the Gol Mountains.

However, east of the Gol Mountains, migrant herbivores also use the Salai plains in the wet season. They are restricted by dense woodland to the north, the rift valley escarpment in the east, and the crater highlands in the southeast (pl. 1). Although the migrants do not move into the crater highlands, there are separate, resident populations within Ngorongoro Crater, which forms part of the highlands. Ngorongoro is a caldera about 18 km in diameter, with an altitude of 2100–2400 m at the rim and 1700 m on the crater floor (Hay 1976). Its vegetation is open grass-land on the floor and northern and western rim, montane forest covering the eastern and southern rim. A detailed description of the area is given by Herlocker and Dirschl (1972).

South of the crater highlands, a dry, rocky woodland along the Lake Eyasi escarpment forms the southern boundary of the Serengeti plains. In the southwest, cultivation by the people of Sukumaland forms a barrier running from the Maswa Reserve to the westernmost portion of the Serengeti park. This narrow western extension of the park reaches nearly to Speke Gulf on Lake Victoria, and includes the corridor between two permanently flowing rivers, the Mbalageti and Grumeti. Cultivation also lies north of the corridor and west of the park boundary and extends north to the Mara River and the Isuria escarpment (pl. 19). The latter acts as the northwest boundary of the system.

The Serengeti, like most of East Africa, is part of a plateau well above sea level with the eastern plains, at 1780 m, sloping to a low point of 1230 m at Lake Victoria. Altitude changes faster in the northeast, provid-ing for a topography dissected by many small gullies. The west is flatter, with fewer tributaries to the main rivers, which meander westward. Low hills (the central range) extend westward along the corridor (pl. 4), and the Itonjo hills are found in the southwest. Other hills are found along the eastern boundary of the park. Most are rounded and stony, rising 100–200 m above the surrounding country; these hills support an open, woody vegetation. They are not grazed much by migratory ungulates, but they form no barrier to the animals' movements. In general, the rest of the Serengeti is composed of gently sloping ridges and valleys.

Climate

The most important aspect of climate in this region is rainfall. Norton-Griffiths, Herlocker, and Pennycuick (1975) have described the rainfall patterns in detail. Rain is seasonal and determined by large-scale weather patterns, modified by local topography. The main weather system in East Africa is associated with the Intertropical Convergence Zone. This zone is a belt of low pressure following six weeks behind the sun as it moves south and north across the equator. The southward movement brings northeasterly prevailing winds, which are relatively dry. It reaches the Serengeti about November and produces the first (short) rains following the dry season. For convenience, November is taken as the beginning of the seasonal year. Rain may sometimes continue until March, thereby merging with the long rains associated with the northward movement of the sun. These rains last from March until May and are the result of the wetter, southeasterly winds coming from the Indian Ocean. On occasion there is a break in the rains in January and February.

This pattern is influenced by the crater highlands, which rise to over 3,000 m. They form a rain shadow immediately to the northwest, so that a gradient of increasing rainfall is produced from the southeast plains to the northwest woodlands on the Mara River. The gradient is present in both the wet season (Nov.-May; fig. 2.1) and dry season (fig. 2.2).

A further modifying influence is associated with Lake Victoria and the wetter Congo weather system to the west. The lake produces its own convergence zone and affects northern and western Serengeti by increasing precipitation between June and November. Furthermore, it also results in a more consistent pattern: the variation from year to year in dry-season rainfall is only 25 percent of the long-term average in the northwest, 50 percent in center and west, and 100 percent in the southeast of the park. This means that migrant herbivores looking for green areas during the dry season will be most likely to find them in the northwest.

As a result of this rainfall gradient, the seasons vary according to area. The first rains start in the north and move south, and the drying sequence at the end of the long rains does the reverse by starting in May on the plains and ending in July in the north. This can be seen in the monthly rainfall distributions for various stations shown in figure 2.2. At Olduvai Gorge, precipitation is low even in the wet season, and the dry season is long and severe. In the west, the wet season has two peaks with much higher precipitation, and the dry season is shorter but still pronounced. In the north, there is as much rain in the dry season as Olduvai receives in the wet season.

Norton-Griffiths et al. (1975) calculated a climatic index (after Thornthwaite 1948) that took into account rainfall, temperature, windspeed, humidity, and solar radiation based on ten-year averages. The isolines of this index were similar to those of rainfall. The far southeast plains are classified as arid, the northwest plains, the western corridor, and the central woodlands (100–200 mm dry-season rain; fig. 2.2) as semi-arid, and the north as dry subhumid. The dry-season rainfall picture, therefore, gives a reasonable indication of the overall climatic conditions.

Pennycuick and Norton-Griffiths (1976) have analyzed the rainfall fluctuations in long-term records. Apart from the annual cycle, there was no very strong indication of longer cycles. Annual fluctuations were correlated over a wide area, however. One significant change has occurred in

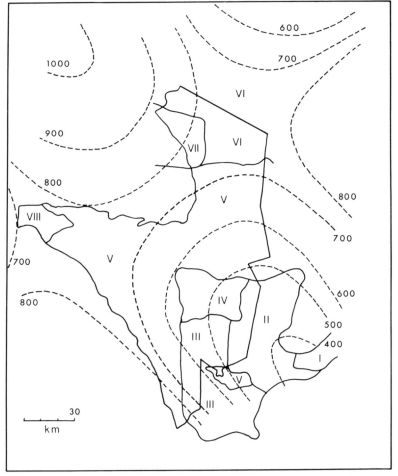

Figure 2.1

Mean wet-season rainfall isohyets (*broken lines*, mm) show a gradient from high in the northwest to low in the southeast. Vegetation zones are indicated by Roman numerals.

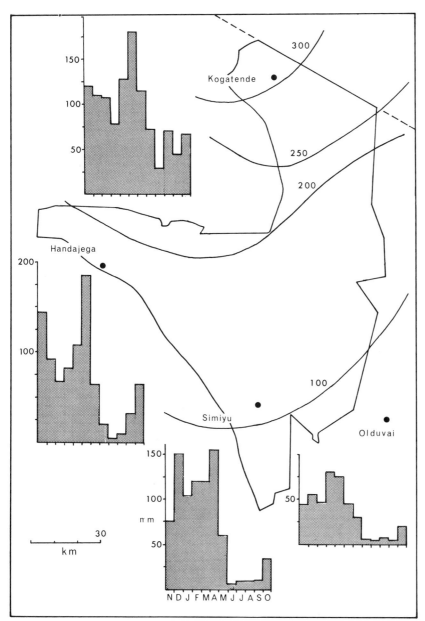

Figure 2.2

Mean dry-season rainfall isohyets (mm) show a similar gradient to that in the wet season. The histograms show the distribution of rainfall through the year at four stations. From Sinclair (1977).

the seasonality of rain: for six years (1971–76), dry-season rainfall was nearly twice as high as the average for the decade earlier (see fig. 10.5), and only in 1977 did it revert to previous values. Such a consistent change has not previously been observed.

Geology

West of a line going through Seronera, the underlying rocks are very old Precambrian (2500 M years) volcanic rocks and banded ironstones forming the Tanganyika Shield. A sedimentary series of late Precambrian quartzites, sandstones, and shales unconformably overlie the shield and form the central and Itonjo hills. East of Seronera, a late Precambrian orogenic event (roughly 475–650 M years ago) distorted the shield. As a result of this event, granitic gneisses and quartzites underlie the plains and eastern woodlands and form the hills along the eastern park boundary and on the plains. Rocky outcrops on the plains (kopjes; pl. 3) are also of this period (Macfarlane 1969; Hay 1976).

The crater highlands are volcanoes of Pleistocene age (Hay 1976). They supplied aerially discharged material which was blown westward and formed the Serengeti plains. The westernmost plains, close to the hills, are fine-grained, those closest to the highlands, such as the Salai plains, are made up of much coarser material. Ngorongoro grew to its maximum height of about 5,000 m as recently as 2 M years ago. It then collapsed to form the caldera (pl. 2). There are older volcanoes dating back to 3.7 M years ago. One volcano, Lengai (pl. 1), is still active, with eruptions recorded in 1917, 1921, 1940, and 1966. Ash deposits from the 1940 eruption were recorded on the Serengeti plains 95 km to the west (Anderson and Talbot 1965).

Soils and Vegetation

The Plains

Soils of the Serengeti plains are highly saline and alkaline as a result of their recent volcanic origin. Anderson and Talbot (1965) and H. De Wit (1978, pers. comm.) have provided a detailed description of these soils and their associated vegetation.

The youngest soils (zone I; fig. 2.1) are in the form of mobile sand dunes derived from Lengai and located on the Salai plains. They are

highly porous, coarse-grained sands, which, combined with the low rainfall, become virtual deserts. Stable dunes support deep-rooted grasses (for
example, *Sporobolus consimilis*), while short grasses grow between the
dunes (Herlocker and Dirschl 1972).

West of the dunes, short grasslands (pl. 3) of the eastern plains (zone
II) are characterized by dwarfed growth forms of *Digitaria macroblephora*, and *Sporobolus marginatus*, among others, and the grasslike
sedges of the genus *Kyllinga*. Basal cover is relatively low, averaging 10
percent, but reaching 30 percent in parts. The dwarf forms are a result of
the peculiarities of the soil (although grazing has some influence; see
chap. 3). There is a high content of potassium, sodium, and calcium, and
low content of magnesium, reflecting the volcanic origin. Rainfall has
leached the salts out of the highly porous sandy top layers and redeposited
them as a calcium carbonate hardpan about 100 cm below the surface
(petrocalcic horizon). The hardpan is impermeable and prevents root
growth. As a result, a highly saline and alkaline soil lies above it, and
only plants that can withstand this and have shallow roots can survive.
The friable top layers are unstable, and erosion steps of up to 50 cm are
common. Grazing accelerates their back erosion (as observed from comparisons of steps inside and outside exclosures), but erosion is not a
significant influence, as soil merely becomes redistributed a few meters
away.

West and south of the short grasslands, the soils are deeper (zone III),
and the rainfall is higher. The hardpan is less continuous and softer, en-
abling roots to penetrate the interstices. Because they are further from
the volcanic source, the soil particles are finer; they hold water longer,
and there is less leaching. Alkalinity and salinity are less, so plants face a
milder environment, and taller growth forms of zone-II grass species can
exist. These are the intermediate grasslands. They include specifically
Andropogon greenwayi, which grows in patches of nearly 100 percent
basal cover. It has a low growth form and is high in productivity and calcium and low in fiber content. Consequently, this area is highly preferred
by wildebeest (Kreulen 1975; chap. 5 below). Average basal cover for
the zone is 30 percent. In both zones II and III, there are extensive areas
of herbs, mainly *Indigofera basiflora* and *Solanum incanum* (pls. 17, 18).

The gradient of soil characteristics described above continues northwest
into zone IV. These are the long-grass plains typified by *Themeda triandra*
and *Pennisetum mezianum*. Basal cover is about 50 percent. The soil is
deeper still (about 2 m), and although calcium carbonate concretions

occur, there is no hard layer. The soil is fine volcanic ash mixed with parent material to form silty clay with low alkalinity and salinity. Herbs are relatively low in abundance, and there are no trees or bushes.

The Termite-Vegetation-Soil Interaction

A noticeable feature of the intermediate and long grasslands is the mosaic of *Cynodon dactylon, D. macroblephora* grass patches set in the surrounding matrix of *A. greenwayi* or *T. triandra*. H. de Wit (1978, pers. comm.) has found that the mosaic is a result of termite activity.

Grasses like *A. greenwayi* or *T. triandra* do not grow in soils with high salinity and alkalinity. When a termite colony starts, it uses easily weatherable subsoil material of high salinity and alkalinity for building the mound (made hard by using salivary cement). Thus, the centers of active mounds are saline and alkaline close to the surface. They are only 30 cm high on the intermediate grasslands, but reach 100 cm in the long grasslands and are probably built by different species. Patches are about 3 m across, and on them grow salt- and alkali-tolerant grasses such as *D. macroblephora* and *C. dactylon.*

When the colony dies, salts are leached outward from the center, down the slope, and, consequently, a ring forms (see also Glover, Trump, and Wateridge 1964). Over time, the mound is reduced to the surrounding level, the salts are washed deeper, and the original grasses return. Since it appears that termite colonies live for several hundred years, the period for a complete cycle is much longer than most other biological interactions considered in the Serengeti. The degree to which ungulates influence this is unknown. But wildebeest, topi, and zebra do use bare areas on the mounds for their territorial activities, dust-bathing and so on, and so could speed up decomposition of termitaria.

The Woodlands

More will be known about soils in the woodlands (zones V, VI, VII; fig. 2.1) when the work of Jager (1979) is complete. These soils are formed from the parent rock of granite or quartzite. Toward the eastern woodlands, there is an increasing volcanic ash content, but, on the whole, salinity and alkalinity are low, and a hardpan is absent, allowing tree growth. Basal cover for grasses varies from 20 to 60 percent, averaging 45 percent (Anderson and Talbot 1965).

Typically, there is a catenary sequence (Milne 1935) of shallow sandy soils on watersheds, with soil becoming deeper and with a greater silt content toward valley bottoms. Grass species are adapted to the different conditions, so that small perennials and annuals (*D. macroblephora, Chloris pycnothrix,* for example) predominate on the sandy soils. On the slopes, deeper soil and good drainage provide optimum conditions for *Themeda triandra* (pl. 19), which is one of the dominant grasses of the woodlands. *Themeda* and associated grasses are fire-tolerant, for these areas are largely fire-induced savannah, that is, grasslands with some trees (Vesey-Fitzgerald 1970).

In drainage swamps, coarser grasses like *Pennesetum mezianum* predominate, but very tall grasses (2 m high), such as *Panicum maximum,* grow in areas that remain flooded for long periods (pl. 8). These are edaphic grasslands, usually small areas close to rivers. However, at the western end of the corridor (zone VIII), there are extensive alluvial soils derived from old beds of Lake Victoria. These support floodplain grasslands with few trees (pl. 8).

Herlocker (1975) has described the woody vegetation. Two broad zones can be recognized: a thorn-tree woodland (zone V) comprising most of the south and center, and a relict forest-bushland (zones VI and VII) that extends from the Grumeti River north to the Loita plains. The thorn-tree woodland is dominated by *Acacia* (38 species in all), with 10 species accounting for 45 percent of the woody vegetation. The most extensive is the small whistling thorn (*A. drepanolobium*) that grows on impeded drainage silt soils. A taller tree, *A. clavigera,* is common on slopes and ridges (pl. 6), and the umbrella tree, *A. tortilis,* is characteristic along the plains border (pl. 5), providing shade for animals. *Commiphora trothae* is the only other tree that occurs commonly with *Acacia* species on ridges and slopes. Regeneration of many of the *Acacia* trees occurs in patches surrounding mature stands (pl. 6). Herlocker noted a general absence of immature trees in the early 1970s. This may be a reflection of the extensive burning that took place in the 1960s, inhibiting the growth of young trees.

North of the Grumeti River, the relict woodlands provide evidence for more radical change. The northwest (zone VII) is mainly open deciduous woodland with the broad-leaved trees *Combretum molle* and *Terminalia mollis.* This is a fire-maintained successional stage. Relict patches of evergreen forest on ridges as well as along valleys (pl. 7) show that it was once the dominant type. The influence of fire is reflected in the abrupt boundary between forest and grassland, with no ecotone. Overstory trees include *Diospyros abyssinica* and *Elaeodendron buchananii,* with a bush layer dominated by *Croton dichogamus* and *Techlea trichocarpa.* Riverine forest also grows along the lower Grumeti, Mbalageti, and Duma rivers (pl. 8).

In the drier eastern part (zone VI), the larger trees are absent, and an evergreen thicket is formed with similar species of *Croton, Techlea,* and *Euclea* (pl. 7). These dense thickets are found on the shallow soils of hillsides at the present time, but there is evidence that they were once more extensive. During the 1960s these thickets were everywhere reduced, and some even disappeared entirely, as can be seen in plates 45 and 46, recorded in 1959 and 1968, respectively. The surrounding grassland is dominated by *A. clavigera* and *A. gerrardii,* both fire-resistant species. From analysis of soil profiles, Schmidt (1975) concludes that thickets were previously of long standing and that changes are recent and unusual. Photographs taken of the same areas in 1978 by myself and M. Norton-Griffiths show much less change during the 1970s than during the previous decade.

The process of thicket degradation involves the activities of elephant, buffalo, and rhino. These species push into the thickets and open up paths. With the extra light, taller grasses grow along the paths and provide fuel for ensuing fires, which spread along the paths. Fires scorch the outer layer of bushes and so widen the path for the next year. Thus, a positive feedback is set in motion. Without fire, paths made by large mammals grow over again, and damage to bushes is minimal, as indicated during the 1970s, when fires were infrequent but animals were abundant.

The evergreen forests and thickets of the northern Serengeti are a special type related to montane forest but now existing in a drier climate. They support a special fauna unlike the rest of the Serengeti, and they clearly require special protection.

Grasslands in the *Terminalia* woodlands are distinct. On ridges and slopes with shallow soil, *Loudetia kagerensis* is associated with *Terminalia* trees. There is a clear boundary contouring the slope where the water table reaches the surface, producing a band of damp soil and springs. Large termitaria are found along this seepage line (pl. 14). Below it, *Themeda* predominate, and no trees or bushes grow. Hence, open grassland extends to the evergreen forest in the valley bottom. Presumably, forest covered the *Themeda* grasslands at one time. On ridges where *Terminalia* trees are not present, other grass species give way to the tall *Hyparrhenia filipendula,* avoided by most grazing ungulates.

The Fauna

Mammals dominate the fauna. There are some 30 species of ungulates and 13 species of large carnivores, with numerous other smaller forms (see appendix B). Numbers of ungulates combined approaches 2.5 million. Detailed species lists can be found in Schaller (1972), Sinclair (1977), and Hendricks (1970). Over 400 species of birds, and an unknown number of other animal groups, occur in the area.

The wildebeest is the most numerous large mammal (pl. 12). Wildebeest perform characteristic seasonal movements, called the "migration," from the short-grass plains in the wet season, where they calve (pls. 9, 10), to the western corridor in the early dry season, and, finally, to the north for the late dry season. Other species, such as zebra (pls. 11, 19), Thomson's gazelle (pl. 17), and eland, perform similar but not identical movements. The other ungulates are more sedentary, but most show some local movement between habitats with change of season.

Paleoecology

The great climatic changes that affected temperate regions during the Pleistocene also affected East Africa, causing forests and savannahs to expand and retreat (Moreau 1966). And such changes may have contributed to the radiation of the ungulates. However, the overriding influence of the pleistocene volcanics in the crater highlands, the creation of alkaline soils in which trees could not grow, provided a relatively stable habitat for Serengeti animals. The essential features of climate, vegetation, and fauna have remained unchanged for millions of years.

Prior to the Miocene, Serengeti drainage was part of an ancient system that drained toward the west coast of Africa. Lake Victoria was formed in the Miocene by uplifting along the western rift valley, causing ponding of the ancient rivers. Hence, the climatic effects of the lake in the Serengeti region presumably date from that time (Kendall 1969).

The ecological data have come from the extensive paleontological and geological studies at two areas on the Serengeti plains (summarized by Hay 1976). The climate has remained semiarid, the prevailing winds have been from the east, the vegetation has been dry scrub or open grassland on alkaline and saline soils, and the fauna has had much in common with that of today, since the earliest evidence from 3.7 M years ago at Laetolil. This site is in volcanic debris just southwest of the crater highlands, which was the source of the material. Fauna was terrestrial and included elephants and other large mammals, and also the earliest footprints of a hominid (Mary Leakey, pers. comm.). The climate was moister than it is today. Evidence points to patches of riverine forest.

More extensive and continuous evidence (2.1 M years ago to the present) is found at Olduvai Gorge. Initially, a lake about 10 km long and 5 km wide occupied a basin there. At that time, the lake was saline and alkaline, with seasonally flooded marshes. Sediments were of volcanic ash from the east; the climate was dry, but locally there were patches of evergreen forest associated with the swamps. Further from the lake, the vegetation was savannah or grassland. The whole area must have been very similar to the present Lake Manyara Park.

As a result of successive faulting and sinking of land just to the east, the lake has become smaller and moved east; today a brackish lake forms intermittently in the Olbalbal depression at the southern end of the Salai plains—in 1977–78, it was full for the first time in several years. Climate appears to have become drier over time: from 2.1–1.7 Mya (million years ago) the climate was moister, perhaps similar to that of the western Serengeti today. Since then, conditions have changed to those resembling Olduvai now. However, there were periods of a few thousand years that were very dry, for example, at 1.65 Mya. During the last glaciation (c. 19,000 years B.P.), the climate was 8°C cooler and was also moister. Thus, there were seasonal fluctuations and longer term ones, but, on the whole, they remained within narrow limits.

In the earlier times (2.1–1.7 Mya), the mammal fauna consisted of elephants, rhinoceroses, horses, pigs, hippos, giraffes, and bovids. In the latter group, there were types resembling reedbuck and waterbuck, which were associated with the local lacustrine habitat, but there were also hartebeest and wildebeest species, which inhabit grasslands. In fact, the wildebeest specimen of that period, shown to me by J. M. Harris, has a broad palate similar to the present-day species, which indicates that it was even then a short-grass plains grazer. The fossil wildebeest did not have the pronounced nasal passages characteristic of modern animals, and resembled the modern topi or kongoni in that respect. Robertshaw and Taylor (1969) have found wildebeest do not sweat, but thermoregulate by panting through the nasal passages. Perhaps the increasing aridity in the past two million years provided the evolutionary pressure to develop water-saving methods of heat dissipation.

Toward the end of the early period (1.7 Mya), gazelles became more numerous in the deposits. These animals are typical of dry, open plains.

The rodent fauna is represented by modern genera and also reflects a trend toward aridity. A list of fossil birds from that period (Mary Leakey, pers. comm.) includes many water birds, but also doves, barbets, mouse-birds, and quail characteristic of savannah, and, more importantly, wheat-ears, which are entirely plains dwellers.

Between 1.7 and 1.15 Mya, conditions were drier than previously, but there were still marshes surrounded by savannah or plains similar to Lake Lagarja now (pl. 13). White rhinoceros (a grazer) was more abundant than the black rhinoceros (a browser). Other types included buffalos, elephants, equids, hippos, and crocodiles, but the most abundant were again wildebeests, hartebeests, and gazelles. Some time in the middle of this period, the modern wildebeest, *Connochaetes taurinus,* appeared (Gentry and Gentry 1977, 1978). Hence, this species has been living in a similar environment with similar associated herbivores for well over a million years. Furthermore, the climatic fluctuations were seasonal, suggesting that herbivores had to migrate in the past.

Also during that period, hominid types were much in evidence. Both *Australopithecus* and *Homo habilis* appeared together about 1.8 Mya, the latter staying to 1.5 Mya, the former extending to 1.25 Mya. At that point, *Homo erectus* appeared and remained to 0.7, or perhaps 0.5 Mya. Occupation sites appear to have been located along the lakeshore, and may have been seasonal.

During the last million years, there have been successive volcanic eruptions, and deposits of ash. The fauna has changed, with white rhinoceros and some buffalo types disappearing. Kendall (1969) has documented climatic changes over the past 15,000 years, based on analysis of Lake Victoria sediments. Periods wetter than present occurred in 12,500–10,500 years B.P. and 9,500–6,500 years B.P., with intervening dry periods. A fossil specimen of *Homo sapiens* in the Serengeti (Hay 1976) dated at 17,000 years B.P. indicates that man has been present in this area for a long time, but his major impact through the use of fire has largely been in the last 3000 years, and the present fire climax vegetation probably developed in this period.

Conclusion

The Serengeti ecosystem appears, on this evidence, to have been very similar to present conditions for a long time, at least a million years and probably much longer. The changes in climate that have occurred were not great and took place gradually. Consequently, the interrelationships between organisms that we see today have probably been highly adapted through a long period of evolution; they are part of a natural ecosystem.

References

Anderson, G. D., and Talbot, L. M. 1965. Soil factors affecting the distribution of the grassland types and their utilization by wild animals on the Serengeti Plains, Tanganyika. *J. Ecol.* 53:33–56.

de Wit, H. 1978. Soils and grassland types of the Serengeti Plain (Tanzania). Thesis (in prep.), Mededelingen Landbouwhogeschool.

Gentry, A. W., and Gentry, A. 1977, 1978. Fossil bovidae (Mammalia) of Olduvai gorge, Tanzania. *Bull. Brit. Mus.* (Nat Hist.), geol. ser. 29:289–446; 30:1–83.

Glover, P. E.; Trump, E. C.; and Wateridge, L. E. D. 1964. Termitaria and vegetation patterns on the Loita Plains of Kenya. *J. Ecol.* 52: 367–77.

Hay, R. L. 1976. *Geology of the Olduvai gorge.* Berkeley: Univ. of California Press.

Hendricks, H. 1970. Schatzungen der Huftierbiomasse in der Dornbusch-savanne nordlich und westlich der Serengetisteppe in Ostrafrika nach einem neuen Vefahren und Bemerkungen zur Biomasse der anderen pflanzenfressenden Tierarten. *Saugetier. Mitt.* 18:237–55.

Herlocker, D. 1975. *Woody vegetation of the Serengeti National Park.* College Station, Texas: Caesar Kleberg Research Program, Texas A & M University.

Herlocker, D. J., and Dirschl, H. J. 1972. *Vegetation of the Ngorongoro Conservation Area, Tanzania.* Canadian Wildlife Service Rep. ser., no. 19. Ottawa: Queens Printer.

Jager, Tj. 1979. The soils of the Serengeti woodlands. Thesis (in prep.), Wageningen.

Kendall, R. L. 1969. An ecological history of the Lake Victoria basin. *Ecol. Monogr.* 39:121–76.

Kreulen, D. 1975. Wildebeest habitat selection on the Serengeti Plains, Tanzania, in relation to calcium and lactation: a preliminary report. *E. Afr. Wildl. J.* 13:297–304.

Kruuk, H. 1972. *The spotted hyena.* Chicago: Univ. of Chicago Press.

Macfarlane, A. 1969. *Preliminary report on the geology of the central Serengeti, northwest Tanzania.* 13th Ann. Rep., Inst. Afr. Geol., Univ. Leeds. Pp. 14–16.

Milne, G. 1935. Some suggested units of classification and mapping, particularly for East African soils. *Soil Res.* 4:183–98.

Moreau, R. E. 1966. *The bird faunas of Africa and its islands.* New York: Academic Press.

Norton-Griffiths, M.; Herlocker, D.; and Pennycuick, L. 1975. The patterns of rainfall in the Serengeti ecosystem, Tanzania. *E. Afr. Wildl. J.* 13:347–74.

Pennycuick, L., and Norton-Griffiths, M. 1976. Fluctuations in the rainfall of the Serengeti ecosystem, Tanzania. *J. Biogeog.* 3:125–40.

Robertshaw, D., and Taylor, C. R. 1969. A comparison of sweat-gland activity in eight species of East African bovids. *J. Physiol.* 203:135–43.

Schaller, G. B. 1972. *The Serengeti lion.* Chicago: Univ. of Chicago Press.

Schmidt, W. 1975. The vegetation of the northeastern Serengeti National Park, Tanzania. *Phytocoenologia* 3:30–82.

Sinclair, A. R. E. 1977. *The African buffalo.* Chicago: Univ. of Chicago Press.

Thornthwaite, C. W. 1948. An approach toward a rational classification of climate. *Geogr. Rev.* 38:55–94.

Vesey-Fitzgerald, D. 1970. The origin and distribution of valley grasslands in East Africa. *J. Ecol.* 58:51–75.

S. J. McNaughton

Three Grassland-Herbivore Dynamics

Evolution of the plant family Gramineae, and the animal family Bovidae, are inextricably linked, with extraordinary adaptive radiation in both groups occurring contemporaneously from late Pliocene to late Pleistocene (Pilgrim 1939; Beetle 1955, 1959; Stebbins 1956; Wells and Cooke 1956; Cooke 1968). Discussions of coevolution at the plant-herbivore interface have placed considerable emphasis on the evolution of novel antiherbivore chemicals by plants and detoxification or avoidance mechanisms by animals (Fraenkel 1959; Ehrlich and Raven 1964; Dethier 1970; Levin 1973, 1976; Freeland and Janzen 1974). Mahadevan (1973) and Fenny (1975) distinguished between specific and general chemical resistance of plants to herbivores and pathogens. Chemicals conferring specific resistance are commonly highly toxic, novel organic molecules, with restricted taxonomic and morphological distribution. Chemicals conferring general resistance are weakly toxic, and are generalized in negative action, taxonomic distribution, and tissue distribution. Specific barriers to herbivory tend to be concentrated in young tissues and to decline with age; general barriers tend to accumulate with tissue age (Levin 1976). Although specific chemical barriers to herbivory are known in the Gramineae (Fraenkel 1959; Roe and Mottershead 1962; Simons and Marten 1971; Marten et al. 1973; Kendall and Sherwood 1975), the general chemical defenses of silicification, lignification, and phenolic accumulation are much more common in grasses (Esau 1960; Minson 1971; Todd et al. 1971; Jambunathan and Mertz 1973). Specific chemical barriers, in contrast, tend to be more widespread in the Dicotyledonae (Fraenkel 1959; Ehrlich and Raven 1964; Levin 1973, 1976).

A fundamental botanical difference between dicots and monocots, including grasses, is that the former grow from terminal meristems, which are highly vulnerable to destruction by large ungulates, while the latter grow from intercalary meristems that are less accessible to large herbivores (Branson 1953; Rechenthin 1956; Langer 1972). It has been known for fifty years that a principal evolutionary response of grasses to herbivory is selection for prostrate, rapidly growing genotypes (Stapledon 1928). Thus, morphological evolution leading to meristem protection by physical isolation and physiological evolution leading to compensatory growth have been major features of grass evolution.

The direct plant responses of compensatory growth and assimilate reallocation are widely studied by agronomists, botanists, foresters, and range managers but are less emphasized in the more fundamental plant-herbivore studies (Parker and Sampson 1931; Robertson 1933; Canfield 1939; Labyak and Schumacher 1954; Stein 1955; Ellison 1960; Brougham 1961; Leonard 1962; Jameson 1963; Pearson 1965; Neales and Incoll 1968; Hutchinson 1971; Vickery 1972; Harris 1974; Mattson and Addy 1975; Ryle and Powell 1975; Dyer and Bokhari 1976). It is well established in this literature that compensatory plant growth under certain conditions may make up for, or more than make up for, tissue reduction by herbivores. Mechanisms of this compensation include enhanced photosynthetic capacity; more efficient light use, due to reductions of mutual leaf shading; hormonal redistributions promoting tillering; leaf cell division and leaf cell expansion; a reduced rate of leaf senescence; nutrient recycling accompanying herbivory; and stimulatory effects of herbivore saliva (McNaughton 1979).

Because grasses have a substantial capacity for vegetative production, grazing-resistant genotypes of perennial species may be perpetuated indefinitely, in spite of the severe decrease in sexual reproduction that commonly accompanies defoliation (Archbold 1942; Sprague 1954; Laude et al. 1957; Roberts 1958; Jameson 1963; Stoy 1965). Harberd (1962) estimated, from clone growth rate and size, that clones of *Festuca rubra* in British grasslands were over 1,000 years old. Vegetative reproduction does not preclude evolution, since grasses are well known for intraplant variations in somatic chromosome number and a consequent potential for genetic modification due to somatic sectoring (Morey 1949; Nielsen and Nath 1961; Nielsen et al. 1962; Nielsen 1968). There is also evidence for somatic inheritance of tillering rate in asexually propagated clones (Breese et al. 1965).

Energy flow through the grazing food chain is higher in undisturbed ungulate-grass ecosystems than in any other terrestrial food web (Wiegert and Evans 1967; Golley 1971). East African ecosystems, in particular, are unique in the high biomass and diversity of large ungulates which they support (Stewart and Talbot 1962; Hendrichs 1970; Eltringham 1974). Since Africa has been a major site of ungulate adaptive radiation (Wells and Cooke 1956; Cooke 1968; Estes 1974), it is a particularly appropriate region for examining the dynamics of the grass-ungulate interface. Specifically, the Serengeti ecosystem is an ideal site for such studies since, as Talbot and Stewart (1964) observed, "The last known great concentrations of mixed species of plains wildlife . . . . in the world are found in the Serengeti-Mara region of Tanzania and Kenya." The objective of my studies is to obtain data on grassland-ungulate dynamics in this unique representative of a formerly widespread type of ecosystem.

Plant-herbivore interactions are characterized by short-term effects, such as modifications of primary productivity by herbivory, and long-term effects, such as changes in plant-community properties and in the genetic properties of plant populations in response to varying grazing regimes. I present data here on both aspects of Serengeti grassland-herbivore dynamics. Four general hypotheses are considered:

H_1: Seasonal rainfall patterns determine the spatial distribution of standing green biomass in the ecosystem.

H_{2a}: Moderate grazing will stimulate aboveground net primary productivity above control (ungrazed) levels.

H_{2b}: Heavy grazing will depress aboveground net primary productivity below control levels.

Taken together, these hypotheses predict an optimum defoliation level (fig. 3.1) at which the balance between residual leaf area and photosynthesis per unit of leaf area maximize net productivity (Dyer 1975; Noy-Meir 1975; Caughley 1976).

H_3: Species composition and structural properties of grassland communities in plots protected from grazing for extended periods will differ considerably from adjacent unprotected stands.

H_4: Genetic properties of plants of the same species from inside and outside protected areas will differ considerably, manifesting selection for traits influencing competitive ability and grazing resistance, respectively.

Research Methods

Two nondestructive methods of vegetation sampling were applied to the Serengeti grasslands: a radiance ratio method of measuring green biomass (Tucker et al. 1973; Pearson et al. 1976a, 1976b; Tucker 1977) and a canopy intercept method of measuring total biomass and the contri-

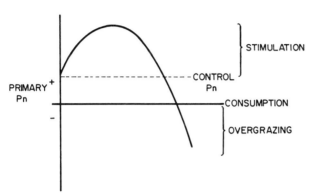

GRAZING MODEL

Figure 3.1   Hypothesis relating consumption by herbivores to net primary productivity of plants. The level of the control line (no grazing) will depend upon soil moisture, plant phenological stage, soil nutrients, and various other factors regulating plant growth.

butions of individual species to biomass. Both methods were calibrated on 0.5 m² oven-dried clip plots. In such a calibration, there are two alternative approaches. One can systematically pursue each possible variable in an attempt to isolate its effect on the method; or one can attempt to maximize the variance in calibration samples. For obvious reasons, I chose the latter approach. In the calibration plots, green biomass as a percentage of total ranged from 36 to 100, foliage height ranged from 4 to 109 cm, number of species in the plot ranged from 1 to 17, background soil color ranged from a pinkish granite to a black silt, measurements were taken in bright sun and under cloud cover and tree canopies, and the percentage of dicots in the sample ranged from 0 to 18 percent. For the radiance technique, a regression was established relating dry green biomass to the ratio of vegetation reflectances at 8000 and 6750 Å (fig. 3.2):

$$\text{green g/m}^2 = 109.8\ R - 83.8$$

where R is R_{8000}/R_{6750}. Open circles above the line represent a potentially

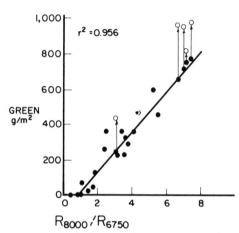

PHOTOMETER CALIBRATION
GREEN g/m² 109.8 R-83.8

Figure 3.2   Relationship between the ratio of vegetation reflectances at 8000 Å and 6750Å and oven-dried green biomasses of calibration clip plots. Open circles are for plots if nongreen stem were to be included in the calibration, a particularly important potential source of error at high biomasses, where the proportion of stems is generally higher.

serious source of error, that is, failure to separate green and nongreen biomass accurately in those species, such as *Pennisetum mezianum,* which have stems with a green cortex and colorless pith, or failure to separate sheaths from underlying nongreen stem. When sorting was careful, however, and only green biomass was measured, the residual error was less than 5 percent. I assume that this is primarily clipping and sorting error, and that the technique is an accurate method of measuring green-plant biomass.

Because nongreen biomass often is a significant portion of available forage in the Serengeti, and because the reflectance method does not separate the contributions of different species, a regression was established relating total oven-dried biomass to the contacts on a wire passed through the vegetation (fig. 3.3):

$$\text{total g/m}^2 = 6.3 + 16.93\, H$$

where H is the mean number of hits per pin in a sample of 5 pins passed through the vegetation at a 53 degree angle from the horizontal. Accuracy was somewhat less for this method, with the residual variance probably a result both of clipping error and error in counting hits; it is apparent that vegetation movement in the wind contributes substantially to the latter. Although a smaller pin angle would increase accuracy, the 53 degree angle was chosen as a compromise between sampling speed and an acceptable level of accuracy. A 15 percent error would be comparable to quadrat replication in my experience. The canopy intercept method also proved accurate in measuring grassland species composition: $\chi^2_{23} = 25.820$ ($P = $ not significant) in testing for whether species relative abundances

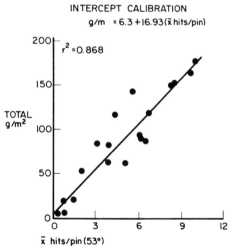

Figure 3.3

Relationship between mean number of hits per pin (using point frame analysis) when 5 pins at 53° angle were passed through calibration clip plots and total above-ground biomass of those plots.

measured by clipped and dried biomass were different from those measured by the number of contacts.

Subsequent to calibration of the reflectance technique on clip plots, a feasibility study was initiated to determine whether the method could be utilized from light aircraft as a means of measuring green standing forage on a total ecosystem basis. A series of twenty-five stands was measured on the ground and from low altitude (60–90 m) in a Piper Supercub aircraft. The area integrated by the photodiode of the photometer increases with altitude, so three ground measurements were taken from separate areas of each stand, and the mean of these was taken as the ground estimate. Aerial measurements explained a significant proportion of the variance in ground measurements (fig. 3.4), with the residual variance probably arising from

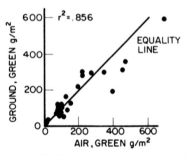

Figure 3.4   Relationship between aerial- and ground-estimated green biomass of grassland stands.

the error attached to estimating biomass of a large area from small sample plots in the ground measurements. Since the slope could not be distinguished from one ($t = .206$), nor the intercept from zero ($t = .034$), the initial calibration equation was applied to aircraft data. Because sample points for mapping seasonal green forage had to be concentrated along surfaces of rapid change, conventional methods of fitting mapping surfaces could not be used. Instead, I plotted values on a map, fitted isolines by eye, tested by correlation with actual values, and readjusted the lines until residual error was minimized.

To determine whether genetic properties of plants differed according to grazing regime, clones were transplanted into a fenced uniform garden at the Serengeti Research Institute. After several months of growth to allow acclimation to the common environment, these samples were scored for several traits.

Plate 1

Lengai, the only active volcano in the crater highlands. Ash from volcanoes produced the Serengeti plains. Edge of the Gregory Rift Valley is in the foreground.

Plate 2

Olduvai Gorge, famous for its fossil pleistocene fauna, including early man, cuts through the eastern Serengeti plains. Rim of Ngorongoro Crater is in the background.

Plate 3

Short-grass plains with herbs (*Solanum incanum*) in the foreground. Occasional outcrops of granitic rocks (*kopjes*) provide shade for predators.

Plate 4   Typical *Acacia* woodland in the central ranges of the Serengeti.

Plate 5   Wildebeest migrating into the woodlands from the plains. The umbrella-shaped *Acacia tortilis* is the dominant tree at the plains edge.

Plate 6

Typical woodland in the center and north. Note the dense patches of young *A. clavigera* trees, with a few mature trees.

Plate 7

Hill thickets in the foreground and background. Riverine forest lies along the Mara River (*center*). Note the abrupt transition between forest and grassland maintained by fire. Kenya Mara Park.

Plate 8

Floodplains on both sides of the lower Grumeti River in the western corridor. During the wet season, the river floods and forms swamps that last well into the dry season.

Plate 9   Wildebeest prefer to graze the short-grass plains.

Plate 10   Wildebeest give birth during a three-week period in February while they are on the plains.

Plate 11

Zebra, like the wildebeest, migrate to the plains and give birth there, but over a longer period.

Plate 12

Wildebeest grazing the long-grass plains. They form one herd numbering one and a-half million.

Plate 13

Wildebeest trying to cross Lake Lagarja in February 1973. Young calves that could not keep up drowned.

Plate 14

The amount of short regrowth grass in the northern Serengeti dry season determines the rate of wildebeest mortality through undernutrition.

Plate 15   Male buffalo in long grass.

Plate 16   Female and calf form the only family
unit in buffalo herds.

Plate 17   Thomson's gazelle browsing herbs on the short-grass plains.

Plate 18   Grant's gazelle browsing the herb *Indigofera basiflora*. Cattle egrets catch insects disturbed by the gazelle.

Plate 19

A family group of zebra in long
Themeda grass, northern Serengeti.
Isuria escarpment in the background.

Plate 20

Topi feeding in long grass. Note the long muzzle used for picking out grass leaves.

Plate 21

Kongoni (*right*) and topi (*left*) often form mixed feeding groups. Mara Park, Kenya.

Plate 22   Dik-dik, a small browser, lives as territorial pairs.

Plate 23   An impala herd responding in complete coordination to the sighting of a leopard.

Plate 24

A territorial male impala deposits secretion from his forehead onto a plant as olfactory marking of his territory.

Plate 25

A subordinate bachelor impala male (*right*) cautiously approaches, and sniffs at the forehead of a higher-ranking male in the bachelor herd. The subordinate is testing the secretion which, in territorial males, is used to mark territory.

Plate 26

The territorial males display during an encounter at their mutual boundary.

Plate 27

A territorial male shows Flehmen—testing for estrus—as he searches among the calmly feeding herd of females for one in estrus.

Plate 28

Lone nomad lion on the plains with a scavenged wildebeest carcass.

Plate 29

Lioness carrying one of her cubs, a few weeks old.

Plate 32

A clan of hyenas on the plains.

Plate 33

A pair of cheetah get a good vantage point while looking for gazelle.

Plate 34

Leopards hunt solitarily, and often
carry their prey into trees.

Plate 35   Wild dogs hunt in packs. Their numbers are declining on the plains.

Plate 36   Wild dogs attacking a zebra.

Plate 37 — Hyenas can steal a carcass from a few wild dogs . . .

Plate 38 — . . . but a larger pack of dogs can reverse the situation. Wildebeest trekking in the background.

Plate 39

A group of Ruppell's griffon and white-backed vultures on a wildebeest carcass.

Plate 40

Female giraffe prefer to browse
immature *Acacia* trees.

Plate 41

Giraffe browsing pressure keeps young trees pruned into rounded bushes. This is not unusual: the trees survive, and they are not "overbrowsed".

Plate 42

Giraffe browsing shapes trees into pyramid and hourglass shapes. Only the center of the tree is out of reach, so that it grows up and spreads out. Tree on left shows a lower line, created by wildebeest standing underneath for shade.

Plate 43

Elephant feeding on an adult *A. clavigera* tree they have pulled over.

Plate 44   Aerial photo of Bologonja spring and hill thickets in northern Serengeti, 1959.

Plate 45   The same area in 1968. Note the disappearance of the thickets that were originally within the white lines.

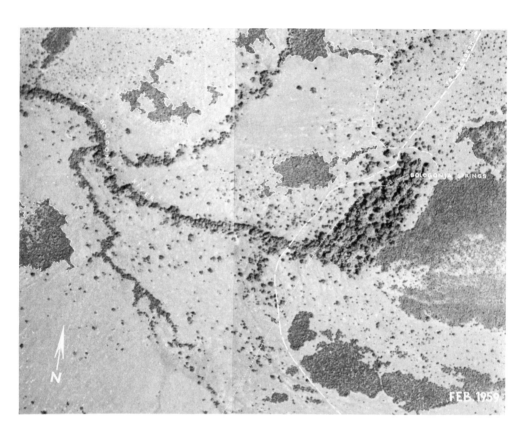

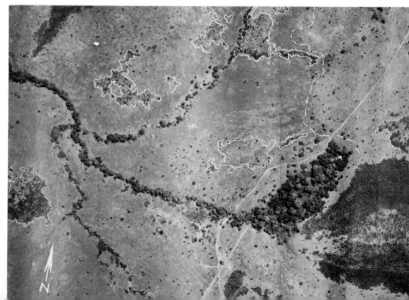

Results

The Seasonal Green Wave

Four aerial surveys of green forage in the Serengeti National Park and Masai Mara Game Reserve provided insight into seasonal and spatial dynamics of the grassland ecosystem. One survey was on 6 September 1974, during the middle of the dry season; one was on 4–5 December 1974, at the dry-wet seasonal transition; one was on 6 March 1975, during the middle of the wet season; and one was on 6 June 1975, at the wet-dry seasonal transition. The number of sample points ranged from 103 to 172, according to spatial complexity. Seasonal dynamics of green forage

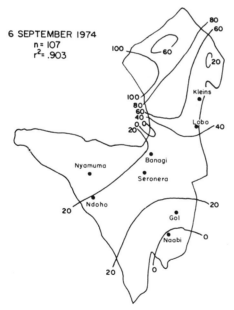

Figure 3.5

Isolines of green biomass on 6 September 1974. r^2 is the proportion of the variance of measured green biomass explained by isolines. Dots are locations of fenced areas sampled in other studies discussed later in this chapter (see table 3.2).

(figs. 3.5–3.8) reflected the general seasonal dynamics of rainfall (compare with figs. 2.1 and 2.2).

During the middle of the dry season (fig. 3.5), green biomass was low everywhere, with a distinct gradient in the northwest, reflecting the last residual high-quality forage available in the Mara River basin. The migratory wildebeest were concentrated in the north at the time of the survey, and the anomalously low values, below 20 g/m², in Masai Mara were associated with very dense wildebeest concentrations. The green wave at the beginning of the 1974–75 wet season was associated with the

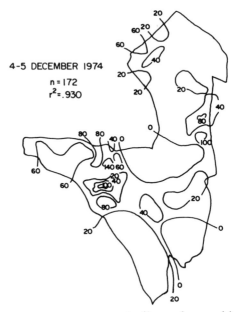

4-5 DECEMBER 1974

n = 172
r² = .930

Figure 3.6   Isolines of green biomass on 4–5 December 1974.

Lake Victoria Convergence Zone, generating a complex pattern from isolated showers centered on the corridor (Ndoho plains and Nyamuma; figure 3.6). The migratory wildebeest were very densely concentrated in the high rainfall centers in the western corridor and in another green patch around Lobo. The Serengeti plains were still extremely dry, and the northwest, which was the only area with significant green biomass in the September survey, was surprisingly dry in December. By the middle of the wet season, in March 1975, green biomass was highest on the Serengeti

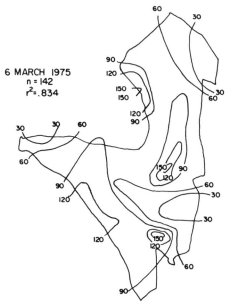

6 MARCH 1975
n = 142
r²=.834

Figure 3.7

Isolines of green biomass on 6 March 1975.

plains, an area east of Banagi Hill, and in the northwest (fig. 3.7). At this time, the migratory wildebeest were concentrated from the high green biomass area near Naabi Hill toward the eastern Serengeti plains. Even at this time, when there is generally high regional rainfall, aerial mapping indicated that the distribution of green forage was surprisingly patchy, with some areas of very low green forage in the northeastern part of the Masai Mara and the extreme western end of the corridor. Only at the wet-dry season transition in June 1975, did the distribution of green biomass generally reflect the overall annual rainfall distribution (fig. 3.8). There was a very steep gradient of increasing green forage from Lobo to Klein's Camp, and the migratory wildebeest were concentrated along this front. The highest green biomasses were, as in the mid–dry-season survey, in the northwest.

In addition to demonstrating the feasibility of monitoring range condition over large areas in a short time period, the periodic grassland monitoring revealed that the spatial and temporal patterns of green forage distribution were much less regular than an examination of mean rainfall

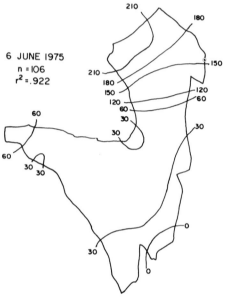

6 JUNE 1975

n = 106

r² = .922

Figure 3.8   Isolines of green biomass on 6 June 1975.

data might have indicated. Even in the middle of the wet season, there were substantial areas in the ecosystem where available high-quality forage was in very low supply. These maps document the highly stochastic nature of forage distribution throughout the ecosystem. The western corridor, in particular, is a highly unpredictable region in both time and space. As would be expected from the seasonal rainfall pattern there, animals are confronted with feast or famine conditions; high green biomass values were recorded there during the wet season, but these values fell essentially to nothing very early in the dry season and remained there for most of the period. The Serengeti plains also can vary from extremely dry, essentially zero green biomass, to the comparatively high values that were recorded in the middle of the wet season. Even the northwest varied radically from survey to survey, with green standing crop ranging from 20 to over 200 g/m². One of the remarkable features of the migratory wildebeest herd, of course, was its members' unerring ability to find these isolated regions of high green biomass. During the June, September, and

December surveys, the animals were always concentrated in the high green biomass regions. This was particularly evident in December, when the two high green biomass patches were approximately 80 km by air apart, and wildebeest were present on both. Although these animals are traditionally referred to as migratory, it seems more appropriate to recognize that they are nomadic, responding to highly stochastic rainfall and subsequent forage availability in an opportunistic fashion. A similar ability to respond to highly stochastic food availability has been reported for an African bird (Ward 1971; Katz 1974).

Short-Term Dynamics

Net aboveground primary productivity. At the beginning of the 1974 dry season, a large concentration of migratory wildebeest built up in Moru Kopjes, and an exclosure was established in a *Themeda-Pennisetum* grassland on one of the major migratory routes to the west (McNaughton 1976). Over the four-day period that wildebeest passed through the area, they had a substantial effect on the plant community (table 3.1), reducing green forage by 85 percent, height by 56 percent, and biomass concentration by 66 percent. Biomass concentration is reduced because of feeding selection for green leaf by the animals, with many culms remaining standing but stripped of leaves. Over the succeeding thirty-two days, however, net aboveground productivity was 2.6 g/m² per day in grazed areas, while green biomass declined by 4.9 g/m² per day in ungrazed areas as leaves continued to senescence in these postflowering grasses (fig. 3.9). Height declined significantly in the grazed areas as culms fell over, and regrowth was a consequence of vigorous tillering. Thus, a dense grazing lawn with a high biomass concentration was produced in grazed areas. At this time, Thomson's gazelles were entering the area in substantial numbers, and over the following thirty days, consumption by gazelle averaged 1.05 g/m² per day in areas previously grazed by wildebeest, but only 0.3 g/m² per day in areas not grazed by wildebeest ($t = 6.916$ for $P < .001$ with $d.f. = 9$). At the end of the dry season, in November 1974, gazelles were still significantly associated with wildebeest-grazed areas and avoided ungrazed stands (McNaughton 1976). The strong association of gazelles with areas previously grazed by wildebeest suggests that coevolution has partitioned grassland exploitation among these two ungulate species during the critical dry season, when food reserves fall below maintenance requirements of the animals (Sinclair 1975). Grazing by wildebeest facili-

Table 3.1   Effect of a four-day passage of
                                             migratory wildebeest on grasslands
                                             dominated by *Themeda* and
                                             Pennisetum on the western border
                                             of the Serengeti plains. The *F*
                                             values are from single contrast tests
                                             built into analyses of variance after
                                             it was determined that there were
                                             significant interaction terms in all
                                             analyses; degrees of freedom are
                                             1,21 in all cases; N.S., not signifi-
                                             cant. From McNaughton (1976).

Time	Biomass (g/m²)	Height (cm)	Biomass concentration (mg/10 cm²)
	Fenced vegetation, wildebeest excluded		
Before	501.9	64	7.9
After	449.2	63	7.1
	$F = 0.217$	$F = 0.010$	$F = 0.123$
	N.S.	N.S.	N.S.
	Vegetation subject to wildebeest grazing		
Before	457.2	66	6.9
After	69.0	29	2.4
	$F = 137.561$	$F = 25.94$	$F = 15.279$
	$P = .005$	$P = .005$	$P = .05$

tates energy flow into the gazelle population by stimulating grass growth,
and spatial partitioning of the grasslands by the two species reduces
competition.

This experiment demonstrated that grazing could stimulate net aboveground productivity; a more careful quantification of the pattern was
obtained in the northern Serengeti and Kenya Mara during the 1974 dry
season (fig. 3.10, top). As these data show, aboveground green biomass
declined in control stands protected from grazing, in spite of the fact that
there was considerable rainfall. During the period wildebeest were in
the area, rainfall was 126 mm, compared with a mean of 39 mm at the
rest of the study sites ($t = 4.989$ for $P < .001$ with $d.f. = 21$). There was
a significant relationship between the proportion of the initial green bio-

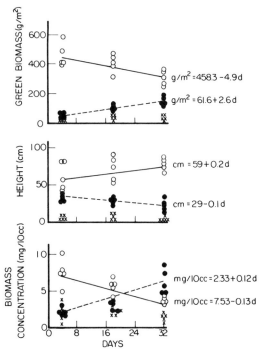

Figure 3.9

Dynamics of vegetation properties after passage of migratory wildebeest through the western Serengeti plains. Data are for enclosed areas protected from wildebeest grazing (*open circles*); unenclosed areas grazed by wildebeest (*solid circles*); and unenclosed areas showing evidence of trampling immediately after wildebeest passage (*crosses*). Lines are best-fit determined by the least squares method when t indicated a significant time trend at $P \leq 0.5$. Biomass concentration, expressed as milligrams of green forage per cubic centimeter, was calculated for a cylinder whose height is foliage height and whose area and biomass are as specified by the reflectrometric technique. From McNaughton (1976).

GRAZING MODEL TEST: WILDEBEEST

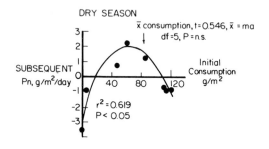

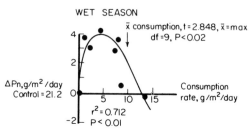

Figure 3.10

Relationship between grazing and net aboveground primary productivity. The top graph represents the dry-season grazing system measured during August and September 1974. Control values are negative because ungrazed grasses are drying out, subsequent to flowering. The bottom graph represents the wet-season grazing system measured during January 1975.

mass eaten by wildebeest and subsequent net aboveground primary productivity, confirming H_2 and the model of figure 3.1. The decline in aboveground green biomass of control stands is probably a consequence of substrate translocation to sites of crown and root storage and leaf senescence that commonly follow seed ripening. Heavy grazing reverses these processes and stimulates growth of shoots.

The wet-season grazing system on the Serengeti plains is decidedly different, since wildebeest enter the area when the plant community already has a substantial productivity potential. Although a superficially similar optimization curve was obtained, the basic form was different (fig. 3.10, bottom). As in the dry season, there was a significant promotion of productivity by moderate grazing, presented here as stimulation above control levels. The fact that the curve is assymetrical, being skewed to the right, indicates that the wet-season grazing areas are stimulated more by light grazing and are more resistant to overgrazing than grasslands in the dry-season concentration areas.

In addition to stimulating growth, wet-season grazing suppresses flowering. At the end of a brief rainy period on the Serengeti plains during January 1975, mean number of flowering culms per m² ranged from 12.5 to 37.5 inside exclosures. Outside, culm density ranged from 1 to 9 per m². All comparisons were significantly different ($P < .05$ to $P < .001$). This suppression of flowering is probably a result of diversion of assimilates from floral differentiation to leaf growth requirements generated by heavy cropping (Jameson 1963; Stoy 1965).

Extremely high short-term productivities were recorded on the Serengeti plains during the wet season. The highest value recorded was 40 g/m² per day, and the values consistently were above 20 g/m² per day. These are higher than any other values I have seen for native or managed grasslands, and suggest that the Serengeti grasslands may be among the most productive in the world. To a considerable extent, this high aboveground productivity is a consequence of grazing stimulation of grass growth.

Long Term Dynamics

Community organization and grazing ecotypes. There commonly is a pronounced diversion of carbohydrates from roots following defoliation (Gifford and Marshall 1973; Ryle and Powell 1975), and it has been observed frequently that defoliation reduces root growth (Crider 1955; Oswalt et al. 1959). Although aboveground growth of Serengeti grasses may be at the expense of stored root reserves to some extent, many of the Serengeti grasses may be described as obligate grazophils. I found, when reexamining the areas fenced in by Watson in 1963, (1967) that species composition had changed radically in many of them. In the one south of Naabi Hill, for instance, *Andropogon greenwayi* and *Sporobolus marginatus* made up 56 percent and 20 percent, respectively, of the biomass outside the exclosure, but were not recorded in samples taken inside. *Pennisetum stramineum* and *P. mezianum,* in contrast, made up 5 percent and 3 percent, respectively, of the biomass outside, but 72 percent and 26 percent inside. Maintenance of this grassland's species composition is clearly grazing-dependent. The species which presently dominate these grasslands are those able to sustain sufficient leaf area under most grazing intensities to prevent substantial root-reserve depletion. Those that come to dominate the protected areas are stemmy species capable of overtopping less erect species and thereby eliminating the latter by competition for light.

Examination of four of these exclosures revealed that grassland organization and structural properties changed radically due to protection from grazing (table 3.2). Diversity had declined significantly, due to an increase in dominance and the associated reduction in equitability. Of course, height was much greater inside the exclosures but, perhaps surprisingly, the amount of green foliage did not differ. Rather, there was a large accumulation of dry foliage and a doubling of the proportion of plant biomass invested in stems inside. I suggest that there is an adaptive trade-off among the species between ability to sustain sufficient leaf area and photosynthetic potential under intense grazing, facilitated by a prostrate growth form; and height growth as a light-competition mechanism,

Table 3.2   Summary properties of grasslands inside and outside exclosures built in early 1960s when resampled in March 1975. Locations of exclosures were Naabi-South; Gol Kopjes-South; Gol Kopjes track-East; and Gol Kopjes track-West.

Property	Location		$t(df = 7)$	P
	Inside	Outside		
Diversity (H')	1.040	1.617	2.544	.05
Richness (S)	7.3	8.0	.728	N.S.
Dominance (DI)	86.8	60.6	2.402	.05
Equitability (J)	.528	.784	2.417	.05
Foliage Height (cm)	38.3	7.2	6.197	.001
Biomass				
green (g/m²)	58.8	64.4	.403	N.S.
dry (g/m²)	345.1	19.8	6.213	.001
proportion stems	.344	.171	2.529	.05

facilitated by investment in stem production. This trade-off in the plants is balanced against the strong selective force that the grazing ungulates constitute. Due to heavy grazing, prostrate, low-grazing species like *A. greenwayi* dominate the grasslands in this area. Species like the *Pennisetums*, with a heavy investment in stems, are very susceptible to damage by grazing, but win at light competition if grazing is reduced.

To determine whether species which occurred both inside and outside these exclosures were represented by genetically distinct individuals, clones of four species were collected inside and outside the south Gol Kopjes exclosure. The species collected were *Sporobolus marginatus* and *Cynodon dactylon,* which were relatively common both outside and inside, and *Themeda triandra* and *Harpachne schimperi,* which were common inside but rare outside. After eight months in the common environment, a major difference between genotypes collected inside and outside the exclosures was that outside genotypes were dwarfed in comparison with those collected from inside. This was quantified by measuring internode and leaf lengths (table 3.3). Only *S. marginatus* was similar regardless of origin. Internodes and leaves of *C. dactylon* genotypes from outside the exclosure were significantly smaller than those of inside genotypes. *T. triandra* genotypes from outside had much shorter internodes, although leaf length was similar. In *H. schimperi,* in contrast, leaves were much shorter in outside genotypes, but the internode lengths were similar. Another conspicuous difference in genotypes of this species was the orientation of culms, which were much more prostrate in genotypes from outside the exclosure. With the exception of *S. marginatus,* then, the general difference between genotypes from protected and unprotected communities was conspicuous dwarfing and prostrate growth forms in the grazed areas. This would be expected from previous studies of the evolution of grasses in different grazing regimes (Stapledon 1928).

Summary Of Results

Returning to the four broad hypotheses stated earlier, research in the Serengeti ecosystem indicates the following:

1. Although green forage reflects the general pattern of rainfall over the ecosystem, the periodic aerial surveys reveal that forage distribution is highly stochastic in time and space. Green regions are commonly of restricted distribution and isolated from one another. Only early in the dry season, as the grasslands are beginning to dry out, did the forage distribution pattern reflect the broad regional rainfall pattern.

Table 3.3   Properties of genotypes of four
                                             grass species collected inside and
                                             outside the exclosure at Gol
                                             Kopjes-South, after eight-months
                                             growth in a uniform garden at
                                             Seronera.

Species	Origin		$t\ (df=4)$	P
	Inside	Outside		
	Internode length (cm)			
Sporobolus marginatus	6.13	6.27	.084	N.S.
Cynodon dactylon	8.12	3.47	7.409	.001
Harpachne schimperi	3.56	2.71	1.659	N.S.
Themeda triandra	17.07	7.87	10.974	.001
	Leaf length (cm)			
Sporobolus marginatus	7.73	9.62	.861	N.S.
Cynodon dactylon	4.26	2.08	3.605	.05
Harpachne schimperi	9.20	4.27	3.656	.05
Themeda triandra	9.54	9.37	.100	N.S.
	Culm angle (° from horizontal)			
Harpachne schimperi	54.4	27.5	3.305	.05

2. Grazing stimulates grassland productivity substantially, and an optimum curve was validated. However, the form of the curve differs in the wet- and dry-season animal concentration areas. In the dry-season range, aboveground green biomass declines spontaneously in the absence of grazing, in spite of heavy rainfall, presumably as a consequence of postflowering senescence. Grazing reverses this process and stimulates aboveground productivity. In the wet-season occupance areas, the grasslands have a substantial control net productivity, and moderate grazing stimulates growth above this level. The wet-season curve is less symmetrical than the dry-season curve, suggesting greater stimulation by moderate grazing and greater resistance to heavy grazing.

3. Releasing the grasslands from grazing by fencing has resulted in conspicuous changes in community species composition and organization. In particular, the prostrate, low-growing species that dominate present grasslands are replaced by taller, stemmier species when grazing is removed. There is an accumulation of dry forage inside the exclosures, and grassland diversity decreases, presumably because the taller-growing spe-

cies "win" in the competition for light, reducing equitability dramatically.

4. Species that occur both inside and outside the permanent exclosures were represented by different genotypes in the two areas. Grazing genotypes were conspicuously dwarfed, a consequence of various combinations of shorter internodes, smaller leaves, and more prostrate culms.

Preliminary Plant Growth and Allocation Model

A simplified model (Ledig 1969; Molz 1971; Evans 1972; Risser 1972; Ares and Singh 1974; Hunt and Parsons 1974; Lewin and Lomas 1974; Noy-Meir 1975; Thornley 1976) of six compartments, relating three plant compartments to principal regulating factors in the Serengeti ecosystem, has been developed for simulation purposes (fig. 3.11).

Shoots are the principal donor compartment, upon which other compartments—roots, grazers, and reproduction—all draw. Stems are not included as a separate compartment, since stems in these grasses are either green and presumably photosynthetic, or they are flowering culms and therefore are put in the reproduction compartment. The system is driven by rainfall through the soil-moisture compartment. Soil moisture promotes shoots, and shoots reduce soil moisture. Roots promote shoots, and shoots either promote or reduce roots, as described below, according to growth stage. Grazing reduces shoots, and shoots promote grazers. Shoots promote reproduction, and reproduction limits shoots. Nutrients promote shoots, roots, and reproduction, depending on phenological stage, soil moisture, and grazing intensity and, in turn, are depleted by growth of

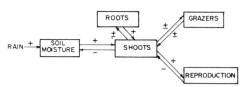

Figure 3.11

Compartmental model of grassland community in relation to grazers and soil moisture. + indicates a promotive effect, − a negative effect.

the plant compartments. All these relationships are qualitatively consistent with the information already discussed. Quantification of the model, for simulation purposes, proceeded as follows:

A. Soil moisture sets the limits on equilibrium shoot biomass by

$$B_t \ (g/m^2) = 1000 \ (S)^{-.9} \tag{1}$$

where S is soil water potential in -bars measured in the major rooting zone. This was established empirically in the 1976 field season. Ungrazed stands were assayed at the beginning of the dry season, when green biomass was driven by declining soil moisture. I found that ($r = -.916$, $P < .001$)

$$B_t \ (g/m^2) = 1090 \ (S+1)^{-.87}$$

Equation 1 was set from this relationship.

B. Net productivity was a skewed function of shoot biomass, of the type suggested in the results discussed above. I used

$$P_n \ (g/m^2 \cdot day) = 4 + 19 \log_e B_t - 3 \log_e B_t^2 \tag{2}$$

as an average relationship from heavily utilized wet-season occupance areas. Note that this is a descriptive approach, not a process-oriented model (Thornley 1976). One of the major differences between these areas and the tall and mid-grasslands that are lightly used during the wet season is that the productivity curve is more symmetrical in the latter and the peak is much lower.

C. Defining the soil moisture limit from equation 1 as

$$1 - B_t/1000 \ (S)^{-.9}$$

and combining this with equation 2 gives a productivity equation of

$$Pn = (4 + 19 \log_e B_t - 3 \log_e B_t^2) \ [1 - B_t/1000(S)^{-.9}] \tag{3}$$

This assumes that productivity may be limited intrinsically by the supportable canopy structure and extrinsically by soil moisture. If the soil is saturated ($S = 1$), equilibrium B_t (in equation 2) is 690 g/m², quite close to values observed in exclosures in the short-grass region after extended periods of heavy rain. If, however, $S = 5$, equilibrium B_t drops to 235 g/m². A particular feature of the dynamics of equation 3 is that soil moisture has a minimal effect upon the initial slope and peak height of the productivity curve, but modifies productivity beyond the peak drastically.

D. I assume that root Pn is a proportion of total Pn of the type

$$p = b \log_e (B_t + 1) - a \qquad (4)$$

where p is the proportion of total Pn translocated to the roots, b is a number less than 1, and a is the maximum root "draw down" if all shoot growth is at the expense of root substrates, that is, when growth starts from $B_t \to 0$, as it would at the beginning of the wet season, or when grazing is severe. I have not yet determined what the values of a and b are, and part of my future research is designed to evaluate them empirically. By approximation to field data from the high occupance areas, I used

$$p = 0.4 \log_e (B_t + 1) - 1.7 \qquad (5)$$

and let

$$P_n r = P_n(p) \qquad (6)$$

where $P_n r$ is root productivity in g/m² per day. Then, applying this partitioning to equation 3, shoot productivity becomes $P_n s = P_n$ if $p \le 0$, $= P_n - P_n r$ if $p > 0$. All shoot growth is at the expense of roots when shoot biomass is low, there is a sharp deletion of roots early in growth, and there is a crossover point beyond which root reserves are replenished (Thornley 1976). I suggest that the form of this curve is a critical determinant of plant species success in these grasslands and their response to grazing.

E. Shoot depletion of soil moisture is a variable dependent upon soil properties, plant properties, wind, canopy geometry, soil moisture, and other variables. For simulation purposes, I let

$$\Delta S = .01 \, B_t \qquad (7)$$

where ΔS is the rate of soil water depletion in bars per day. Although I assume a strict linear function, a decelerating exponential might be more realistic for obvious reasons; as written, it probably accentuates soil moisture depletion relative to reality. In addition, a soil moisture feedback to stomatal resistance would be required for meaningful mechanistic description.

F. Finally, I let animals graze in relation to the crossover point of equation 6: $C = 0$ if $p < .1$, $= P_n s$ if $p \ge .1$, where C is consumption in g/m² per day. This essentially argues that the grazers, on average, graze at a point above where root reserves are seriously depleted and that, beyond this point, they keep B_t constant.

G. At this stage, reproduction is ignored, since these grasses flower only sporadically under the existing grazing regime. A reasonable addition would be to set an allocation of P_n to reproductive structures relative to B_t of the type proposed for root reserves. Field data are available to approximate this term.

H. Also ignored at this time were potential nutritional limitations.

Time dynamics of three runs at three different constant soil-water potentials suggest that the ability of these grasses to replenish root reserves may be strongly dependent upon equilibrium soil-water potentials (fig. 3.12). Root reserves were rapidly replenished and the shoot compartment grew rapidly when the soil was water-saturated. At a water potential of -5 bars, root reserves were replenished after twenty-five days, and the shoot compartment was approaching a steady-state size after forty days. At the -10 bars simulation, however, shoot biomass reached a steady state after about thirty days, and root reserves were not replenished.

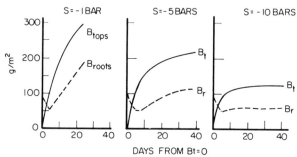

Figure 3.12   Simulation of shoot (B tops) and root biomass (B roots) dynamics for forty-day periods at constant soil-water potentials in the absence of grazing (B = biomass)

A more reasonable simulation is to let soil moisture vary; I used a five-day rain interval when S went to 1, and used equation 7 as a water-depletion term. Simulation dynamics were strikingly different in grazed and ungrazed treatments (fig. 3.13). Root reserves were replenished much more rapidly under grazing because net productivity was arrested at a relatively high level, creating a larger assimilate flow to roots. In the ungrazed simulation, soil moisture was depleted so rapidly by the higher shoot biomass that roots approached an equilibrium value lower than that at the beginning of growth. In the grazed simulation, consumption kept

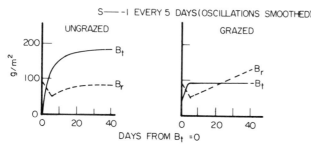

Figure 3.13   Simulations of shoot (B_t) and root (B_r) biomass dynamics for forty-day periods of alternate dry and rainy periods specified by making soil-water potential go to -1 bar every five days. On the left is the simulation without grazing; on the right is the simulation with grazing.

shoot biomass lower, reducing the rate of soil-water depletion and allowing root reserves to be replenished after about twenty-five days.

More interesting than the time dynamics of such simulations, however, is their ability to predict field situations they were designed to model. One such prediction is of particular interest because it summarizes much of the dynamics of the system, that is, grazing intensity under different rainfall sequences. I here define grazing intensity as $1\text{-}g/ug$, where g is plant biomass in exclosures, and ug is biomass in unfenced areas. I have data on six study sites where the interval between sequential rainfalls varied. For five of the stands, all relatively intensely grazed, observed values were close to those from the simulation (fig. 3.14). For another, relatively lightly grazed stand, however, the observed grazing intensity was significantly lower than the predicted value. This suggests the perfectly reasonable modification of making consumption variable in relation to shoot biomass, soil water, and the consequent net productivity of the grassland.

This is a comparatively simple model of limited scope, designed more to provide a framework for formulating hypotheses about grass growth dynamics than to provide a predictive model for the Serengeti ecosystem. One thing which it does emphasize, however, is an important gap in our understanding of grassland dynamics in the Serengeti: our total lack of

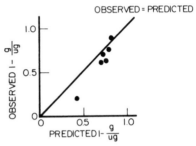

Figure 3.14   Relationship between observed and
predicted values of the depression
of green biomass below control
levels when the interval between
rains varied.

knowledge of root dynamics. I suggest that the form of the root draw
down–replenishment curve (equation 4) is a critical feature of grass
physiology determining species and genotype success in these grasslands.
Species like those of *Pennisetum,* which were of low abundance under
grazing, but became very abundant when grazing was reduced by fencing,
probably have higher crossover points for equation 4 and suffer fatal root-
reserve depletion under the intense grazing of unfenced grasslands. In
addition, selection for prostrate, low-growing genotypes would produce
a growth form less accessible to grazers, allowing the crossover point to
be reached even under the prevailing grazing regime. The model also
suggests an important interaction between soil moisture and grazing which
influences the balance between roots and shoots. Grazing, so long as it is
above the crossover point for root-reserve repenishment, may actually
facilitate substrate flow to roots by maximizing net productivity and
reducing the rate of soil-water depletion.

Dynamics of the model are reasonable, but insufficient data are presently in hand to allow it to be regarded as a realistic description of the
Serengeti grasslands. Among the data needed to develop such a predictive
model, I identify the following as particularly significant: (1) the root
draw down–replenishment curve for different grassland types; (2) the
relationship between soil-moisture depletion and shoot biomass for different grassland types, including a feedback from soil moisture to depletion rate mediated by plant stomates; (3) evaluation of herbivore
consumption rate as shoot biomass and soil moisture balances vary; and

(4), information on nutrient cycling and depletion rates in relation to consumption and plant compartment growth, respectively.

Plant-Herbivore Strategies

The fundamental paradox of grassland-herbivore dynamics in the Serengeti ecosystem is the extent to which coevolution of grasses and ungulates has superimposed a deterministic functional system on an inherently stochastic environment. Rainfall is the principal extrinsic variable driving system function, but it is extremely variable in time and space, as chapter 2 documents. In the grasslands, this is reflected in extreme temporal and spatial patchiness of green forage. Green areas commonly are widely separated in space at any one time, and are highly unpredictable at any one place from time to time. In the face of this stochastic variation, it is apparent why wildebeest so dominate the grazing fauna: their nomadic behavior allows them to exploit widely separated bursts of grassland productivity wherever they occur, and the dense herding behavior allows them to crop the grasslands in a way that increases forage yield. Their ability to do this, however, also is clearly a consequence of selection for grazing resistance in Serengeti grasses; few cultivated grasses would be able to sustain significant levels of net productivity under the severe defoliation regimes characteristic of this grazing system (Milthorpe and Davidson 1965; Davidson and Milthorpe 1966a, b). Thus, mobility of the major grazing animals, and ability of the grasses to sustain themselves under the present grazing regime, account for the ability of the Serengeti to support an unparalleled grazing fauna. A critical feature of grazing resistance undoubtedly is ability to maintain a high enough shoot biomass to maintain root reserves.

Various mechanisms may be adduced to explain why productivity of these grasslands is stimulated by grazing. First, the rate of photosynthesis commonly is inhibited by the accumulation of photosynthetic storage products, particularly starch, in actively photosynthesizing leaves. Increasing the drain on leaf substrates by promoting translocation to sinks elsewhere in the plant commonly results in increased photosynthetic rates (Neales and Incoll 1968; Thorne and Koller 1974). Partial defoliation, by increasing assimilate demand in meristems of remaining shoots, usually stimulates photosynthetic rate per unit of remaining leaf tissue (Wareing et al. 1968; Gifford and Marshall 1973). In addition to the direct stimulatory effect of defoliation upon photosynthetic capacity of the remaining canopy, defoliation may allow more efficient light utilization by reducing mutual leaf shading (Hughes 1969; Heslehurst and Wilson 1971; Robson 1973). In particular, since large ungulates graze upon grasses from above, and growth is primarily from basal intercalary meristems, the older and less efficient tissues are preferentially removed, leading to greater light intensities on younger, previously shaded tissues (Rechenthin 1956; Jameson 1963; Langer 1972).

There often is a pronounced diversion of carbohydrates from roots following defoliation (Gifford and Marshall 1973; Ryle and Powell 1975), and it has been observed frequently that defoliation reduces root growth (Crider 1955; Oswalt et al. 1959). However, it also is apparent that the Serengeti grasses must sustain sufficient leaf area to prevent fatal depletion of root reserves, and the simulation model suggests that root productivity might be enhanced by moderate grazing. Clearly, root dynamics are a critical aspect of Serengeti grassland ecology about which we know nothing. In addition to influencing the balance between shoots and roots, flowering and fruit set are commonly inhibited following defoliation (Archbold 1942; Sprague 1954; Laude et al. 1957; Roberts 1958; Stoy 1965), and my studies indicate that such inhibition occurs in the wet-season occupance areas subject to heavy grazing whenever it rains.

Defoliation substantially modifies hormonal balance within the plant, particularly in the balance of growth regulating hormones produced in the root and translocated to the shoot (Weiss and Vaadia 1965; Meidner 1967; Pallas and Box 1970; Torry 1976). A greater flow of growth-promoting hormones to residual meristems following defoliation promotes cell division and enlargement and activity in quiescent meristems—additional mechanisms of compensatory growth accompanying herbivory. A longer-term effect of modifications of hormonal equilibria following defoliation is a reduction in the rate of photosynthesis decline with leaf aging, a factor that insures the maintainance of the assimilatory capacity of residual leaf tissue at higher levels over a longer time period (Richmond and Lang 1957; Woolhouse 1967; Neales et al. 1971; Gifford and Marshall 1973).

Additional longer-term effects of defoliation may account for productivity increases following grazing. First, soil water may be conserved, due to reduction of the transpiration surface (Jameson 1963) and because photosynthetic rate increases are associated primarily with reductions in mesophyll resistance rather than stomatal resistance (Gifford and Marshall 1973; Thorne and Koller 1974). Second, growth stimulation by nutrients recycled from dung and urine is so well known that it warrants little additional comment (Peterson et al. 1956; Lotero et al. 1966; Weeda

1967). Here again, however, is an extremely important aspect of eco-system function, particularly in an area like the Serengeti, where ungulate biomass is high and soil nutrient pools are low, about which nothing is known. Finally, one potential direct stimulatory effect of grazing rumi-nants upon grass productivity may arise out of plant-growth-promoting agents in saliva (Vittoria and Rendina 1960; Reardon et al. 1972, 1974; Dyer and Bokhari 1976). Yield increases of up to 50 percent above con-trol levels have been recorded when ruinant saliva was added to clipped plants.

Conclusion: A Botanical Preservationist Coda

The Serengeti is universally recognized as the world's premier national park. However, this recognition is based almost solely on its unparalleled large-mammal fauna. And although I never saw migratory wildebeest walking single file silhouetted against the sky without feeling sadness for mankind's failure to preserve fragments of the once-great North American and Eurasian grazing ecosystems, I am convinced that one of the great resources of this ecosystem is its pool of grazing-adapted plant genotypes.

Considerable emphasis has been placed recently on the importance of maintaining stocks of plant germplasm in the face of rapid destruction of natural ecosystems throughout the world (Fraenkel and Hawkes 1975; Harlan 1975; Myers 1976). Much of the high-quality protein in the human diet comes from grazing ecosystems, and although substantial stocks of crop plants, such as small grains, are maintained, less effort has been made to identify and preserve forage genotypes. The Serengeti is one of the few remaining vestiges of a naturally operating grazing system. My studies document the astonishing productivity potential and grazing re-sistance of the plant genotypes there, and suggest how closely their preser-vation is tied to maintenance of the present grazing system. I believe conservation of the plant genotypes at the base of this remarkable food web will be one of the important benefits for future generations to be derived from preserving this great natural area. The genetic storehouse of grasses presently in the Serengeti may one day be a major resource for increasing agricultural productivity of the earth's semiarid tropical regions.

The author is grateful to the board of trustees and the director of Tanzania National Parks for allowing him to live and work in the Seren-geti National Park, to Dr. T. Mcharo, former director of the Serengeti

Research Institute, for inviting him to work there, and to the governments of Kenya and Narok District for permission to work in Masai-Mara Game Reserve. Valuable assistance was provided by Margaret, Sean, and Erin McNaughton. R. H. V. Bell convinced me to do the aerial green biomass surveys and he, J. J. R. Grimsdell, and R. Pellew kindly served as pilots. This research was supported by grants BMS 74–02043, DEB 74–20043 (A02), and DEB 77–20360 from the U.S. National Science Foundation Ecosystem Studies Program to Syracuse University.

References

Archbold, H. K. 1942. Physiological studies in plant nutrition. Part 13. Experiments with barley on defoliation and shading of the ear in relation to sugar metabolism. *Ann. Bot.* 6:487–531.

Ares, J., and Singh, J. S. 1974. A model of the root biomass dynamics of a shortgrass prairie dominated by blue grama (*Bouteloua gracilis*). *J. Appl. Ecol.* 11:727–43.

Beetle, A. A. 1955. The four subfamilies of the Gramineae. *Bull. Torrey Bot. Club* 82:196–97.

———. 1958. *Piptochaetium* and *Phalaris* in the fossil record. *Bull. Torrey Bot. Club* 85:179–81.

Branson, F. A. 1953. Two new factors affecting resistance of grasses to grazing. *J. Range Manage.* 6:165–71.

Breese, E. L.; Hayward, M. D.; and Thomas, A. C. 1965. Somatic selection in perennial ryegrass. *Heredity* 20:367–79.

Brougham, R. W. 1961. Factors limiting pasture production. *Proc. N. Z. Soc. Anim. Prod.* 21:33–46.

Canfield, R. H. 1939. The effect of intensity and frequency of clipping on density and yield of black grama and tobosa grass. *USDA Tech. Bull.,* no. 681.

Caughley, G. 1976. Plant-herbivore systems. In *Theoretical ecology: principles and applications,* ed. R. M. May, pp. 94–113. Philadelphia: W. B. Saunders.

Cooke, H. B. S. 1968. Evolution of mammals on southern continents. Part 2. The fossil mammal fauna of Africa. *Quart. Rev. Biol.* 43: 234–64.

Crider, F. J. 1955. Root-growth stoppage resulting from defoliation of grass. *USDA Tech. Bull.,* no. 1102.

Davidson, J. L., and Milthorpe, F. L. 1966a. Leaf growth in *Dactylis glomerata* following defoliation. *Ann. Bot.* 30:173–84.

————. 1966b. The effect of defoliation on the carbon balance in *Dactylis glomerata. Ann. Bot.* 30:185–98.

Dethier, V. G. 1970. Chemical interactions between plants and insects. In *Chemical ecology,* ed. E. Sondheimer and H. B. Simeone, pp. 83–102. New York: Academic Press.

Dyer, M. I. 1975. The effects of red-winged blackbirds (*Agelaius phoeniceus* L.) on biomass production of corn grains (*Zea mays* L). *J. Appl. Ecol.* 12:719–26.

Dyer, M. I., and Bokhari, U. G. 1976. Plant-animal interactions: studies of the effects of grasshopper grazing on blue grama grass. *Ecology* 57: 762–72.

Ehlich, P. R., and Raven, P. H. 1964. Butterflies and plants: a study in coevolution. *Evolution* 18:586–608.

Ellison, L. 1960. Influence of grazing on plant succession of rangelands. *Bot. Rev.* 26:1–78.

Eltringham, S. K. 1974. Changes in the large mammal community of Mweya Peninsula, Rwenzori National Park, Uganda, following removal of hippopatamus. *J. Appl. Ecol.* 11:855–66.

Esau, K. 1960. *Anatomy of seed plants.* New York: Wiley.

Estes, R. D. 1974. Social organization of the African Bovidae. In *The behaviour of ungulates and its relation to management,* ed. V. Geist and F. Walther, N. S. no. 24, pp. 166–205. Morges, Switzerland: I.U.C.N.

Evans, G. C. 1972. *The quantitative analysis of plant growth.* Oxford: Blackwell.

Feeny, P. P. 1975. Biochemical coevolution between plants and their insect herbivores. In *Coevolution of plants and animals,* ed. E. Gilbert and P. H. Raven, pp. 3–19. Austin, Texas: Univ. of Texas Press.

Fraenkel, G. S. 1959. The *raison d'etre* of secondary plant substances. *Science* 129:1466–70.

Fraenkel, O. H., and Hawkes, J. C., eds. 1975. *Plant genetic resources for today and tomorrow.* Cambridge: Cambridge Univ. Press.

Freeland, W. J., and Janzen, D. H. 1974. Strategies in herbivory by mammals: the role of plant secondary compounds. *Amer. Nat.* 108:269–89.

Gifford, R. M., and Marshall, C. 1973. Photosynthesis and assimilate distribution in *Lolium multiflorum* Lam following differential tiller defoliation. *Aust. J. Biol. Sci.* 26:517–26.

Golley, F. B. 1971. Energy flux in ecosystems. In *Ecosystem structure and function,* ed. J. A. Wiens, pp. 69–80. Corvallis, Ore.: Oregon State Univ. Press.

Harberd, D. J. 1962. Some observations on natural clones in *Festuca ovina*. *New Phytol.* 61:85–100.

Harlan, J. R. 1975. Our vanishing genetic resources. *Science* 188:618–21.

Harris, P. 1974. A possible explanation of plant yield increases following insect damage. *Agro-ecosystems* 1:219–25.

Hendrichs, H. 1970. Schatzungen der Huftierbiomasse in der Dornbusch-Savanne nordlich und westlich der Serengeti in Ostafrica nach einem neuen Verfahren und Bemerkungen zur Biomass der anderen pflanzen-fressenden Tierarten. *Saugetier. Mitt.* 18:237–55.

Heslehurst, M. R., and Wilson, G. L. 1971. Studies on the productivity of tropical pasture plants. III. Stand structure, light penetration, and photosynthesis in field swards of *Setaria* and green leaf *Desmodium*. *Aust. J. Agric. Res.* 22:865–78.

Hughes, A. P. 1969. Mutual shading in quantitative studies. *Ann. Bot.* 33:381–88.

Hunt, R., and Parsons, I. T. 1974. A computer program for deriving growth-functions in plant growth-analysis. *J. Appl. Ecol.* 11:297–307.

Hutchinson, K. J. 1971. Productivity and energy flow in grazing-fodder conservation systems. *Herb. Abs.* 41:1–10.

Jager, T. 1975. *Soils research in Serengeti woodlands.* Ann. Rep. Serengeti Res. Inst. (1974–75), pp. 42–51. Arusha: Tanzania National Parks.

Jambunathan, R., and Mertz, E. T. 1973. Relationship between tannin levels, rat growth and distribution of proteins in sorghum. *J. Agric. Food Chem.* 21:691–96.

Jameson, D. A. 1963. Responses of individual plants to harvesting. *Bot. Rev.* 29:532–94.

Katz, P. L. 1974. A long-term approach to foraging optimizations. *Amer. Nat.* 108:758–82.

Kendall, W. A., and Sherwood, R. T. 1975. Palatability of leaves of tall fescue and reed canarygrass and some of their alkaloids to meadow voles. *Agron. J.* 67:667–71.

Labyak, L. F., and Schumacher, F. 1954. The contribution of its branches to the main-stem growth of loblolly pine. *J. Forestry* 52:333–37.

Langer, R. H. M. 1972. *How grasses grow.* London: Arnold.

Laude, H. M.; Kadish, A.; and Love, R. M. 1957. Differential effect of herbage removal on range species. *J. Range Manage.* 10:116–20.

Ledig, F. T. 1969. A growth model of tree seedlings based on the rate of photosynthesis and the distribution of photosynthate. *Photosynthetica* 3:263–75.

Leonard, E. R. 1962. Inter-relations of vegetative and reproductive growth, with special reference to indeterminate plants. *Bot. Rev.* 28: 353–410.

Levin, D. A. 1973. The role of trichomes in plant defense. *Quart. Rev. Biol.* 48:3–15.

———. 1976. The chemical defenses of plants to pathogens and herbivores. *Ann. Rev. Ecol. Syst.* 7:121–59.

Lewin, J., and Lomas, J. 1974. A comparison of statistical and soil moisture modeling techniques in a long-term study of wheat yield performance under semi-arid conditions. *J. Appl. Ecol.* 11:1081–90.

Lotero, J.; Woodhouse, W. W.; and Petersen, R. G. 1966. Local effect on fertility of urine voided by grazing cattle. *Agron. J.* 58:262–65.

Mahadevan, A. 1973. Theoretical concepts of disease resistance. *Acta Phytopath. Acad. Sci. Hung.* 8:391–423.

Marten, G. C.; Barnes, R. F.; Simons, A. B.; and Wooding, F. J. 1973. Alkaloids and palatability of *Phalaris arundinacea* L. grown in diverse environments. *Agron. J.* 65:100–201.

Mattson, W. J., and Addy, N. D. 1975. Phytophagous insects as regulators of forest primary production. *Science* 190:515–22.

McNaughton, S. J. 1976. Serengeti migratory wildebeest: facilitation of energy flow by grazing. *Science* 191:92–94.

McNaughton, S. J. 1979. Grazing as an optimization process: grass-ungulate relationships in the Serengeti. *Amer. Nat.* 113:691–703.

Meidner, H. 1967. The effect of kinetin on stomatal opening and the rate of intake of carbon dioxide in mature primary leaves of barley. *J. Exp. Bot.* 18:556–61.

Milthorpe, F. L., and Davidson, J. L. 1965. Physiological aspects of regrowth in grasses. In *The growth of cereals and grasses,* ed. F. L. Milthorpe and J. D. Ivins, pp. 241–55. London: Butterworths.

Minson, D. J. 1971. Influence of lignin and silicon on a summative system for assessing the organic matter digestibility of *Panicum. Aust. J. Agr. Res.* 22:589–98.

Molz, F. J. 1971. Interaction of water uptake and root distribution. *Agron. J.* 63:608–10.

Morey, D. D. 1949. The extent and causes of variability in Clinton oats. *Iowa Agr. Expt. Sta. Res. Bull.,* no. 363.

Myers, N. 1976. An expanded approach to the problem of disappearing species. *Science* 193:198–202.

Neales, T. F., and Incoll, L. D. 1968. The control of leaf photosynthesis rate by the level of assimilate concentration in the leaf: a review of the hypothesis. *Bot. Rev.* 34:107–25.

Neales, T. F.; Treharne, K. J.; and Wareing, P. F. 1971. A relationship between net photosynthesis, diffusive resistance, and carboxylating enzyme activity in bean leaves. In *Photosynthesis and photorespiration,* ed. M. D. Hatch, C. B. Osmond, and R. O. Slatyer. New York: Wiley-Interscience.

Nielson, E. L. 1968. Intraplant morphological variation in grasses. *Amer. J. Bot.* 55:116–22.

Nielson, E. L.; Drolsom, P. N.; and Jalal, S. M. 1962. Analysis of F_2 progenies from *Bromus* species hybrids. *Crop Sci.* 2:459–62.

Nielson, E. L., and Nath, J. 1961. Somatic instability in derivatives from *Agroelymus turneri* resembling *Agropyron repens. Amer. J. Bot.* 48: 345–49.

Noy-Meir, I. 1975. Stability of grazing systems: an application of predator-prey graphs. *J. Ecol.* 63:459–81.

Oswalt, D. L.; Bertrand, A. R.; and Tell, M. R. 1959. Influence of nitrogen fertilization and clipping on grass roots. *Proc. Soil Sci. Soc. Amer.* 23:228–30.

Pallas, J. E., and Box, J. E. 1970. Explanation of the stomatal response of excised leaves to kinetin. *Nature* 227:87–88.

Parker, K. W., and Sampson, W. W. 1931. Growth and yield of certain Gramineae as influenced by reduction of photosynthetic tissue. *Hilgardia* 5:361–81.

Pearson, L. C. 1965. Primary production in grazed and ungrazed desert communities of eastern Idaho. *Ecology* 46:278–85.

Pearson, R. L.; Miller, L. D.; and Tucker, C. J. 1976a. Hand-held spectral radiometer to estimate graminaceous biomass. *Appl. Optics* 15:416–18.

Pearson, R. L.; Tucker, C. J.; and Miller, L. D. 1976b. Spectral mapping of shortgrass prairie biomass. *Photogrammetric Eng. and Remote Sensing* 42:317–23.

Peterson, R. G.; Woodhouse, W. W.; and Lucas, H. L. 1956. The distribution of excreta by freely grazing cattle and its effect on pasture fertility. Part 2. Effect of returned excreta on the residual concentrations of some fertility elements. *Agron. J.* 48:444–49.

Pilgrim, G. E. 1939. The fossil Bovidae of India. *Mem. Geol. Surv. India, Palaeont. Indica.* 26:1–356.

Reardon, P. O.; Leinweber, C. L.; and Merrill, L. B. 1972. The effect of bovine saliva on grasses. *J. Anim. Sci.* 34:897–98.

————. 1974. Response of sideoats grama to animal saliva and thiamine. *J. Range Manage.* 27:400–401.

Rechenthin, C. A. 1956. Elementary morphology of grass growth and how it affects utilization. *J. Range Manage.* 9:167–70.

Richmond, A. E., and Lang, A. 1957. Effect of kinetin on protein content and survival of detached *Xanthium* leaves. *Science* 125:650–51.

Risser, P. G. 1972. Systems analysis of a tall-grass prairie. Proc. 3d Midwest prairie conf. (unpaged reprint). Manhattan, Kans.: Kansas State Univ.

Roberts, H. M. 1958. The effect of defoliation on the seed-producing capacity of bred strains of grasses. Part 1. Timothy and perennial rye-grass. *J. Brit. Grassl. Soc.* 13:225–61.

Robertson, J. H. 1933. Effect of frequent clipping on the development of certain grass seedlings. *Plant. Physiol.* 8:425–47.

Robson, M. J. 1973. The growth and development of simulated swards of perennial ryegrass. Part 1. Leaf growth and dry weight changes as related to the ceiling yield of a seedling sward. *Ann. Bot.* 37:487–500.

Roe, R., and Mottershead, B. E. 1962. Palatability of *Phalaris arundinacea* L. *Nature* 193:255–56.

Ryle, G. J. A., and Powell, C. E. 1975. Defoliation and regrowth in the Graminaceous plant: the role of current assimilate. *Ann. Bot.* 39:297–310.

Simons, A. B., and Marten, G. C. 1971. Relationship of indole alkaloids to palatability of *Phalaris arundinacea* L. *Agron. J.* 63:915–19.

Sinclair, A. R. E. 1975. The resource limitation of trophic levels in tropical grassland ecosystems. *J. Anim. Ecol.* 44:497–520.

Sprague, M. A. 1954. The effect of grazing management on forage and grain production from rye, wheat, and oats. *Agron. J.* 46:29–33.

Stapledon, R. G. 1928. Cocksfoot grass (*Dactylis glomerata* L.) ecotypes in relation to the biotic factor. *J. Ecol.* 16:71–104.

Stebbins, G. L. 1956. Cytogenetics and the evolution of the grass family. *Amer. J. Bot.* 43:890–905.

Stein, W. I. 1955. Pruning to different height in young Douglas fir. *J. For.* 53:352–55.

Stewart, D. R. M., and Talbot, L. M. 1962. Census of wildlife on the Serengeti and Loita Plains, *E. Afr. Agric. For. J.* 28:58–60.

Stoy, V. 1965. Photosynthesis, respiration, and carbohydrate accumulation in relation to yield. *Physiol. Plantarum,* supp. 4, pp. 1–125.

Talbot, L. M., and Stewart, D. R. M. 1964. First wildlife census of the entire Serengeti-Mara region, East Africa. *J. Wildl. Manage.* 28: 815–27.

Thorne, J. H., and Koller, H. R. 1974. Influence of assimilate demand on photosynthesis, diffusive resistances, translocation and carbohydrate levels of soybean leaves. *Plant Physiol.* 54:201–7.

Thornley, J. H. M. 1976. *Mathematical models in plant physiology.* London: Academic Press.

Todd, G. W.; Getchum, A.; and Cress, D. C. 1971. Resistance in barley to greenbug, *Schizaphis graminum* L.: toxicity of the phenolic and flavonoid compounds and related substances. *Ann. Entomol. Soc. Amer.* 64:718–22.

Torrey, J. G. 1976. Root hormones and plant growth. *Ann. Rev. Plant Physiol.* 27:435–59.

Tucker, C. J. 1977. Asymptotic nature of grass canopy spectral reflectance. *Appl. Optics* 16:1151–56.

Tucker, C. J.; Miller, L. D.; and Pearson, R. L. 1973. Measurement of the combined effect of green biomass, chlorophyll, and leaf water on canopy spectroreflectance of the short-grass prairie. *Proc. 2d Annual Remote Sensing of Earth Resources Conf.,* ed. I. F. Shahrokhi, pp. 601–27. Tullahoma, Tenn.: Univ. of Tennessee.

Vickery, P. J. 1972. Grazing and net primary production of a temperate grassland. *J. Appl. Ecol.* 9:307–14.

Vittoria, A., and Rendina, N. 1960. Fattori condizionanti la Tanzionalita tiaminica in piante superiori e cenni sugli effetti dell bocca dei ruminanti sull erbe pascolative. *Acta Med. Vet.* 6:279–405.

Ward, P. 1965. Feeding ecology of the black-faced dioch (*Quelea quelea*) in Nigeria. *Ibis* 107:173–214.

———. 1971. The migration pattern of *Quelea quelea* in Africa. *Ibis* 113:275–97.

Wareing, P. F.; Khalifa, M. M.; and Treharne, K. J. 1968. Rate-limiting processes in photosynthesis at saturating intensities. *Nature* 220: 453–57.

Watson, R. M. 1967. The population ecology of the wildebeest in the Serengeti. Ph.D. dissertation, Cambridge University.

Weeda, W. C. 1967. The effect of cattle dung patches on pasture growth, botanical composition, and pasture utilization. *N.Z. J. Agric. Res.* 10: 150–59.

Weiss, C., and Vaadia, Y. 1965. Kinetin-like activity in root apices of sunflower plants. *Life Sci.* 4:1323–26.

Wells, L. H., and Cooke, H. B. S. 1956. Fossil Bovidae from the lime-
    works quarry, Makapansgat, Potgietersrus. *Paleont. Afr.* 4:1–55.
Wiegert, R. G., and Evans, F. C. 1967. Investigations of secondary pro-
    ductivity in grasslands. In *Secondary productivity of terrestrial ecosys-
    tem,* ed. K. Petrusewicz, pp. 499–518. Warsaw: Polish Acad. Sci.
Woolhouse, H. W. 1967. The nature of senescence in plants. *Symp. Soc.
    Exp. Bot.* 21:179–213.

A. R. E. Sinclair

Four   The Eruption of the
        Ruminants

Although rapid increases in animal populations—eruptions—have been observed on a number of occasions, the causes of their initiation and termination are rarely understood. With large mammals, most such instances have been associated with introduced species in new habitats (Caughley 1970; Scheffer 1951; Klein 1968), where the cause of the increase—a superabundance of suitable habitat—is simple and relatively well understood. The same cannot be said for indigenous species in their natural habitats. There are few well-documented cases, and understanding is limited to guesses, after the event, in the form of well-worn but still untested beliefs. Increases are often attributed to the removal of predators (Murie 1944; Bergerud 1971; Keith 1974), and crashes in population attributed to "overutilization" of resources (Murie 1944; Buechner 1960; Klein 1970; Jordan et al. 1971; Keith 1974). In very few cases has the available food been measured accurately. Some of these hypotheses have been reiterated so often as "explanations" of observed events that in wildlife management circles they have come to be regarded as truths. Caughley (1970), for example, has exposed the classic case of the Kaibab deer. Keith (1974, p. 49), in discussing ungulate fluctuations in North America, noted a paucity of reliable quantitative information on both ungulate and predator populations but nevertheless concluded: "The attending circumstances are not particularly well known, but I believe some logical inferences can be drawn which strongly implicate a lack of predation as the chief cause of these irruptions."

In the Serengeti, we have observed the eruption of two ruminant species —the wildebeest and buffalo—in the presence of predators. The cause has been traced to the disappearance of a major mortality factor, the viral disease known as rinderpest. The disease was removed by veterinarians through vaccination of surrounding domestic livestock and hence constitutes an experiment, albeit unintentional, in population dynamics. The subsequent monitoring of these populations and the factors impinging upon them has allowed an analysis of the role of food in the regulation of the populations. These events have also thrown some light on the degree to which interspecific competition is taking place and whether or not it remains constant. With some ten species of grazing ungulates in this area, the study of competition is not only interesting in itself, but has implications for conservation and management. Basically, the question is: If a dominant species changes in number, how much does this affect the numbers of other species?

In this chapter I shall first document these eruptions and then present the evidence for their cause. The way the wildebeest population responds to these causal factors through changes in recruitment and mortality is followed with a discussion of how the wildebeest may affect other species, notably zebra and buffalo, through interspecific competition.

Changes in Ungulate Populations

Wildebeest

The first photographic census of the main migratory population took place in 1961 (Talbot and Stewart 1964). Prior to this, an aerial survey had been conducted in January 1958 by Grzimek and Grzimek (1960). These authors were then only beginning to develop the techniques for aerial census. They used low-level (100 m height) transects and made estimates of what they saw. It has not been possible, therefore, to compare the results of this initial survey with later censuses in which photography was used and where counting errors could be calculated. However, since their estimate of 94,000 animals is less than half that found in 1961, it is reasonable to assume that the population in 1958 was not very much greater than that in 1961, certainly not of the size found in the 1970s.

Further total counts were conducted in 1963, 1965, 1967, and 1971 (Watson 1967; Sinclair 1973), by which time the population had grown so large that an aerial photographic sampling technique was developed by Norton-Griffiths (1973) and subsequently adopted for censuses in 1972 and 1977. Details of the methods, and the analysis of errors involved can be found in Norton-Griffiths (1973) and Sinclair (1973, 1977).

The results of the censuses are shown in figure 4.1. The estimate from the 1977 census gave a figure of 1,440,000 ± 400,000 (95% C.L.). To

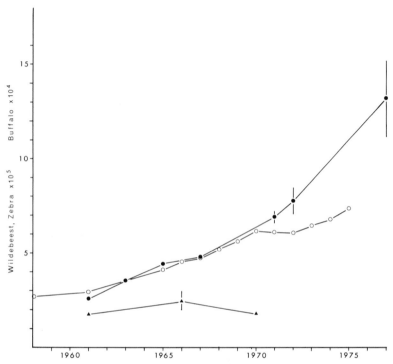

Figure 4.1

Changes in the populations of migratory wildebeest (*solid circles*) and buffalo (*open circles*), compared to zebra (*triangles*). Vertical bars represent one standard error.

compare this with the total counts, the estimate was multiplied by the ratio of total count to sample count conducted simultaneously in 1971. This procedure (also used for 1972) gave a figure of 1,320,000.

The results show that the migratory population has increased fivefold over a sixteen-year period since 1961. From the survey of 1958, and other more subjective descriptions (Pearsall 1957), it appears that the population had remained low for an indefinite period prior to 1961; no major collapse of the population appears to have taken place, at least during the 1950s. Between 1961 and 1967, the population increased toward an equilibrium point. However, a second period of rapid increase then took place and continued until 1978.

Buffalo

The methods of census are described in Sinclair (1973, 1977). For this species, a more definite figure for population size in 1958 was calculated from a survey by Darling (1960). Total counts of various areas were conducted in 1961, and then annually from 1965 onward. The most recent figures for 1974 and 1975 have been provided by J. J. R. Grimsdell (1975; and pers. comm.). The results are presented in figure 4.1. As suspected in the case of wildebeest, there was little, if any, increase between 1958 and 1961. Thereafter the population increased at a rate similar to that of wildebeest (about 10% per year) until 1970. Then there was a period of three years when the population remained nearly constant, followed by a second increase. These events seem to parallel, after a two-year delay, those of the wildebeest.

Zebra

The Serengeti zebra live in small family groups scattered over the woodlands; these groups move long distances in response to changes in rainfall and food distribution (chap. 5). During the wet season, part of the population gathers in large groups on the plains. As a result of this distribution, the population is difficult to census, and only a few crude estimates are available. Talbot and Stewart (1964) give the first total count, taken in 1961. I have corrected this for undercounting errors by the same proportion as I did for wildebeest (Sinclair 1973) to give a figure of 179,000. This is a conservative correction, and the result is certainly under the true total.

A second estimate obtained from a sample count in 1966 by R. M. Watson is given in Jolly (1969), this being 259,000 ± 88,180 (95% C.L.). A third estimate is derived from a series of total counts in 1970 by R. O. Skoog (1970) and estimates from the systematic aerial transects organized by M. Norton-Griffiths and described by L. Maddock (chap. 5). This gave an approximate total of 180,000. In May 1978 an aerial transect census gave a preliminary estimate of 215,000 zebra.

These estimates show that zebra did not increase to any great extent during the 1960s (fig. 4.1). Despite the crude data and the different errors involved in each census, it is safe to say that if an increase did take place, it was nothing like that evidenced by the wildebeest, which nearly tripled its numbers during the same time period. Moreover, I suggest that zebra have not increased at all during the period of study.

Topi and Other Grazers

In 1971, topi, impala, kongoni, waterbuck, and eland were censused by random aerial transects (Sinclair 1972), and this was repeated in 1976 by J. J. R. Grimsdell (appendix A) for strata in the western and central woodlands.

Topi occurs in small family groups over most of the central woodlands. The censuses show that in this area the population has remained the same during the seventies. In the west, there has been a considerable increase. Herds on these open western plains are large, often comprised of several hundred animals, and they are clearly influenced by different factors than those affecting topi in the central woodlands. Intensive studies of the western herds by Duncan (1975) between 1971 and 1973 confirm the increase: in one study area, he found a population of 1800 in 1971, increasing to 2400 in 1972, and to 3200 in 1973. Grimsdell notes that a decrease in human settlement around these western plains has probably allowed the topi to increase there without affecting those animals in the central woodlands. Grimsdell shows that all other grazers have remained more or less constant in population size in both the west and center between 1971 and 1976.

Grant's gazelle numbers in 1970–71 were about 30,000, based on estimates from systematic aerial surveys. In 1978, a similar census gave an estimate of 52,000—approximately double that of the first. The majority of this population inhabits the plains and feeds on small herbs.

It is clear that wildebeest and buffalo have shown a remarkable increase between 1961 and 1977. These two species showed similar rates of increase in the sixties, and for buffalo there have been two periods of increase, with an intermediate pause. The wildebeest and buffalo populations show similar characteristics but the wildebeest appears to be about two years ahead of the buffalo. Populations of other grazing species did not change during the seventies, at least in the central woodlands. The zebra population showed little, if any, increase during the sixties. These observations lead to the question: What caused this increase of some species while others remained stationary?

Rinderpest and the Increase of Ruminants

Rinderpest is a viral disease of ruminants. It is a kind of bovine measles and is normally associated with cattle, the usual host. Of the wild ungulates, buffalo and wildebeest are particularly susceptible, giraffe are moderately susceptible, and most others hardly susceptible at all. Zebra, because it is a nonruminant, is not affected (Carmichael 1938). The disease is an exotic to Africa, having been introduced to Ethiopia in 1889. The subsequent devastating epizootic in cattle, buffalo, and a number of other ruminants, including wildebeest, suggests that the disease had probably not entered the continent previously on a wide scale. But after its introduction, it remained in Tanzania, and probably in Serengeti (see chap. 1), for the first sixty years of this century.

Between 1958 and 1960, and probably before, rinderpest was killing calves of buffalo and wildebeest annually, and, in fact, was termed "yearling disease" at that time (Talbot and Talbot 1963). Serum collected from Serengeti wildebeest (table 4.1; fig. 4.2) showed that most animals had developed antibodies to the disease, indicating that it was constantly present and challenging the population. Only young wildebeest over seven months old were susceptible to the disease and suffered mortality. Younger calves obtained immunity derived from the mothers' colostrum (Plowright and McCulloch 1967; Taylor and Watson 1967; Plowright, pers. comm.). Similar evidence was obtained for buffalo (Sinclair 1977), as can be seen in figure 4.2.

Wildebeest calves born in 1961 and 1962 still developed antibodies to the disease, but little mortality was observed (Talbot and Talbot 1963), and veterinarians have suggested that a less-virulent strain had evolved (Plowright and McCulloch 1967). Calves born in 1963 and all years sub-

Table 4.1		Incidence of rinderpest antibody in the migratory wildebeest.	
Year	Number Positive	Total	% positive
1958	38	44	86
1959	72	84	86
1960	38	48	79
1961	61	91	67
1962	48	94	51
1963	0	21	0
1964	0	13	0
1965	0	25	0
1966	0	31	0
1969	0	24	0

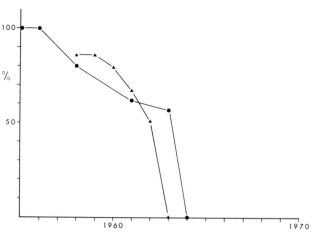

Figure 4.2   The incidence of rinderpest antibodies found in wildebeest (*triangles*) and buffalo (*circles*). No wildebeest have been affected by the disease since 1963, nor buffalo since 1964.

sequently have not developed antibodies to the disease, indicating that within one year, it had died out of the population. The same picture has emerged for buffalo, except that it died out one year later, in 1964 (Sinclair 1977). Why it should have disappeared so quickly remains uncertain. There was at the time an intensive campaign to immunize cattle in the areas surrounding the Serengeti, and cattle are the major carriers for the disease (Atang and Plowright 1969). It is possible that wild ruminants were able to avoid the disease once an attenuated strain had evolved, and they were further protected by the ring of immunized cattle.

Whatever the reasons for the disappearance of rinderpest, there is no doubt that it disappeared rapidly, thereby suddenly releasing the ruminant populations from an important yearling mortality. During the rinderpest period, Talbot and Talbot (1963) recorded that one-year-old wildebeest comprised 8 percent of the population. Since then, this figure has always been higher, usually ranging between 14 percent and 17 percent, indicating that yearling survival doubled, from about 25 percent of those born, to 50 percent.

The timing of this release from rinderpest mortality suggests it was the cause of the increase in wildebeest and buffalo populations. Furthermore, population models of the buffalo simulate the observed census figures only if a rinderpest mortality is included in the years 1958–63 (fig. 4.3). This demonstrates that rinderpest could have been a cause of the increase. The most conclusive evidence for this can be seen from the lack of response by the zebra population, which was not affected by rinderpest. Others factors, such as rainfall, burning, predation, and human

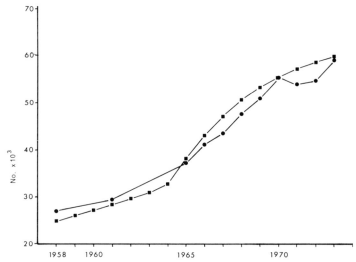

Figure 4.3   The population model for buffalo (*squares*) incorporating rinderpest mortality before 1964, provided a close fit to the observed data (*circles*). From Sinclair (1977).

interference, would have affected both wildebeest and zebra simultaneously if they had been important. Finally, the one-year delay in the disappearance of rinderpest from the buffalo population compared with wildebeest, in combination with the slower population response of buffalo (longer gestation period, lower fertility and recruitment rates) could explain why the buffalo population curve was two to three years behind that of the wildebeest.

Rainfall and the Food Supply

The single most important environmental variable affecting the ungulates is rainfall, for it determines the amount of food available to the animals in the dry season. And it is in the dry season that lack of food and mortality become most severe. The relationship between rainfall and grass production is illustrated in figure 4.4. The four ungulates considered in this chapter are all grass feeders, preferring to eat green leaves, although zebra tend to eat more grass stems than the others (see chap. 6 below; Gwynne and Bell 1968; Sinclair 1977; Duncan 1975). Hence, the more dry-season rainfall, the more available food.

Dry-season rainfall showed no significant trend through the sixties. However, 1971 produced a record high dry-season rainfall, and 1972 had even more rain (table 4.2). Since then, the dry season has remained wetter than in the sixties. There is evidence to suggest that this has been the cause of the second period of population increase evident in buffalo and wildebeest.

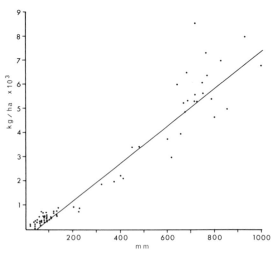

Figure 4.4   The relationship between accumulated rainfall and grass production in the woodlands. From Sinclair (1975), with permission of Blackwell Scientific Publications, Ltd.

Table 4.2

Wildebeest density in relation to average dry-season grass production. The production per month (y) was calculated from the average monthly dry-season rainfall in the central and northern woodlands (x), by the regression $y = 7.6719x - 201.85$. Population estimates were interpolated from figure 4.1.

Year	Population estimate	Average monthly dry-season rainfall (mm)	Grass production (kg·ha^{-1}·day^{-1})	Animals per food (indiv· kg^{-1}·ha^{-1}·day^{-1})
1961	263,362			
1962	309,743	38.75	3.2	96,794
1963	356,124			
1964	397,624	54.25	7.1	56,003
1965	439,124	32.50	1.6	274,452
1966	461,208	38.00	3.0	153,736
1967	483,292	29.25	0.8	604,115
1968	535,663	33.50	1.8	297,590
1969	588,034	32.50	1.6	367,521
1970	640,406	40.75	3.7	173,082
1971	692,777	57.50	8.0	86,597
1972	773,014	68.75	10.9	70,919
1973	882,411	52.75	6.8	129,766
1974	991,808	60.75	8.8	112,705
1975	1,101,205	69.75	11.1	99,207
1976	1,210,603	62.75	9.3	130,172

From the relationship seen in figure 4.4, I calculated the amount of grass grown per hectare per day in each dry season, and then divided this estimate into the wildebeest population figure for that year to obtain a density of individuals with respect to their food (table 4.2). Figure 4.5 shows that when dry-season food is taken into account, the density of animals (relative to their food) peaked in the late sixties, but then declined and has remained low ever since. Thus, there was more food per individual in the seventies than in the late sixties, despite the increase in population.

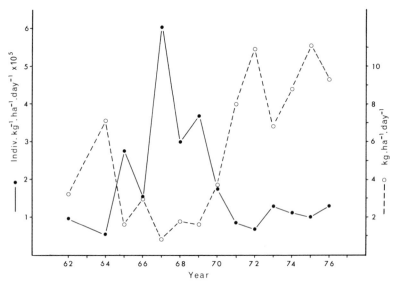

Figure 4.5

Daily dry-season grass production calculated in table 4.2 (*broken line*). Wildebeest population divided by this grass production (*solid line*) shows a decline in the early seventies, despite the population increase, because of increased rainfall.

If the density of animals relative to their food is related to the degree of mortality in the dry season, then the change in density could have been the cause of the second population increase seen in the 1970s.

Mortality and the Food Supply

Using vehicles to traverse this region, total counts of wildebeest carcasses on the dry-season range (pl. 14) were conducted in 1967, 1968, and 1969. An estimate of search error was obtained by H. Kruuk, who put out a number of marked skulls and found that 60 percent of them were recovered; this information was used to correct the estimate of mortality. In 1971 and 1972, systematic ground transects provided an estimate of carcass number in a given area using Jolly's method (1969) for unequal-sized transects (table 4.3).

The average number of wildebeest using the area in which the carcasses were found was calculated from aerial photographic censuses carried out every ten days to two weeks throughout the time the wildebeest were present (in 1967, visual estimates were later corrected from an estimate/photograph calibration). With this density and the number of carcasses, I calculated the proportion of the population that died per dry-season month (table 4.3). The average monthly mortality rate, starting in May of the year when the population censuses were conducted, was calculated by subtracting the adults and yearlings of one year from the adults, yearlings, and calves of the previous year in May. This gave the annual mortality, from which the average monthly mortality was calculated (table 4.3).

Comparing the measured dry-season mortality with the average over the same year, it is clear that in the drier years of 1967–69, when densities in relation to food were relatively high (fig. 4.5), the dry-season mortality was above average (fig. 4.6). Conversely, in the wetter years of 1971 and 1972, the dry-season mortality was either the same, or even slightly below,

Table 4.3		Wildebeest dry-season mortality, food requirements, and available food.	

Year	Carcasses km^{-2}	% mortality per dry-season month	% mortality per month (whole year)
1967	2.43	1.63	1.09
1968	2.49	1.77	1.06
1969	2.71	1.24	0.74
1971	1.02	0.78	0.88
1972	1.21	0.44	0.72

Year	Population food requirement $kg \cdot ha^{-1} \cdot day^{-1}$	Available food $kg \cdot ha^{-1} \cdot day^{-1}$	Available food per animal $kg \cdot day^{-1} \cdot indiv^{-1}$
1967	—	—	—
1968	2.3–4.6	0.82	1.139
1969	3.0–6.1	2.20	2.683
1971	3.07	6.88	6.985
1972	6.74	26.04	12.050

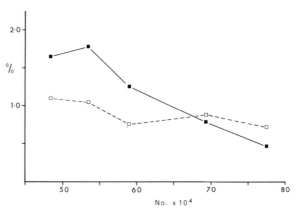

Figure 4.6
The proportion of wildebeest dying per dry-season month (*solid squares*), compared with the average monthly mortality rate during the year (*open squares*), plotted against population size.

the average for the year. This indicates that dry-season mortality is sensitive to changes in rainfall and food supply. It also suggests that undernutrition is the probable cause of the mortality. Indeed, the energy condition of wildebeest that had died in the dry season was, on average, low (Sinclair 1975).

Figure 4.6 also shows that dry-season mortality was inversely related to the size of the population. However, this does not take into account the changes in rainfall and available food in the dry season. The relationship between mortality rate and the total amount of available food is curvilinear because there is still some mortality even when there is a large excess of food. Figure 4.7 shows that the mortality rate is inversely related to the natural log of total amount of food available to the population. This could result either from changes in the environment (through rainfall) or from changes in population density affecting the population's own food supply. If the latter were taking place (through intraspecific competition), then there should have been a stronger inverse correlation between mortality rate and natural log of food per individual than between mortality rate and total food. The slope of the regression lines conforms with this expectation. Although the data are too few to make definite statements, it

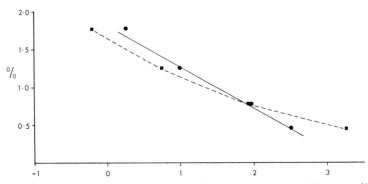

Figure 4.7

The dry-season monthly mortality rate of wildebeest is related to the natural log of total available food (*squares*). Mortality rate (y) is better related to the log of food per individual (x), (*circles*), shown by the regression line $y = 1.8241 - 0.5525\,x$.

does appear that wildebeest may be directly depleting their dry-season food supply through grazing, resulting in competition and in mortality for the weaker individuals.

An average wildebeest weighs about 130 kg and needs to eat for maintenance about 2.4 percent of its weight per day (Sinclair 1977), which amounts to 3.12 kg of dry grass. The product of this figure and the density provides an estimate of the population requirement (table 4.3). When these requirements are compared with the total available food measured from the exclosure plots, it is noticeable that the animals needed more than was available to them during 1968 and 1969, whereas in 1971 and 1972, there was an excess of food throughout the dry season. Presumably during the former years, the difference between requirement and available food was made up by use of fat reserves. Those animals with insufficient reserves—old adults and calves (Sinclair 1977)—became susceptible to disease and died. This accounts for the higher mortality rate seen in those years. In 1971 and 1972, food stress was entirely absent, and mortality was reduced to a residual amount caused by predation, accident, and diseases unaffected by body condition.

In the northern Serengeti, where the wildebeest are found in the dry season, the density of predators, particularly hyenas, is very much lower than on the plains (Kruuk 1972). Hence, dry-season predation rates are probably lower than those for the wet season. This factor, combined with the good food conditions in 1971 and 1972, could explain why the dry-season mortality rates in those years were below the average for the year. Indeed, mortality was higher in the wet season than in the dry.

One of the most conspicuous forms of accidental mortality was not measured in these surveys. Some animals drowned when groups of wildebeest moved from one grazing area to another. Usually this took place in the woodlands when the animals were crossing rivers. The leaders of a large group would be pushed into the water by the oncoming masses behind them. These first animals would either be swept away by the current or often would be trampled and drowned. The leaders were usually males, so this sex predominated in the mortality, but calves were also drowned because they were weaker than the rest of the population. Drownings like this occurred two or three times every year, and anywhere from twenty to a maximum of five hundred animals perished. Although these catastrophic events were mostly associated with rivers, two drownings were observed at Lake Lagarja, on the plains. They occurred in the wet season, when groups of females with newborn calves took a shortcut across the lake (pl. 13). Normally this lake is shallow and the animals can walk across, but on these occasions higher rainfall had filled the lake and the animals had to swim. The small calves, weaker than the rest, dropped behind and lost contact with their mothers. They either drowned or reached the shore and died later. In early 1969, a few hundred calves were found (Norton-Griffiths, pers. comm.), and in February, 1973, more than 3,000 calves perished this way. Despite the high losses on some of these occasions, drowning forms a small proportion of the total annual mortality and, hence, has an insignificant effect on the population.

To conclude this section, the evidence suggests a strong relationship between the amount of food available per individual wildebeest and the degree of mortality in the dry season. When there was insufficient food, mortality rates were above average for the year; conversely, in 1971 and 1972, when there was excess food, mortality was average or below average. This evidence points to intraspecific competition as the cause of the mortality. Furthermore, I suggest that the excess food in the dry season, a result of abnormally high rainfall, allowed the further increase of the wildebeest population. I should mention that moderate grazing by wildebeest actually stimulates grass production (chap. 3), and this, in turn, may have contributed to the wildebeest increase when they were at low density.

The Mechanism of
Population Response

The increase in wildebeest population is probably traceable to two separate events: the removal of rinderpest and an increase in rainfall. What segment of the population is most sensitive to these causes, so producing the increase?

Considering fertility first: the proportion of wildebeest that were pregnant was determined directly from samples of animals collected in the field. The earliest information for both adults and yearlings covers the period 1959–60 (Talbot and Talbot 1963). Yearlings can conceive in the rut of their second year, that is, when they are one year, four months old. In the late fifties, 83 percent were pregnant. Information given by Watson (1967) shows that only about 34 percent of this age-group was pregnant between 1964 and 1965, and samples I collected in 1968, 1970, and 1971 produced even lower figures, ranging between 4 percent and 11 percent. Clearly, there has been a steady decline in fertility of yearlings (fig. 4.8).

However, the same sources of information show that adult wildebeest have maintained the same high fertility rate (90–100%) throughout this

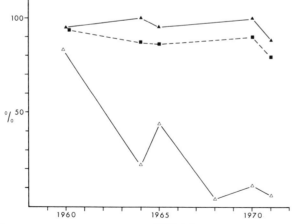

Figure 4.8   Wildebeest pregnancy rate in yearlings (*open triangles*), adults (*solid triangles*), and the total population (*squares*).

period. Although yearlings contribute more than any other age-group to overall population fertility, their influence on this fertility has so far been small, as can be seen in figure 4.8; population fertility remained high and relatively constant throughout the period of wildebeest increase, and I have found similar results in buffalo (Sinclair 1977). Perhaps one reason why fluctuations in yearling fertility contribute so little is that yearlings amount to only 8 to 20 percent of the population. Thus, even large changes in fertility within this group have a small overall effect. This is not to say that such changes have no effect on the intrinsic rate of increase in the long run.

Wildebeest calf survival, on the other hand, has shown considerable fluctuation. Between 1959 and 1960, the calves comprised 18 percent of the population when four months old, and 8 percent when one year old (Talbot and Talbot 1963). In 1963, Watson (1967) estimated a much higher proportion of both calves and yearlings. Until then, estimates had been obtained by counting samples of animals from vehicles. The samples were small and nonrandom, because at that time scientists found it difficult to sample the whole population evenly. Although there is a possibility of bias and sampling error accounting for these large fluctuations, both the lowest and highest figures were obtained by the same method, indicating that bias may not be important. From 1964 onward, sampling was carried out via low-altitude aerial photography, and the population was sampled more evenly. Calves up to ten or eleven months old can be readily identified on the photographs, and I have classified this age-group as equivalent to yearlings.

Figure 4.9 shows the proportions of the four-month and yearling age-groups plotted against population size. The four-month class shows a certain amount of fluctuation but no consistent change with increasing population, and, on average, remains at about 17 percent. Since population fertility was also constant over the same period, this means that early calf survival was independent of density and fairly constant. At birth, calves would comprise 30 percent of the population if they were all born at once. Hence, by four months of age, an average of 43 percent of them had died.

In contrast to early calf survival, the proportions of yearlings appeared to show some definite changes as the population grew. In the early sixties, there was a marked increase in yearling survival; the proportion of yearlings in the population rose from 8 percent to 23 percent. Although this latter figure is probably an overestimate (small sample size), estimates in succeeding years were also high, suggesting that the improvement in survival was real. This change took place in 1962, the year when rinder-

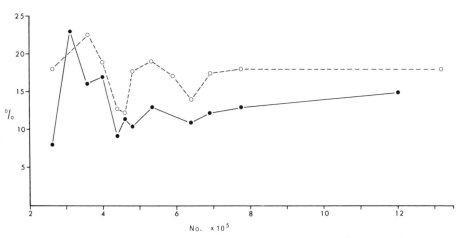

Figure 4.9

The survival of wildebeest calves at four months old (*open circles*) and one year old (*solid circles*), as reflected by their proportion in the population, plotted against population size.

pest died out of the population. Since early calf survival was approximately constant, these changes in yearling survival were due to changes in mortality occurring between four and twelve months of age, a time spanning the dry season. Calves obtain immunity to rinderpest from their mothers, but they lose this at seven months of age (Plowright and McCulloch 1967), during the middle of the dry season, August-September. Talbot and Talbot (1963) report that rinderpest was most noticeable in wildebeest calves during the dry season. Hence, the presence of rinderpest accounts for the initial low yearling survival, and when it disappeared, the major effect on the population was increased survival of yearlings—mortality in the first year dropped from 74 percent to 24 percent.

After the high point, yearling survival gradually declined as the population increased, reaching low points at intervals between 1965 and 1970. This was the period when the population appeared to be leveling off, and the evidence suggests that low yearling survival may have been one of the contributory mechanisms. Consistent with this idea, survival improved slightly after 1970, when the population appeared to undergo a second increase. Survival of yearlings, however, is not due solely to dry-season mortality but is also influenced by survival in the first few months. In fact, figure 4.9 shows that the difference between the four-month-old and yearling proportions (reflecting dry-season mortality) is unrelated to density.

A more detailed analysis of the buffalo population (Sinclair 1977) showed features similar to those of the wildebeest. Yearling mortality fluctuated greatly and caused disturbances in the population, while adult mortality was lower but density-dependent, and so acted as a stabilizing influence on buffalo numbers. Although the data for wildebeest are less reliable because of necessary interpolations between censuses, the information on adult mortality also suggests a density-dependent relationship. When data on recruitment rates in recent years become available, a more reliable analysis can be attempted.

Conclusion

It is clear that there has been a major population increase in wildebeest and buffalo populations, and the timing of this increase coincides with the disappearance of rinderpest. In buffalo, population models require the presence of rinderpest in the early years, followed by its removal, to account for observed census figures. Perhaps the best evidence to support this comes from the observation that zebra, a nonruminant, was not affected by rinderpest and did not increase during the sixties. Any other factor, such as an increase in habitat or food through change in environment or decrease in predation or human interference, would have affected both ruminants and nonruminants alike. Rinderpest is the only factor that differentiates between these groups, and, therefore, the only factor that can account for the observed population changes.

Rinderpest kept the wildebeest and buffalo populations well below the limit imposed by food. When the disease was removed, the populations increased steadily to this new limit, the buffalo lagging two to three years behind the wildebeest. By the late sixties, both populations were leveling off, as a result of mortality induced by the lack of food in the dry season. Food appeared to be limited both by low dry-season rainfall and by grazing. However, in 1971, there began a series of years of high dry-season rainfall, which produced a superabundance of food for the animals and allowed the populations to increase a second time. But other grazers have not responded to this climatic change. During these years, wildebeest mortality in the dry season, normally higher than the average monthly rate, fell below it. These responses to climate, food, and grazing show that intraspecific competition was probably important in allowing the increase and leveling out once rinderpest disappeared.

Through two separate periods, in the early sixties and again in the seventies, food supply was not limiting the wildebeest and buffalo. Therefore, interspecific competition between wildebeest and zebra would have been at a low level during those periods. If interspecific competition had been important as an evolutionary pressure in determining resource partitioning of the ungulates, then one would have expected that the wildebeest, artificially released from the exotic rinderpest and increasing toward a new level, would have a negative effect on the zebra population. The evidence does indicate a decline in zebra between 1966 and 1970, suggesting that interspecific competion was operating, but the crude zebra population estimates preclude any firm conclusion on this. The fact that other grazers such as topi, kongoni, and impala did not increase significantly in the central woodlands during the seventies suggests that the expanding wildebeest and buffalo populations were removing most of the extra food and were preventing the increase of these other species through competition. In chapter 6, we show that it is the larger species, such as wildebeest and buffalo, that dominate the herbivore species complex.

An independent measurement on the magnitude of interspecific competition between wildebeest and buffalo shows that the latter population could be depressed by as much as 18 percent by the wildebeest (Sinclair 1977), while the reverse effect was negligible, due to the tenfold difference in the size of their populations. If interspecific competition was taking place, then one would expect the buffalo to be held in check by the wildebeest, which itself could continue to increase. The buffalo, in fact, leveled off for four years in the early seventies, a longer period of time than the leveling out exhibited by wildebeest, which was already increasing again by 1972. This supports the hypothesis that interspecific competition was taking place.

Most of the studies on ecological separation of herbivores in the Serengeti (Watson 1967; Bell 1970, 1971; Gwynne and Bell 1968) took place at a time when food was abundant, by virtue of the rinderpest effect; interspecific competition, then, was at a low level, and this may well have been the case throughout the period when rinderpest was present in the ecosystem (some seventy years). Hence, the observed species differences can be maintained by interspecific competition acting intermittently (with gaps as long as seven generations); or it may be that the differences are fortuitous and caused by factors quite unrelated to competition. Between wildebeest and buffalo, it appears that the former is the case.

Finally, the fact that the wildebeest can increase its numbers fivefold and not cause immediate serious declines in the numbers of other species is important for park management. This observation eliminates the immediate need for artificial control of the wildebeest population in order to save other species; but the long-term impact on the vegetation and, indirectly, on other animals is not yet clear and needs further monitoring.

References

Atang, P. G., and Plowright, W. 1969. Extension of the JP-15 rinderpest control campaign to Eastern Africa: the epizootiological background. *Bull. Epizoot. Dis. Afr.* 17:161–70.

Bell, R. H. V. 1970. The use of the herb layer by grazing ungulates in the Serengeti. In *Animal populations in relation to their food resources,* ed. A. Watson, pp. 111–23. Oxford: Blackwell.

————. 1971. A grazing ecosystem in the Serengeti. *Sci. Am.* 224 (1): 86–93.

Bergerud, A. T. 1971. *The population dynamics of Newfoundland caribou.* Wildl. Monogr. no. 25. The Wildlife Society.

Buechner, H. K. 1960. *The bighorn sheep in the United States: its past, present, and future.* Wildl. Monogr. no. 4. The Wildlife Society.

Carmichael, J. 1938. Rinderpest in African game. *J. Comp. Path.* 51: 264–68.

Caughley, G. 1970. Eruption of ungulate populations, with emphasis on Himalayan thar in New Zealand. *Ecology* 51:53–72.

Darling, F. F. 1960. *An ecological reconnaisance of the Mara plains in Kenya Colony.* Wildl. Monogr. no. 5. The Wildlife Society.

Duncan, P. 1975. Topi and their food supply. Ph.D. dissertation, Nairobi University.

Grimsdell, J. J. R. 1975. Serengeti Research Institute, Annual Report, 1973–74, p. 21. Arusha: Tanzania National Parks.

Grzimek, M., and Grzimek, B. 1960. Census of plains animals in the Serengeti National Park, Tanganyika, *J. Wildl. Manage.* 24:27–37.

Gwynne, M. D., and Bell, R. H. V. 1968. Selection of grazing components by grazing ungulates in the Serengeti National Park. *Nature* (Lond.) 220:390–93.

Jolly, G. M. 1969. Sampling methods for aerial censuses of wildlife populations. *E. Afr. Agric. For. J.* 33:46–49.

Jordan, P. A.; Botkin, D. B.; and Wolfe, M. L. 1971. Biomass dynamics in a moose population. *Ecology* 52:147–52.

Keith, L. B. 1974. *Some features of population dynamics in mammals.* 11th Int. Congr. Game Biol. (Stockholm 1973), pp. 17–58.

Klein, D. R. 1968. The introduction, increase, and crash of reindeer on St. Matthew Island. *J. Wildl. Manage.* 32:350–67.

———. 1970. Food selection by North American deer and their response to overutilization of preferred plant species. In *Animal populations in relation to their food resources*, ed. A. Watson, pp. 25–44. Oxford: Blackwell.

Kruuk, H. 1972. *The spotted hyena.* Chicago: Univ. of Chicago Press.

Murie, A. 1944. *The wolves of Mount McKinley.* U.S. Nat. Park Serv., Fauna Ser. no. 5.

Norton-Griffiths, M. 1973. Counting the Serengeti migratory wildebeest using two-stage sampling. *E. Afr. Wildl. J.* 11:135–49.

Pearsall, W. H. 1957. Report on an ecological survey of the Serengeti National Park, Tanganyika. *Oryx* 4:71–136.

Plowright, W., and McCulloch, B. 1967. Investigations on the incidence of rinderpest virus infection in game animals of N. Tanganyika and S. Kenya, 1960–63. *J. Hyg.* (Camb.) 65:343–58.

Scheffer, V. B. 1951. The rise and fall of a reindeer herd. *Sci. Monthly* 73:356–62.

Sinclair, A. R. E. 1973. Population increases of buffalo and wildebeest in the Serengeti. *E. Afr. Wildl. J.* 11:93–107.

———. 1975. The resource limilation of trophic levels in tropical grassland ecosystems. *J. Anim. Ecol.* 44:497–520.

———. 1977. *The African buffalo.* Chicago: Univ. of Chicago Press.

Skoog, R. O. 1970. Serengeti Research Institute, Annual Report, pp. 28–31. Arusha: Tanzania National Parks.

Talbot, L. M., and Stewart, D. R. M. 1964. First wildlife census of the entire Serengeti Mara region, East Africa. *J. Wildl. Manage.* 28: 815–27.

Talbot, L. M., and Talbot, M. H. 1963. *The wildebeest in western Masailand.* Wildl. Monogr. no. 12. The Wildlife Society.

Taylor, W. P., and Watson, R. M. 1967. Studies on the epizootiology of rinderpest in blue wildebeest and other game species of northern Tanzania and southern Kenya, 1965–67. *J. Hyg.* (Camb.) 65:537–45.

Watson, R. M. 1967. The population ecology of the wildebeest (*Connochaetes taurinus albojubatus* Thomas) in the Serengeti. Ph.D. dissertation, Cambridge University.

Linda Maddock

Five The "Migration" and Grazing
Succession

The "migration," one of the best-known phenomena of the Serengeti area, is the term generally applied to the movements of the wildebeest population (consisting of over a million animals) between their wet-season range on the open plains and their dry-season range in the wood-lands (pl. 12). Zebra (pl. 19) and Thomson's gazelle (pl. 17) populations are involved in similar movements, although these movements do not involve such enormous numbers of animals.

Modern accounts of the wildebeest migration date back to the mid-1950s. More detailed data, available from 1960 onward, have been ana-lyzed by Pennycuick (1975) to give a picture of the average pattern of the migration and of the variations from year to year. The main factor deter-mining both the annual movements and the variations in them was found to be rainfall, through its effect on the food supply. In this chapter, I shall review this information and compare it with data on the distributions of zebra and Thomson's gazelle between 1969 and 1972 to see how they affect each other.

It is clear that the migratory populations represent a large proportion of the total biomass of the Serengeti ecosystem; by migrating to the plains, which provide only seasonal grazing, they avoid competition with other ungulates for a large part of the year. Since zebra, wildebeest, and Thom-son's gazelle occupy similar areas, it is probable that they are separated by differences in food requirements. Vesey-Fitzgerald (1960) shows how a number of grazing species, far from competing, may form a grazing suc-cession and be dependent on one another. Initially, larger species stimulate new grass growth by their grazing and trampling, thus enhancing the food supply for the smaller species which follow them. In the Serengeti, Bell (1971) has described the grazing succession of zebra, followed by wilde-beest and then by Thomson's gazelle, both in the migratory populations and in the small resident populations of these species in the western part of the national park.

Previous Accounts
of the Migration

Since Pearsall (1957) made a preliminary survey of the Serengeti area, many workers have described the migrations, but most concentrate on wildebeest and mention zebra and Thomson's gazelle only in passing. These accounts agree that wildebeest use the plains in the wet season and move into the woodlands in the dry season, but differ in some details, depending on when the data were collected and the extent of the area studied.

Pearsall (1957), in one of the earliest surveys, done at the end of the dry season in 1956, described large concentrations of wildebeest and zebra in the lower reaches of the Mbalageti and Grumeti rivers of the western corridor, in the north around Bologonja, and in Ngorongoro Crater (fig. 5.1). The north and west groups moved onto the plains in November, as had been observed during the previous five years. Previously it had been thought that the Ngorongoro animals left the crater, but this was not seen in the wet season of 1956–57.

Swynnerton (1958) described similar movements, but does not specify the years when his observations were made. He stated that the timing of movements was dependent on the onset and cessation of the rains, and that the wildebeest moved further north in dry years. Fraser Darling (1960) also noted that the wildebeest migrate to find grass, but found no marked movement to the north in 1958, and so concluded that the wildebeest in the Kenya Mara were a separate population.

Grzimek and Grzimek (1960) observed the migration between December 1957 and January 1960, and found that the wildebeest tended to follow the rain and new grass growth, which led them to move onto the plains around the middle of December. They left the plains in May or June, when the rain ceased, and moved to the western corridor; some moved from there to the north later in the dry season.

Talbot and Talbot (1963) noted that some Serengeti wildebeest, after leaving the corridor, moved to the Kenya Mara, where there was another population with its own migratory pattern. They also observed a resident population in the west of the corridor. The Ngorongoro population was found to make irregular movements, with most wildebeest remaining in the crater. The Serengeti migration was different every year; the wilde-

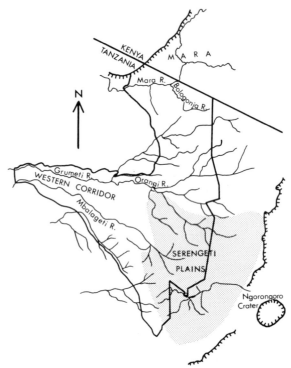

Figure 5.1   The Serengeti area, showing the boundary of the national park and the main geographical features. The plains are shaded.

beest preferred fresh green grass and avoided muddy ground, but tended to move through areas where there was likely to be permanent water. Talbot and Stewart (1964) state that the wildebeest were on the plains between January and May, and sometimes also in November and December, but when a census was taken in May, 1961, many had already moved into the corridor. For this census, the Kenya Mara and Serengeti were treated as a single ecosystem for the first time.

Watson (1967) showed that the wet- and dry-season ranges were of approximately equal size and that half the year was spent in each. Although the wildebeest generally left the plains at the end of the wet season, in May or early June, they never stayed on the plains later in the year, even if the rain continued, as it did in 1962. He suggested that rutting had to take place during the movement off the plains, when all the wildebeest are concentrated into a relatively small area (pl. 12). If rutting were delayed, young would be born at a less favorable time the following season, or the births would be less well synchronized. After moving west in the early dry season, the animals continued north, in the wake of zebra, who had grazed the grass down, following rain showers that signaled the regrowth of green grass. They then moved back to the west, or sometimes returned directly to the plains in the early wet season.

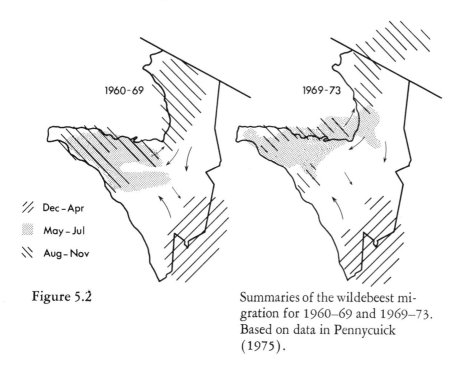

1960-69

1969-73

/// Dec – Apr

May – Jul

\\\ Aug – Nov

Figure 5.2   Summaries of the wildebeest migration for 1960–69 and 1969–73. Based on data in Pennycuick (1975).

Between 1960 and 1973, Pennycuick (1975) found that, on average, wildebeest moved from the plains to the west, but only some went on to the north. They returned to the plains either directly or via the west at the start of the wet season (fig. 5.2). The timing of movements to and from the plains depended on the rainfall: if there were more than 50 mm of rain in a month between November and June, the wildebeest could be found on the plains. If there was a dry month in the middle of the wet season, the animals moved off the plains to the west and southwest. There has also been a long-term change in the pattern of the migration, in that the northern parts of the area, in particular the Kenya Mara, have been utilized to a greater extent in recent years (fig. 5.2). This utilization also occurred in dry years, but there has been no evidence of a corresponding long-term change in annual rainfall. It has been suggested, therefore, that the long-term change in the migration is associated with the increase in numbers of wildebeest in the 1960s (Sinclair 1973; Norton-Griffiths 1973).

Bradley (1977) shows that, as in wildebeest, there are several sub-populations of Thomson's gazelle in the Serengeti area—a main migratory group, and smaller resident or semimigratory groups. The main migratory gazelles were found on the eastern plains in the wet season (pl. 17), and moved steadily westward as the dry season proceeded. They preferred areas receiving monthly rainfall of between 50 and 65 mm.

Seasonal Distributions

This account of migration in zebra, wildebeest, and Thomson's gazelle is based mainly on data obtained during systematic survey flights carried out by members of the Serengeti Research Institute at the end of each month (five months were omitted) from August, 1969, to August, 1972. The area covered by survey flights roughly corresponds to the whole Serengeti ecosystem, defined as the total range of the migratory wildebeest population. This area was divided into a grid of 10 x 10 km squares, and these are the basic units used in the analysis. However, as there are 300 squares in all, corresponding to a total area of 30,000 km², they have been grouped into land regions.

Land regions are separated according to their topography, geology, soils, and vegetation (Gerresheim 1974). Figure 5.3 shows the nineteen land regions. Region 14 represents the Serengeti plains, and, as it is by far the largest of the regions, it has been divided into five subregions (a-e). Region 15 is also mainly grassland. In general, the grass is short on the eastern plains and becomes progressively longer to the west. The other regions are mainly woodland, although there are some fairly large areas of grassland or open, wooded grassland in the western corridor (regions 9, 12, and 13), while the northerly parts of the park have thicker wood-land or bush (Herlocker 1975).

Watson (1967) described the distribution of wildebeest in terms of occupance, expressed as biomass, or numbers of animals per square of ⅛ degrees. A similar index was used here, applied to the 10 km grid squares, but, as the data were qualitative rather than quantitative, the occupance

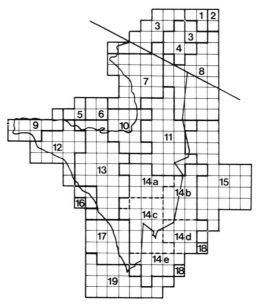

Figure 5.3

Map of the Serengeti ecosystem, showing the area covered by the survey flights and its division into 10 x 10 km grid squares. The boundary of the Serengeti National Park is indicated, and the grouping of the grid squares into land regions 1–19. The plains region, 14, is divided into five subregions, *a-e.*

was expressed as a percentage: a square was counted as occupied if more than about one hundred animals were observed in it during a survey flight. The number of times a square was occupied over a given time period was expressed as a percentage of the total number of observations for all squares. Thus, for any one month, season, or year, the occupances of all squares taken together equaled 100 percent, and the occupance of a region, or group of regions, was the sum of the occupances of its constituent squares.

Mean Migratory Patterns

Occupances of the three species are summarized in figure 5.4, with land regions grouped (as in Pennycuick, 1975) into north (regions 1, 2, 3, 4, 6, 7, 8), central (10, 11), west (5, 9, 12, 13), and plains (14, 15, 17, 19). Figures 5.5 (a-c) show further details of the geographical distributions with total occupance in a 10 km square indicated by the size of the circle, and occupance during the wet season (November-May) indicated by the shaded part of the circle. Thus, areas in which circles are mainly black were occupied during the wet season, while those in which circles are mostly white were occupied during the dry season. Areas with half-shaded circles are those through which the animals moved between their wet- and dry-season ranges. These movements occurred mainly in October-November and May-June but the exact timing depended on the rainfall.

In general, migratory species spend the wet season on the plains and the dry season in the woodlands, but there are differences in the areas that the three species occupy (figs. 5.4–5.6). Zebra and wildebeest use the

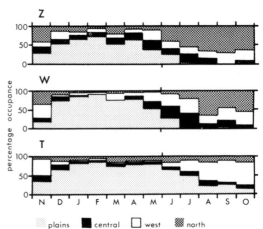

Figure 5.4   Summary of the mean monthly occupance data for zebra (Z), wildebeest (W), and Thomson's gazelle (T), between 1969 and 1972, for land regions grouped into four areas.

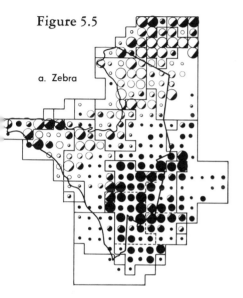

a. Zebra

Mean occupance maps for (*a*)
zebra, (*b*) wildebeest, and
(*c*) Thomson's gazelle. Size of the
circles indicates the percentage
occupance for a grid square.
Shaded portion of the circles shows
the occupance during the wet season (November-May); and the
open part, the occupance during
the dry season (June-October).
Boundaries of the Serengeti
National Park and the land regions
are also shown.

Figure 5.5

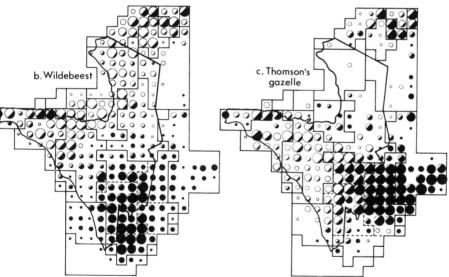

b. Wildebeest

c. Thomson's
gazelle

plains between November and June; while there were few wildebeest in
other areas from December to April, zebra were more widely dispersed.
Toward the end of the wet season, a greater proportion of both zebra
and wildebeest were found in the western and central areas, and from June

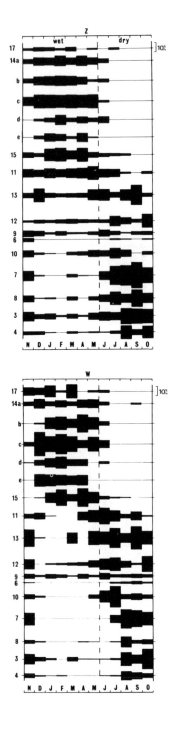

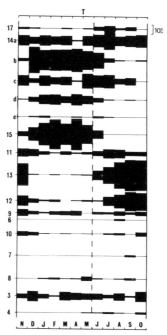

Figure 5.6

Mean occupance of each land region for each month for zebra (Z),
wildebeest (W), and Thomson's
gazelle (T).

to October many zebra were located in the north. Wildebeest, on the other
hand, did not occupy the north to a significant extent until August. A
fairly constant proportion of both these species was found in the west for
most of the dry season, but there were more wildebeest there in November
before they returned to the plains.

The Thomson's gazelle showed a somewhat different pattern of migration: they seldom occupied the north or central areas, their main movements being between plains and west. As with wildebeest, most Thomson's
gazelles were on the plains during the wet season. But they remained
longer than wildebeest or zebra, and some were found there throughout
the dry season. From June to October, the proportion of gazelles increased
in the west; in November, they started the return to the plains.

There are small resident populations of all three species in the Kenya
Mara, and in the far west of the corridor near Lake Victoria. They are distinct during the wet season, but in the dry season they mingle with the
migrants (except the northern Thomson's gazelle).

The migratory movements of the three species can be seen in more detail from the histograms for individual land regions in figure 5.6. During the wet season, the wildebeest were found on the central and southern parts of the plains, and a large proportion of their time was spent outside the Serengeti National Park (regions 14e, 14c). In May and June, they moved mainly to the west (regions 12, 13, 10, 11). As the dry season progressed ,they moved north through region 10, and into Kenya (regions 1, 2, and 3). Some wildebeest remained in the western regions throughout the dry season. In October, the return movement toward the plains began; this was either straight south through region 11, or back via the western regions, 12 and 13, especially if the onset of the rains was delayed.

During the wet season, zebra occurred further to the north and northeast of the plains than the wildebeest, their concentration being centered around the northern half of region 14c, and there was less occupance of areas outside the Serengeti National Park. They moved north to region 11 if there was a dry period during the wet season, whereas wildebeest moved to the west and the southwest (regions 13, 17, and 19). Toward the end of the wet season, zebra moved west, taking a more northerly track than wildebeest, and many zebra also went directly north to regions 3 and 7. Some zebra remained in the west throughout the dry season: although Talbot and Stewart (1964) thought wildebeest and zebra wandered throughout the woodlands at this season, the low occupance of regions 10 and 11 between August and October shows there is little movement between the northern and western groups. There appeared to be little difference in the northern areas occupied by the two species, except that zebra occurred further east in the drier region 8.

The Thomson's gazelle spent longer on the plains than either of the other two species; they used the far eastern parts (regions 14b, 15) during the wet season and then moved west in May and June to occupy regions 14a and 14c after the zebra and wildebeest had left. Few Thomson's gazelles occurred in the far southern plains (14d, 14e), where the wildebeest concentration was found, and few used the northern plains until after zebra had moved. Some gazelles remained in region 14a on the edge of the long-grass plains throughout the dry season, but most were found in the western corridor (regions 13 and 12), there being little northward movement beyond regions 10 and 11. Figure 5.7 summarizes the seasonal distributions of zebra and Thomson's gazelle for comparison with wildebeest distributions in figure 5.2.

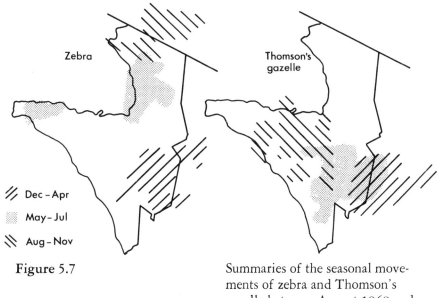

Figure 5.7

Summaries of the seasonal movements of zebra and Thomson's
gazelle between August 1969 and
August 1972, for comparison with
wildebeest distributions in fig. 5.2.

Association between Species

Association between the three species was indicated by the degree of
correlation of their densities. During the survey flights, a measure of
density in each square was recorded on an order-of-magnitude scale from
1 (under 10 animals) to 4 (over 1,000 animals). The logarithmic scale
had the effect of normalizing an otherwise highly skewed distribution.

Figure 5.8 shows the correlation coefficient between pairs of species
for each month. Wildebeest and zebra were highly correlated throughout
the year, the lowest correlations occurring at the beginning of the dry
season, when most of the wildebeest were in the west and zebra in the
north, and at the beginning of the wet season, when zebras moved to the
plains earlier than wildebeest.

Correlations of both wildebeest and zebra with Thomson's gazelle were
generally quite similar, but lower than those between the wildebeest and

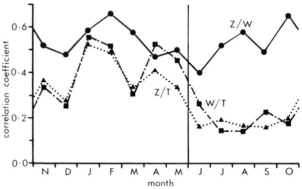

Figure 5.8   Correlation of density between
pairs of species for each month.
Z = zebra; W = wildebeest;
T = Thomson's gazelle.

zebra, with highest values in the wet season and lowest values in the dry
season. This pattern is expected, as gazelles remained on the plains longer
than the other two species and did not go north.

   The grazing succession of zebra followed by wildebeest and then by
Thomson's gazelle has been described by Gwynne and Bell (1968), Bell
(1970, 1971), and McNaughton (1976). The present data do not show
this clearly in the monthly distributions for the land regions, except at the
end of the wet season, when Thomson's gazelle occupied those parts of
the plains previously used by zebra and wildebeest. To investigate whether
the grazing succession occurred, I calculated overall correlation coefficients
with lags of up to two months. However, highest correlations were obtained with no time lag between species, both for the entire three-year
period covered by the data and for separate seasons. It is likely that our
sampling technique accounts for this lack of evidence, since succession
probably occurs on a smaller scale than can be observed from the 10 km
grid squares we used. McNaughton (1976) found that Thomson's
gazelle followed the wildebeest after about three weeks, but that the
spatial distribution of areas grazed by wildebeest was patchy. Bell (1971),
using transects covering a much smaller area than our 10 km grid squares,
found there was a two-month lag between zebra and wildebeest and

between wildebeest and Thomson's gazelle of the western resident populations.

Effect of Rainfall on Movements

The major features of the rainfall distribution have been described by Norton-Griffiths et al. (1975) and by Pennycuick and Norton-Griffiths (1976). In general, there is a gradient of increasing rainfall from southeast to northwest, shown both by the isohyets for mean annual rainfall and by the climatic index. The seasonal distribution of rainfall also changes along this gradient: the drier southeastern areas have a single peak of rainfall about December, whereas the wetter northwestern areas have two peaks, in November and April. In the drier areas, a smaller percentage of total annual rain falls during the dry season, so that the dry season is more severe in the southeast than in the northwest. The drier areas and drier months also have more variable rainfall.

The year can conveniently be divided into two seasons—wet and dry. A wet month is here defined as one whose mean rainfall exceeds the mean for all months; for the area as a whole, then, the wet season extends from November to May, and this also holds for nearly all the individual rain-gauge sites.

Mean Annual Rainfall

I have already shown that migratory species spend the wet season on the plains, where the annual rainfall is lower, and the dry season in the woodlands, where the annual rainfall is higher. I have quantified this by calculating the mean annual rainfall of the occupied regions, weighted according to the percentage occupance, for each month of the year for each species (fig. 5.9). All species occupied drier regions in the middle of the wet season and wetter regions in the middle of the dry season (fig. 5.9a). There are, however, some differences between the species: during the wet season, Thomson's gazelles were found in regions with intermediate annual rainfall, while zebra and wildebeest were found in wetter and drier areas, respectively. In the dry season, zebra and wildebeest occupied regions of similar rainfall, while the gazelles occupied much drier areas. This is shown by the fact that zebra and wildebeest both moved from drier-than-average to wetter-than-average areas in June and July, but Thomson's gazelles remained in drier regions throughout the year. All three species moved from wetter to drier regions in November and December.

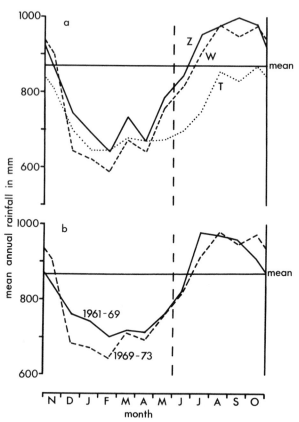

Figure 5.9

The relationship between mean annual rainfall and occupance over all regions. *a* shows zebra (*Z*), wildebeest (*W*), and Thomson's gazelle (*T*) between 1969 and 1973. *b* shows wildebeest during 1961–69 and 1969–73.

The mean annual rainfall in regions occupied by wildebeest in 1961–69 and 1969–73 is plotted in figure 5.9b. This graph emphasizes the tendency for wildebeest to utilize the wet northern regions to a greater extent in the latter period (see fig. 5.2). The basic trends for wildebeest shown in figures 5.9a,b are similar; the curves in the latter are somewhat smoother, due to the inclusion of more data.

Monthly Rainfall

Correlations between monthly values of percentage occupance of each species and rainfall are given in table 5.1. There is a positive relationship for regions occupied in the wet season and a negative one for regions occupied in the dry season, as would be expected. The most significant correlations are for regions occupied during the wet season, although some of the main dry-season areas also show significant correlations.

Pennycuick (1975) showed that in all areas, the frequency distributions of rainfall for all months were significantly different from those for months when wildebeest were present. In general, the wildebeest preferred the plains in months that had between 30 and 110 mm of rain-

Table 5.1

Correlation coefficients for monthly rainfall against occupance for each region. The regions are arranged as in figure 5.6. A value exceeding ± 0.35 is significant at the 5% level ($N = 32$).

Region	Zebra	Wildebeest	Thomson's gazelle
17	0.01	−0.04	−0.47
14 a	0.40	0.56	−0.42
b	0.58	0.73	0.56
c	0.67	0.45	−0.16
d	0.42	0.61	0.58
e	0.41	0.45	0.45
15	0.61	. 0.55	0.64
11	0.14	0.11	−0.44
13	−0.46	−0.48	−0.46
12	−0.16	−0.40	−0.22
9	−0.21	−0.17	0.18
6	−0.03	−0.12	−0.14
10	0.01	−0.49	−0.09
7	−0.38	−0.20	−0.23
8	−0.33	−0.17	0.38
3	−0.32	−0.37	0.07

fall, and other areas when there was a monthly rainfall of less than eighty mm.

Differences between Years

Since data were obtained for zebra and gazelle for only three years, no quantitative assessment could be made of the effects of changes in annual rainfall on their migration patterns. Some individual instances of unusually wet or dry periods did, however, give an indication of possible effects. The movement of all species to wetter areas in March (fig. 5.9a), for example, was due to the exceptionally dry March in 1971, when animals left the plains. This movement was typical for dry periods in the middle of the wet season. Most animals moved to areas that they normally occupied at the beginning of the dry season; wildebeest, however, moved more to the southwest (region 17). If the rains started earlier or were delayed, the animals responded accordingly; for example, in 1972, the rains continued until June, and the animals stayed longer on the plains.

Year-to-year variations have been examined in more detail for wildebeest (Pennycuick 1975), using data from 1961 on. It was found that wildebeest generally moved onto the plains in the first month after September, when more than 40 mm of rain fell. When the plains dried up, wildebeest left, but returned if there was sufficient rainfall before June to produce fresh green grass.

The relationship between annual wildebeest occupance and rainfall for plains, west, central, and northern areas, is shown in figure 5.10. Only data for plains and the north showed significant correlations. In wetter years, wildebeest used the plains more and the north less. Hence, the occupance pattern paralleled that of monthly rainfall fluctuation (Table 5.1).

Variations in annual and monthly rainfall probably affect zebra and Thomson's gazelle in the same way, so that if there is a positive correlation between monthly rainfall and occupance in a particular region, then that region will also be occupied to a greater extent in a wetter year. Pennycuick (1975) showed this was true for wildebeest in the sixties, for they moved into Kenya in dry years. Wildebeests' increased use of the north in latter years was not due to a decrease in rainfall, and probably resulted from the population increase (chap. 4 above; Sinclair 1973, 1977; Norton-Griffths 1973), which Sinclair suggests was caused by eradication of rinderpest from cattle in the surrounding areas.

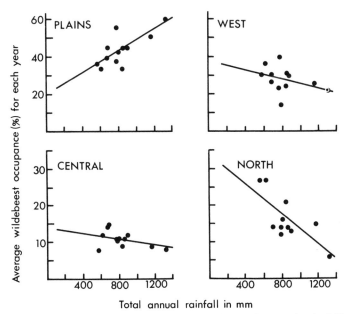

Figure 5.10   The effects of annual rainfall on the percentage wildebeest occupance in the four areas for 1961–73. Regression lines are significant for plains and north only. From Pennycuick (1975), with permission of Blackwell Scientific Publications, Ltd.

The Grazing Succession

The distribution of biomass over the Serengeti ecosystem is shown in Table 5.2. The three migratory species represent 60 percent of the total ungulate biomass over the whole ecosystem, but 94 percent of that on the plains. Therefore, it is only those species which move off the plains in the dry season which are able to make use of the total area to any great extent. Because it does not take migration into account, the equation of Coe, Cumming, and Phillipson (1976) for the relationship between rain-

Table 5.2   Biomass in kg · km^{-2} of ungulates
                                             in different parts of the Serengeti
                                             (Norton-Griffiths, pers. comm.).

	North	Central/West	Plains	Total
Zebra	1493	1400	1973	1600
Wildebeest	2066	2583	4330	2952
Thomson's gazelle	164	368	982	491
Others	6874	3501	434	3225
Total	10597	7852	7719	8268
% migratory species	40	55	94	60

fall and biomass in East Africa predicts a much smaller biomass for the plains than actually occurs there.

Although the seasonal distributions of zebra, wildebeest, and Thomson's gazelle are not identical, there is considerable overlap between them, and ecological separation must be due largely to differences in feeding. Gwynne and Bell (1968) and Bell (1970, 1971) have shown that, while all grazing ungulates prefer green growing grass, which is the stage with highest protein content, the tolerance to less ideal conditions depends on body size and digestive system. Thus, nonruminants, such as the zebra, can tolerate poorer-quality food than can ruminants because the former are able to process a large quantity. Similarly, a smaller ruminant requires food with a higher protein content than a larger one because of the former's higher metabolic rate. Zebras and wildebeest eat grass almost exclusively, the zebra taking more of the coarser parts. Thomson's gazelles take a high proportion of dicotyledon material, particularly the protein-rich fruits. Figure 5.11, based on data from Gwynne and Bell (1968), shows the mean percentages of the different grass parts eaten by wildebeest, zebra, and Thomson's gazelles at different times of the year.

Bell (1971), in a study of the resident populations in the western corridor, describes how differences in food requirements can be satisfied by the grazing succession. This he relates to the catena, which is the sequence of soil types characterized by dry conditions on the tops of ridges and wetter conditions with dense vegetation in the hollows. The grazing species tend to congregate on the higher ground in wet weather when there is abundant short green grass. The effect of the animals' grazing is to maintain the grass in a growing state, ideal for all grazers.

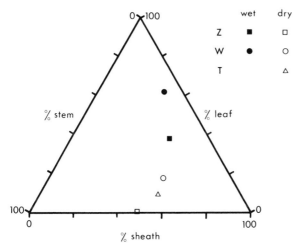

Figure 5.11

Partitioning of food resources among the three migratory species in different seasons. The mean percentage of different grass parts (sheath, stem, and leaf) in the western corridor. Redrawn from Gwynne and Bell (1968), with permission of MacMillan Journals, Ltd.

Only when conditions become drier and the grass stops growing do the animals move down the slopes. The larger species, the buffalo and zebra, move first, followed by topi and wildebeest and finally by Thomson's gazelle. Zebra are able to eat longer grass, which is of too poor a quality to support smaller ruminants. This enables wildebeest to reach parts of the grass which have higher protein content, and also stimulates new, high-quality growth. When wildebeest have removed the next layer, Thomson's gazelle can reach high-protein grass parts and dicotyledon material at the lowest levels of the sward (see chap. 6). This catenary grazing succession is analogous to the migratory sequence: the movement of migrants from the short-grass plains to the long grass north is similar to movements down the catena but on a larger scale (Bell 1971). In the migration, buffalo and topi are not important, as few are found on the plains, and most remain in the long grass west and north.

McNaughton (1976) found that wildebeest on the western edge of the plains removed 85 percent of the standing crop of grass. This stimulated new growth, which was eaten by Thomson's gazelles. There was a strong correlation between the areas grazed by wildebeest and those subsequently utilized by gazelles, and this was maintained throughout the dry season.

Watson and Kerfoot (1964) had noted earlier that the long-grass parts of the plains were grazed by gazelles after the wildebeest and zebra had been feeding there. In a different habitat with a different range of herbivores, Vesey-Fitzgerald (1960) also found a grazing succession, with new grass growth stimulated by grazing, trampling, and burning. In fact, Bell (1971) suggests that the grazing succession is a general feature of the savanna ecosystem, with different ranges of species utilizing different areas.

The Advantages of Migration

Seasonal distributions of migrant zebra, wildebeest, and Thomson's gazelle in the Serengeti can be explained largely in terms of their food requirements. The migratory movements appear to be a compromise strategy, balancing the average conditions and those prevailing at any one time. The main factors affecting the food supply are rainfall and grazing; fire also has an important effect, but it is more variable and less well quantified.

There are probably a number of factors other than food which contribute to the animals' preference for the plains. For example, the fact that they dislike wet, muddy ground (Talbot and Talbot 1963; Bell 1971), as is commonly found in the woodlands during the wet season, may be related to the avoidance of parasites and diseases. Darling (1960) also notes that wildebeest avoid thick cover; certainly, predators can be seen from a far greater distance on the plains, where visibility is better. This is particularly important during the period when the young are born, as this period is the most vulnerable stage in their lives. Wildebeest have a marked birth peak, with nearly all calves being dropped in about three weeks in January and February. Zebras also have a birth peak about this time, although their foals appear over a more extended period. Thomson's gazelles breed twice a year (Kruuk 1972; Bradley 1977), so that birth peaks are less marked, but, even so, many of the young are born while gazelles are on the plains.

Leuthold and Leuthold (1975) have shown that in Tsavo National Park, grazers tend to have a more marked breeding peak than browsers, and this they relate to seasonal changes in food quality. Jarman (1974) suggests that type of food, feeding style, body size, group size, and anti-predator behavior of antelopes are interrelated, so that species whose social systems involve large herds have more synchronized breeding than those that live in smaller groups. The advantage of this is obvious, in that predators are overwhelmed with prey and can take only a small fraction of the newborn. Solitary species have less need for such measures, as they can rely on being relatively inconspicuous.

The January-February birth peaks in the Serengeti migratory species correspond to the time when the protein content of grass is greatest and is most readily available for lactating females. And lactation is the time of maximum stress. Anderson and Talbot (1965) did not consider the supply of nutrients as an important factor in the distribution of animals on the plains, but Kreulen (1975) has suggested that the wildebeest's preference for the eastern plains may be related to the availability of calcium, needed particularly during lactation. He also suggests that the grass is shorter in these areas only because of heavy grazing. This is undoubtedly a contributory factor, for McNaughton (1976) has shown that a large proportion of the grass may be removed by grazing, and Anderson and Talbot (1965) state that the species composition is also altered by grazing and burning. However, there are also gradients of rainfall and soil type corresponding to gradients in grass length, which suggests that these also influence ungulate distribution and timing.

Furthermore, the timing of wildebeest calving may be related to the timing of the rut in May-June. Watson (1967) noted that the wildebeest left the plains at this time, even when there was still suitable grazing; he suggested that by moving, they concentrated in a smaller area, which was necessary to synchronize the rut. Rutting is the time of maximum stress for male wildebeest; between rounding up females and chasing off other males, there is little time for feeding. It is essential, therefore, that males be in peak condition at the start of the rut, as they are at the end of the wet season (Sinclair 1977), and that they should be able to regain that peak condition before the worst of the food shortages occur during the dry season. Similar factors influencing optimum breeding time also apply to zebra and Thomson's gazelle populations.

The migratory species in the Serengeti appear to avoid heavy predation by their movements. Kruuk (1970, 1972) has shown that in Ngorongoro Crater, where the herbivore populations are resident throughout the year, the relative number of hyenas is much greater than it is in the Serengeti, and there is a higher rate of turnover of both predator and prey populations. The predators are limited by the numbers of prey and, at the same time, have a significant effect on the prey populations. In the Serengeti, however, hyenas and lions seem to be limited by their ability to supply food to their nonmobile young (pl. 29) from the nonmigratory prey populations (see chaps. 9, 10 below; Bertram 1975). The predator-prey ratio is so low that it seems unlikely that the predators could have any controlling effect on the migratory prey populations.

Assuming that the Serengeti wildebeest population was previously controlled by rinderpest, as has been suggested by Sinclair (chap. 4 above, 1973, 1977), the eradication of this disease has probably had several effects on both the wildebeest population and on the other migratory populations. It has already been shown that there is facilitation, rather than competition, between the three migratory species; an increase in one, then, does not necessarily imply a decrease in the others, and can, in fact, cause them to increase as well. Another important factor appears to be the increased use by wildebeest of the northern parts of Serengeti National Park and Kenya Mara (Pennycuick 1975). This has been accompanied by a decrease in the woody vegetation of these areas, a decrease which Herlocker (1975) attributes to the influence of fire. It has been assumed that elephants are also having an important effect on the trees of the Serengeti, but Watson and Bell (1969) have found no evidence of any increase in the elephant population in the area. Changes in the amount of rainfall may have affected the vegetation, and although Pennycuick and Norton-Griffiths (1976) show that there have not been any great changes in annual rainfall in recent years, there is abundant evidence from other areas, where lake levels have risen (Morth 1967; Western and van Praet 1973), that the water table could well have changed drastically. Finally, changes in the boundaries of the Serengeti National Park and in the use of the land surrounding it may have affected the migratory patterns.

The way in which all these factors are related is not clear at present. It does seem, however, that if the wildebeest numbers were previously controlled by rinderpest, then the removal of this disease could cause the population to increase to a new level, probably determined by the food supply.

Conclusion

Migratory populations of wildebeest, zebra, and Thomson's gazelle, comprising over half the total ungulate biomass, move between plains in the wet season and woodlands in the dry season. Migration patterns are not identical, for wildebeest move from the plains to the west and then to the north, while more of the zebra move directly from the plains to the north. Thomson's gazelles spend the early dry season on the western plains before moving into the corridor and do not utilize the north at all.

These differences can be related to the different food requirements of the animals, which is also the chief factor keeping them ecologically separate. The nonruminant zebras can tolerate the poorest-quality food and, hence, are the first to move into an area of long grass. Wildebeest are next in the grazing succession, moving into an area after zebra have already grazed it down. Finally, the smaller Thomson's gazelle, which requires high-quality food, grazes an area once wildebeest have made high-protein dicotyledons accessible and stimulated growth of new grass by removing another layer of the sward. Rather than competing, each species is thus partly dependent upon the others for its food supply.

The migration enables these populations to utilize the plains area, which provides grazing only during the wet season. This has the effect of removing them from competition with other grazers for much of the year and also reduces pressures from the less-mobile predators. In addition, it means that the wet, fly-infested woodlands are avoided. All migratory species have well-marked birth peaks while they are on the plains; the young, then, are born in a more favorable habitat at the time of year when most food is available.

The main migratory patterns are thus determined by food supply, which is largely dependent upon rainfall. Yearly variations in the migrations are also related to rainfall. Animals use the plains only while there is sufficient rain to produce green grass. Wildebeest, at least, use the wetter areas more in drier years. This flexibility is essential in an environment with widely fluctuating rainfall.

References

Anderson, G. D., and Talbot, L. M. 1965. Soil factors affecting the distribution of the grassland types and their utilisation by wild animals on the Serengeti Plains, Tanganyika. *J. Ecol.* 53:33–56.

Bell, R. H. V. 1970. The use of the herb layer by grazing ungulates in the Serengeti. In *Animal populations in relation to their food resources,* ed. A. Watson. Oxford: Blackwell. Pp. 111–23.

———. 1971. A grazing ecosystem in the Serengeti. *Sci. Am.* 224:86–93.

Bertram, B. C. R. 1975. The social system of lions. *Sci. Am.* 232(5): 54–65.

Bradley, R. M. 1977. Aspects of the ecology of the Thomson's gazelle in the Serengeti National Park, Tanzania. Ph.D. dissertation, Texas A&M University.

Coe, M. J.; Cumming, D. H.; and Phillipson, J. 1976. Biomass and production of large African herbivores in relation to rainfall and primary production. *Oecologia* 22:341–54.

Darling, F. F. 1960. *An ecological reconnaissance of the Mara Plains in Kenya Colony.* Wildl. Monogr. no. 5. The Wildlife Society.

Gerresheim, K. 1974. *The Serengeti landscape classification.* Serengeti Ecological Monitoring Programme. Nairobi: African Wildlife Leadership Foundation.

Grzimek, M., and Grzimek, B. 1960. Census of plains animals in the Serengeti National Park, Tanganyika. *J. Wildl. Manage.* 24:27–37.

Gwynne, M. D., and Bell, R. H. V. 1968. Selection of vegetation components by grazing ungulates in the Serengeti National Park. *Nature* (Lond.) 220:390–93.

Herlocker, D. 1975. Woody vegetation of the Serengeti National Park. College Station, Texas: Texas A&M University.

Jarman, P. J. 1974. The social organisation of antelope in relation to their ecology. *Behaviour* 48:215–67.

Kreulen, D. 1975. Wildebeest habitat selection on the Serengeti Plains, Tanzania, in relation to calcium and lactation: a preliminary report. *E. Afr. Wildl. J.* 13:297–304.

Kruuk, H. 1970. Interactions between populations of spotted hyenas (*Crocuta crocuta* Erxleben) and their prey species. In *Animal populations in relation to their food resources,* ed. A. Watson. Oxford: Blackwell. Pp. 359–74.

————. 1972. *The spotted hyena.* Chicago: Univ. of Chicago Press.

Leuthold, W., and Leuthold, B. M. 1975. Temporal patterns of reproduction in ungulates of Tsavo National Park, Kenya. *E. Afr. Wildl. J.* 13:159–69.

McNaughton, S. J. 1976. Serengeti migratory wildebeest: facilitation of energy flow by grazing. *Science* 191:92–94.

Morth, H. T. 1967. *Investigations into the meteorological aspects of variations in the level of Lake Victoria.* East African Meterological Department Memoir, vol. 4, no. 2. Nairobi.

Norton-Griffiths, M. 1973. Counting the Serengeti migratory wildebeest using two-stage sampling. *E. Afr. Wildl. J.* 11:135–49.

Norton-Griths, M.; Herlocker, D.; and Pennycuick, L. 1975. The patterns of rainfall in the Serengeti ecosystem, Tanzania. *E. Afr. Wildl. J.* 13:347–74.

Pearsall, W. H. 1957. Report on an ecological survey of the Serengeti
National Park, Tanganyika. *Oryx* 4:71–136.

Pennycuick, L. 1975. Movements of migratory wildebeest population in
the Serengeti area between 1960 and 1973. *E. Afr. Wildl. J.* 13:65–87.

Pennycuick, L., and Norton-Griffiths, M. 1976. Fluctuations in the rainfall of the Serengeti ecosystem, Tanzania. *J. Biogeog.* 3:125–40.

Sinclair, A. R. E. 1973. Population increases of buffalo and wildebeest in
the Serengeti. *E. Afr. Wildl. J.* 11:93–107.

————. 1977. *The African buffalo.* Chicago: Univ. of Chicago Press.

Swynnerton, G. H. 1958. Fauna of the Serengeti National Park. *Mammalia* 22:435–50.

Talbot, L. M., and Stewart, D. R. M. 1964. First wildlife census of the
entire Serengeti-Mara region, East Africa. *J. Wildl. Manage.* 28:
815–27.

Talbot, L. M., and Talbot, M. H. 1963. *The wildebeest in western Masailand, East Africa.* Wildl. Monogr. no. 12. The Wildlife Society.

Vesey-Fitzgerald, D. F. 1960. Grazing succession among East African
game animals. *J. Mammal.* 41:161–72.

Watson, R. M. 1967. The population ecology of the wildebeest (*Connochaetes taurinus albojubatus* Thomas) in the Serengeti. Ph.D. dissertation, Cambridge University.

Watson, R. M., and Bell, R. H. V. 1969. The distribution, abundance and
status of elephant in the Serengeti region of Northern Tanzania. *J.
Appl. Ecol.* 6:115–32.

Watson, R. M., and Kerfoot, O. 1964. A short note on the intensity of
grazing of the Serengeti plains by plains game. *Z. Saugetierk.* 29:
317–20.

Western, D., and van Praet, C. 1973. Cyclical changes in the habitat and
climate of an East African ecosystem. *Nature* (Lond.) 241:104–6.

P. J. Jarman
A. R. E. Sinclair

Six Feeding Strategy and the
Pattern of Resource
Partitioning in Ungulates

The Serengeti's ungulates impress the visitor by their huge numbers, the vast extent of country they occupy, and the numbers of species that occur together. These combined impressions pose the classic ecological question How do so many individuals of so many similar species coexist in such an apparently homogeneous plant community?

Studies have now provided some evidence that interspecific competition is taking place (chap. 4) and is likely to be the process leading to the pattern of resource partitioning shown by the ungulates. In this chapter, we describe some feeding strategies that have evolved to reduce competition and that produce the observed pattern. We deal first with three species, impala, topi, and buffalo, that are sedentary in the woodlands. We then compare features observed in these three with those of three migrant species, Thomson's gazelle, wildebeest, and zebra, to show that they all face the same problems of obtaining sufficient quantity of good-quality food, and that they overcome these problems by methods dictated by body size, metabolic rate, and shape of mouth parts.

An understanding of resource partitioning is of direct relevance to conservation and management of the area, particularly since the species composition of the herbivore community is dynamic. It is only through such understanding that predictions about the response of the ecosystem to external or internal pressures will eventually be possible.

Grzimek and Grzimek (1960) were the first to recognize that not all areas were being equally grazed by herbivores, nor all grass species equally used. From chemical analyses of preferred and rejected species, they found that higher protein and digestibility characterized the former. At that time Vesey-Fitzgerald (1960) also put forward his suggestion of "interspecific facilitation," whereby one herbivore provides a feeding opportunity for another, through the former's grazing of the vegetation. Subsequent studies have elaborated on these ideas, detailing the methods by which animals select their habitats and food, and revealing patterns of association between herbivores.

The Ungulate's View
of Vegetation

The vegetation provides food, water, minerals, shade, and cover to hide in; of these requirements, food is the most important. The vegetation also provides cover for predators, obstruction to visual communication, and a lot of inedible material among which the animal must search for usable food; of these, the last is particularly important.

These features of vegetation vary in time and space, and the importance of each varies for different ungulate species, and even between individuals of one species. The ungulate responds to these features in choosing where to live and what to eat.

Physical attributes of vegetation may affect habitat use. For instance, an impala is untroubled by the precise water content of the plants that it eats, provided that this exceeds 30 percent of the plants' dry weight (Jarman 1973). Once water content falls below that value, the impala must drink, and, hence, must live and feed within reach of water. Water requirements are influenced by the availability of shade when the sun is high. Impala may be restricted to habitats where there is shade, such as mature *Acacia tortilis* woodland, and, more precisely, to small patches of those habitats. The animal is restricted to sedentary activities, like resting and ruminating, when confined to shade. This, in turn, may produce a shift in the daily activity schedule, so that more mobile activities occur predominantly in the cool of the day or at night. And nocturnal activity may then restrict the animal to open habitats with good visibility.

An individual ungulate may choose a vegetation type for cover. Alternatively, it may avoid cover that could contain predators. For example, a dik-dik may always remain near dense bush because it uses cover to hide from predators. But we have no test of this explanation, because dik-dik are rarely found away from bush, and, consequently, differential predation rates (near versus far from bushes) are unobtainable. Impala at Banagi avoided areas where grass was higher than 40 cm, despite the abundant food there. The risk of predators being hidden in such tall grass was probably one of the factors influencing their choice; but we can only guess at this.

Timing is an important factor in the plant communities' production of food for ungulates. Plant growth is limited by soil moisture, rather than by temperature. Soils differ in their water-holding capacities: in general, those on ridgetops hold the least amount of water, those on sumps and valleys the greatest amount. Adjacent soil types are described as an inter-linked sequence, or "soil catena," from the eluvial soils on the ridges, from which materials are removed, through colluvial soils of the slopes, to alluvial soils of the valleys, to which materials flow. Plant growth starts when the soil moisture is recharged by periodic rain, continues while moisture remains available at root level, and ceases when the soil dries out. Plants on eluvial soils grow for a shorter time after rains have finished than those on lower alluvial soils, where soil moisture persists longer. Year-round plant growth may be supported by permanently available soil moisture next to lakes, rivers, or springs, but such areas make up a relatively small fraction of the Serengeti.

Plants with deeper roots, such as shrubs and trees, obtain soil moisture for longer than those with superficial roots and, hence, are active for longer after the end of the rains than annuals. Plant productivity, particu-larly that of grasses, is directly related to the amount of rainfall (chap. 4 above; Braun 1973). Thus, annual production is least in the south and east of the Serengeti, and greatest in the north and west. Taking all these facts together, an ungulate should find plant production most restricted in time and quantity on ridgetop soils of the southeast, and greatest on the alluvial soils of the northwest.

Physical structure and the timing of growth are not the only character-istics relevant to ungulates. Where soil moisture is abundant, tall grass, shrubs, and trees flourish. This has two effects. Firstly, the vegetation may grow out of reach of most ungulates. Secondly, much of the production will be of structural tissues, rich in insoluble lignin, cellulose, and hemi-cellulose, which are of low value as food for ungulates. By contrast, ridge-top soils of low rainfall areas support only a brief burst of growth from small grasses and herbs. They have relatively little structural matter, and are, consequently, high-quality ungulate food.

High-quality foods contain easily digestible soluble carbohydrates and, especially, proteins. Low-quality foods are typically old, tough, woody or fibrous, and indigestible; those of high quality are either young, soft, and green, or are storage organs, such as tubers, fruits, and seeds. Plant com-munities contain mostly low-quality food. This is especially true in peren-nial communities, particularly forest. It is least true in communities of annuals and in early growth stages of communities in the rains.

Average food value in plant communities available to ungulates changes from the start of the rains: high-quality young growth matures with the development of low-quality structural tissue. In the dry season, annuals die, and soluble constituents are withdrawn from accessible parts of perennials. Brief reversals in this seasonal decline may benefit some animals, as, for example, when trees drop high-quality fruit in the dry season. Individual plants follow the same seasonal pattern. Thus, as plant material increases in abundance during the wet season, it gradually decreases in quality, a decrease which continues after growth has stopped.

Changes in quality are accompanied by changes in the dispersion of food items. A perennial grass plant, which was burned or eaten to the ground in the previous season, will consist of a tuft of new leaves of uniformly high value early in the rains. As differentiation proceeds, high-value leaves become scattered among lower-quality stems. Similar changes occur within communities, as plants grow and mature at different rates. This can be seen in table 6.1, where some typical values are presented. Perennial dicots, especially shrubs and trees, are more differentiated in food quality between stems and leaves than are monocots. The more dif-

Table 6.1

Examples of crude protein content (expressed as %-age of dry matter) of grass parts harvested in the Serengeti. Data are from Duncan (1975), from swards at different stages of maturity. (For other examples of seasonal changes, see Sinclair 1977, fig. 13.)

Growth stage and species	Green leaf %	Dry leaf %	Stem and Sheath %
Early green growth			
Pennisetum mezianum	20.0	—	15.0
Digitaria macroblephora	16.6	—	10.5
Heteropogon contortus	14.8	—	5.5
Mature, but growing			
Pennisetum mezianum	9.8	4.3	4.4
Digitaria macroblephora	10.0	4.0	3.8
Dry-season sward			
Pennisetum mezianum	6.8	3.9	3.1
Digitaria macroblephora	11.1	4.9	2.5
Heteropogon contortus	6.8	3.8	2.9

ferentiated the adacent parts within a plant are, the more a feeding animal will have to distinguish between them.

Feeding Strategies

The seasonal and spatial differences between plant communities in species composition, production, food quality, and food-item dispersion all have a bearing on how an individual ungulate sets about feeding. So, too, do the animal's intrinsic characteristics: whether it is large or small, short or tall, and the specific adaptations of its mouth and gut. In general, among related species with similar digestive capabilities, smaller species require better-quality food, than do large species because of the former's higher metabolic rate. On the other hand, individuals of the larger species require larger absolute quantities of food (Bell 1969; Jarman 1968, 1974). Over the course of evolution, these differences have tended to produce, at one end of the spectrum, small ungulates with mouths adapted for carefully selecting discrete, high-quality food items, and, at the other end, large species with mouths adapted for rapid ingestion of large quantities of undifferentiated items, possibly of low quality. The social consequences of these differences are discussed in chapter 8.

An ungulate might exist by taking random bites from the nearest plant in whatever vegetation community it found itself; but it would likely die, as the quality of its diet would almost certainly be too low, at least for part of the year. If the animal were more selective, if it obtained the right quality and type of food at the fastest rate, within habitats that afforded it sufficient protection from the climate and predators, it would have a better chance of survival. Selection can occur at a number of levels, and these are sequential to some extent. Firstly, in a given season, an animal can choose a vegetation community in which to feed. Secondly, it can choose among plant types or species to eat. Thirdly, it may choose which parts of those plants to eat. Each refining step in this process of selection should increase the rate at which it obtains high-quality food. Not all species emphasize each step equally. The most highly selective may need to go through the whole sequence, while more generalist species might be sufficiently rewarded by the first level of selection. Differences between individuals in feeding strategies has received scant attention, so we present, where possible, evidence for heterogeneity within a population.

To demonstrate nonrandomness in a species' feeding, it is necessary to measure where and what that species eats and to compare these with what is available for it to eat. There is some confusion in the literature over what is meant by "available." When considering seasonal choice of habitat, we take all areas within the animal's annual range as "available"; when considering diet, we consider all plants within physical reach of the animal within the habitats used at that time as "available." We start with impala, the smallest of the three sedentary species, and then compare the other two species with it.

Impala in the Serengeti

Impala (pls. 23–27) are medium-sized antelope (40–55 kg), widely distributed through the woodlands of Africa, from Swaziland in the south to northern Kenya in the north. In all regions they avoid the habitat extremes of desert, open grassland, montane heath, and closed forest. Based on several studies of their diet, they are known to be mixed feeders, taking varying proportions of grass and browse (Lamprey 1963; Talbot and Talbot 1962; Stewart 1971; Jarman 1971; Azavedo and Agnew 1968). In nearly all regions, individuals move seasonally between habitats in a small or medium home range (a few hectares to a few square kilometers, depending on region). In the Serengeti, impala occur in herds throughout the woodlands; they are especially numerous in the center and north of the park, and have annual home ranges of 100 to 1000 ha.

Selection for feeding habitat. Within their home range, impala show clear-cut seasonal preferences for feeding habitats. Figure 6.1 illustrates two years' seasonal use of habitats in the central woodlands. Impala preferred *Acacia senegal* woodlands in the wet season, particularly toward the end of that season, and *Acacia drepanolobium* savanna, grassland, and drainage-line vegetation in the dry season. In the immediate vicinity of Banagi, they occupy *A. senegal* and *A. clavigera* woodlands in the wet, and mature *A. tortilis* woodland in the dry season; elsewhere in the park they alternate between other vegetation types with season. In general, vegetation communities higher on the local catena are preferred in the wet season, those on alluvial soils low on the catena in the dry season; and this pattern is similar elsewhere in Africa (Jarman 1972).

Within this seasonal pattern, classes of impala may differ in their habitat selection. Females with newborn young often seek areas with bushes. At the start of the rains, territorial males may occupy territories high on the catena long before other impala have started to use such habitat. Bache-

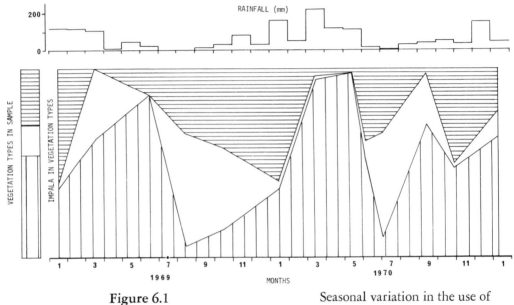

Figure 6.1 Seasonal variation in the use of habitats by impala between Banagi and Ikoma. The proportions of impala using *Acacia senegal* woodland high on the catena (*vertical shading*); tall *A. tortilis* woodland (*no shading*); and grassland, drainage-line vegetation, and *A. drepanolobium* communities low on the catena (*horizontal shading*), during thirteen sample periods, are shown. The relative proportions of the three habitat types are shown to the left of the figure, and the rainfall at Banagi above.

lor males may use vegetation communities peripheral to those favored by females, thereby avoiding harassment by territorial males (Jarman and Jarman 1973). Hence, different individuals identify suitable vegetation communities for different reasons not always connected with food.

Selection for plant species. Impala are able to select the species of plants
on which to feed. When two captive impala were offered either browse or
grass species two or three at a time, a hierarchy of preference for plant
species became apparent. Field observations of feeding impala recorded
those plants that were eaten and all others within 10 cm radius to give a
direct measure of choice. At that time, impala ate predominantly grasses,
with nine out of fourteen available grass species being selected (in 177
recorded bites). Two species, *Digitaria macroblephora* and *Panicum
maximum,* were significantly preferred, (P < 0.001), and two others,
Heteropogon contortus and *Themeda triandra,* were significantly avoided
(P < 0.001 and P < 0.01, respectively; Jarman unpublished).

The effect of selection for species could be seen by comparing the
proportions of grass and browse in the average impala's diet and in the
plant communities. In the wet season, impala preferred grass, but in
the dry season changed to browse, selecting this component at a rate above
its availability, even though grass remained the major dietary constituent.
This can be seen in the comparison between pasture composition and
impala stomach contents (table 6.2). The trend from grazing in the wet
season to browsing in the dry is also seen in the analysis of mouth contents

Table 6.2   Seasonal variation in the proportions of grass and forbs in pasture samples at Banagi, and of grass and dicot material in the stomach contents of impala. Data are from Jarman and Gwynne (unpub.).

Month of sample (1970)	Average pasture		Impala stomach contents	
	Grass %	Forbs %	Grass %	Dicot material %
January	78.3	21.7	97.9	2.1
February	70.2	29.8	—	—
March/April	77.2	22.8	98.8	1.2
June	86.4	13.6	86.4	13.6
July/August	91.0	9.0	87.0	13.0
September	92.9	7.1	—	—
October	80.8	19.2	96.8	3.2
December	82.4	17.6	98.4	1.6

(fig. 6.2), which is not biased by differential digestion of plants. Further-more, feeding observations of impala over a year (fig. 6.3) showed that browse from bushes and small trees was eaten more frequently as rainfall declined. This negative relationship is seen clearly in figure 6.4.

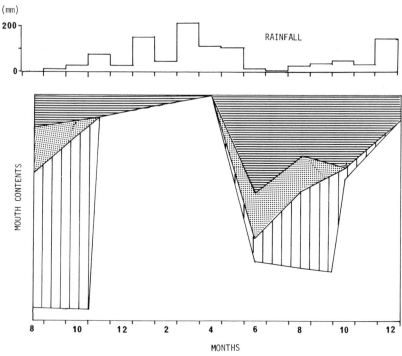

Figure 6.2   The relative proportions of green grass (*no shading*), dry grass (*vertical shading*), fruits and seeds (*dotted shading*), and browse (*horizontal shading*) in the mouths of impala shot during eight sample periods ($N > 10$ in all samples) are shown, with the rainfall at Banagi during the same period (1969–70).

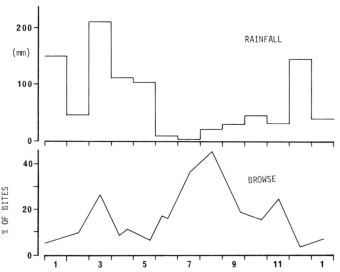

Figure 6.3   Seasonal use of browse by impala.
The proportion of bites that
observed impala took from browse
plants above the herb layer, and
Banagi rainfall over the same
period (1970–71).

Selection for Plant Part

Fruits, flowers, and leaves were the only parts of woody browse species
that impala ate; stems were avoided completely. Impala also differentiated
between parts of grasses. A comparison between the proportions of grass
leaf, sheath, and stem in pastures and in stomach contents (table 6.3)
showed that in all seasons impala ate significantly ($P < 0.001$) lower proportions of stem and higher proportions of sheath than occurred in the
pasture; the proportion of leaf eaten was either the same as, or less than,
that in the pasture. Because no allowance has been made for differential
digestion of these parts, the analyses probably underestimate the extent of
selection against stem and for leaf and sheath.

Causes and effects of feeding selection. Impala chose for feeding not only
the plant community, but also the plant type and species, and the plant

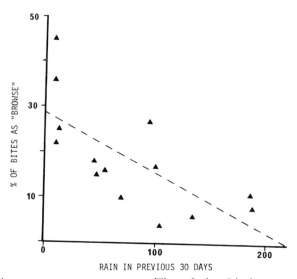

Figure 6.4   The relationship between use of
browse by impala and recent rain.
The proportion of observed bites
that were of browse plants (y)
plotted against rainfall summed
over the preceeding thirty days
(x). The broken line indicates the
relationship: $y = 28.3 - 0.129 \, x$
($r = 0.689$, $d.f. = 12$, $P < 0.01$)
where y is a percent, and x is in mm.

parts. These levels of selection combine to form a coherent strategy. Grass
was always the major component of the diet, so aspects of the strategy
that improved the quality of grass eaten were most important. By avoiding stem, impala avoided the most fibrous, lowest-quality fraction of the
plant. The preferred grass species (especially *D. macroblephora*) were
usually green, short, soft, and leafy; the avoided species, such as *H. con-*
tortus and *T. triandra,* grow into tall, tough plants. Selection for species
is probably brought about in the same way as selection for plant part: by
avoiding tough, fibrous material.

The proportion of green grass leaf in the stomach contents increased
in the wet season and decreased in the dry. Moreover, green leaf was

Table 6.3

Seasonal variation in the proportions of grass parts found in the
stomach contents of impala collected near Banagi. Data are from
Jarman and Gwynne (unpub.).

Sampling Month	Leaf %	Sheath %	Stem %	Number of impala sampled
August 1969	34.1	45.3	20.6	12
January 1970	53.9	39.8	6.3	25
March/April 1970	32.4	53.6	14.0	23
June 1970	27.7	49.7	23.3	12
August 1970	24.2	53.1	22.7	20
October 1970	30.1	54.1	15.8	20
December 1970	50.7	40.5	8.8	19

closely correlated with the proportion that total leaf formed of the grass fraction in the stomach contents (fig. 6.5). This suggests that the impala's selection for leaf was governed by its quality, as reflected by how green it was. Their avoidance of dry grass, and preference for green material, are other facets of their selection for low-fiber food items, as may be their seasonal use of browse. We lack data on relative fiber content (or its inverse, nutritive value) of browse compared with green or dry grass in each season, but the implication is that the still-green browse becomes relatively more attractive as grasses mature in the dry season. Further, as the grass crop diminishes after growth ceases, browse on trees and shrubs becomes relatively more available because its standing crop tends to be more constant through the seasons than is that of grass.

Fruits and seeds, mainly of trees (*Acacia* spp., *Balanites aegyptica*) and the herb *Solanum incanum,* form a minor, highly seasonal part of the impala's diet. They are eaten as soon as they become available. With their high protein (up to 27%; Gwynne 1969) or sugar content, they form valuable food sources at a season when the quality and availability of other foods are falling.

The seasonal choice of feeding habitat also seems to favor the quality of the impala's diet. When we compared the stomach contents of impala collected in preferred and nonpreferred habitats, we found that those collected in the former had, on average, a higher proportion of green grass leaf in their stomachs than had those from nonpreferred habitats. On six out of the seven occasions, those in preferred habitat had eaten a

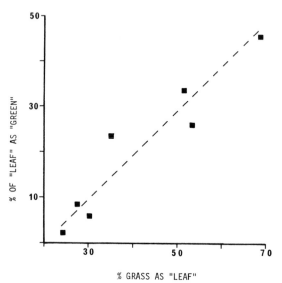

Figure 6.5   Selection for greenness of leaf by
impala in relation to the proportion
that "leaf" formed of the grass
component of rumen contents.
Average values for large seasonal
samples. From Jarman and
Gwynne (unpub.).

higher proportion of total grass leaf than had the others. On the seventh,
although the grass eaten in the preferred habitat was less leafy, impala
there had eaten much higher proportions of protein-rich *Acacia* seeds
than had those animals elsewhere (P. J. Jarman and M. D. Gwynne,
unpublished). Thus, there appear to be consistent advantages, in terms of
food quality, in occupying preferred habitats in all seasons. For those animals remaining on nonpreferred habitats, there may be compensatory
advantages in terms of quantity or dispersion of food items, for instance.

The whole strategy depends upon individuals switching habitats, food
types, and plant parts in an attempt to delay an inevitable seasonal decline
in food quality. One result is that the impala's diet includes a wide array
of species, at least seventy in the Serengeti, with a range of palatability.
Thus, some species are eaten readily, while others, such as the evergreen
shrub *Salvadora persica,* are eaten only in extremely dry conditions,

despite being available year-round. It may be important for an impala to learn what and where to eat. Two captive impala had to be taught to eat some foods that wild impala ate avidly.

Analyses of stomach contents revealed minor differences in diet quality between classes of individuals. In one dry-season sample, impala over eight years old ate a consistently higher proportion of *Acacia* seeds and pods than younger adults. In all seasons, lactating females obtained slightly better quality grass (in terms of greenness and leafiness) than did pregnant females (P. J. Jarman and M. D. Gwynne, unpublished). While the advantages of this are apparent, the cause is obscure. Why should pregnant females not be as successful as lactating ones in finding high-quality food?

Impala feeding selection appears nutritionally advantageous compared to random feeding. There is some evidence to support this. The physical condition of a large number of impala females was judged visually every month for a year. We knew from post-mortem inspection of females that the visual criteria used to judge condition correlated closely with the ratio of muscle to bone, which is a measure of stored protein, and less closely with the amount of fat reserves. The average condition of the impala females showed a marked seasonal variation, which reflected rainfall but with a considerable time lag. Whereas aspects of the feeding strategy, such as the amount of green grass or browse eaten, were responsive to very recent rain (within the past 11–30 days), female condition was most strongly correlated with average rainfall over the preceding 31–90 days. Also, at the start of the rains, females responded to recent rainfall by improving their condition, while in the dry season, their falling condition best correlated with rainfall between the previous 61–90 days (Jarman, unpublished data). The difference between the rapid response of feeding behavior at the end of the rains and the delayed response of physical condition implies that the feeding strategy was having some success in postponing a decline in the sufficiency of the diet.

Topi in the Serengeti

Topi are large (100–120 kg) antelope, with elongate, narrow faces (pls. 20, 21), distributed widely but irregularly through Africa—from Botswana, where they are called *tsessebe,* to Sudan in the north and Nigeria in the west, where they are called *tiang.* Their habitats are all mesic, perennial grassland, or savanna communities. The Serengeti contains one of the last great populations of topi in East Africa.

Their home ranges include anywhere from a few to many hundreds of square kilometers, since in many areas they are extensively nomadic, or even migratory. Several studies have demonstrated that they are almost exclusively grazers (Talbot and Talbot 1962; Gwynne and Bell 1968; Stewart and Stewart 1970; Field 1972). Major studies of topi in the Serengeti have been undertaken by Bell (1969, 1970, 1971) and Duncan (1975), supplemented by information from the Kenya Mara (Jarman 1976).

Selection for feeding habitat. Within the Serengeti, topi avoid the eastern plains and the southeast, occur sparsely on the central plains, in fair numbers through the central woodlands to the Kenya Mara, and in high numbers on the western plains. They avoid dense woodland, forest, and thicket, and steep, rocky hillsides. The western plains have low relief, with each catena level or distinct vegetation community covering an area many kilometers wide. In the woodlands, a rolling topography of hills and river valleys carries a diversity of closely packed vegetation communities. In the latter case, one day's wandering may carry a topi through several vegetation types; in the former, many days or weeks could be spent in one community, except for excursions to drink.

Despite these regional differences in diversity of vegetation communities, topi display a consistent seasonal pattern of habitat use. In the wet season, they move toward the top of the local catena, and are concentrated there in the late wet season, on communities of short or medium-length (5 to 15 cm) grass (pl. 20). They avoid areas where grass is either very short or tall, mature, and flowering. In the dry season, they move down to vegetation on alluvial soils. They seem to select seasonally for categories of plant communities, distinguishable by plant structure rather than by species composition (Duncan 1975).

In the woodlands, these shifts from upper to lower catena levels involved short movements of the same magnitude as normal daily movements, and for individuals amounted to little more than a change in the proportions of time they spent in different parts of their home range. On the western plains, however, changes in seasonally selected habitat required movement of several kilometers and a shift in home range.

Selection for plant species. Since topi eat grass almost exclusively, they clearly select against dicot components of the sward. Duncan (1975) took tame topi to natural-grass communities and found they selected certain species over others. However, their selection was not consistent; a species strongly favored in one season would be rejected in another. For example, *D. macroblephora* was preferred over *P. mezianum* in the dry season, but early in the wet season, as the sward was regenerating after being burned, *P. mezianum* was preferred. When the sward matured, *D. macroblephora* returned to favor. It seems that topi were guided by growth rather than by species.

Selection for plant part. Gwynne and Bell (1968) showed that in the dry season, zebra, wildebeest, and topi differed in the proportions of grass leaf, sheath, and stem in their stomachs. Since the animals were collected at the same time and in the same area, at least two of the species were, by inference, selecting from among the available grass components.

Duncan (1975) carried this investigation further, and found marked seasonal fluctuations in the relative proportions of the grass parts in the rumen (fig. 6.6). Using topi with esophageal fistulae to graze natural

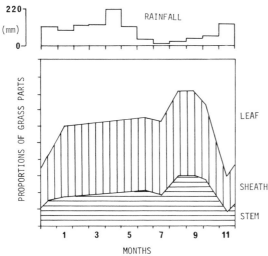

Figure 6.6   Seasonal variations in the relative proportions of grass parts in the rumina of topi, and rainfall over the same period. Data, from Duncan (1975), have not been corrected for differential digestion.

swards of known composition, Duncan found that in all seasons, topi selected for green leaf and avoided stem and sheath. In the wet season, they ate a diet of up to 86 percent green grass leaf, and took more than 50 percent of this component in nearly all seasons, except the late dry season, when it fell to 25.5 percent.

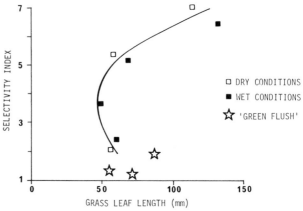

Figure 6.7   Changes in the selectivity index of topi with leaf length of grass. Data from Duncan (1975).

Causes and effects of feeding selection. During the wet season, the amount of food that fistulated topi obtained in each bite was related linearly to the biomass of available grass. In the dry season, topi obtained smaller bite-sizes than they would have from the same quantities in the wet season. Leaf length between 4 and 7 cm influenced bite-size strongly, there being a fourfold linear increase in bite-size over this range. Above 7 cm leaf length made no further difference to bite-size. In general, intake rates were depressed by dry-season conditions on all except the highest-standing crops of grass, and by grass leaves shorter than 7 cm.

Using the ratio of proportion of green leaf in the diet to its proportion in the sward as an index of selectivity, Duncan found that selectivity by his captive animals varied from 7.0 to 1.2:1. Selectivity increased linearly as the quantity of grass available increased; but selectivity bore a complex curvilinear relationship to leaf length (fig. 6.7). Grass with long leaf produced high selectivity values, that with short or medium leaf, especially

young regrowth of burned sward, low selectivity. Figure 6.8 shows that
selectivity rose sharply when green leaf fell below 15 percent to 20 percent
of the sward. Selectivity also rose as the crude protein content of the stem
and sheath component of the grasses fell, but there was no relationship
with the protein content of green leaf. In summary, selectivity increased as
stem and sheath quality declined, and as available biomass increased.

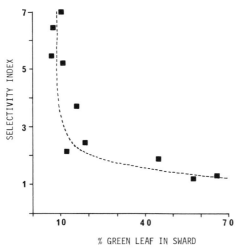

Figure 6.8   Selectivity shown by topi in relation to the proportion of green
leaf in the sward. Data are from
Duncan (1975).

The advantages of selecting for green grass leaf are clear, for it contains
one and a-half to three times as much protein as dead leaf or stem and
sheath. These differences in quality are greatest in dry grass and in mature,
growing swards and least in young growth after fires. Accordingly, feeding selection by topi is greatest and most advantageous with mature or dry
grass, and least with young grass. Duncan proposed that the primary
function of selection for plant species and habitats was to increase the
protein quality of the diet. Chemical analyses showed that topi usually
selected those species with the highest protein content. Since grass species
grow, mature, and dry at differing rates, each species seasonally changes
its value for a feeding topi relative to other species, which accounts for the
inconsistency in species preference shown by topi.

The seasonal use of habitat reflected the topi's wet-season choice of areas characterized by fairly short grasses (since during the wet season, higher quantities depress quality) with high proportions of green leaf; and their dry-season choice of areas with large quantities of grass with long leaves. Topi in all regions effectively moved to keep themselves in grasslands of similar growth stage, height, and proportions of leaves to maintain an adequate quality of diet.

Under Duncan's experimental conditions, his topi obtained a diet that was 9 percent to 110 percent richer in crude protein than a random sample of the sward, which reflects the success of their feeding strategy. The diet gave the animals an abundant supply of energy (carbohydrates) in all seasons, but dietary protein intake in the dry season fell to barely maintenance level. In the dry season, all ages and both sexes of topi lost protein reserves (muscle bulk) as each animal drew upon its bodily stores. This would happen sooner and with more severe consequences without the observed feeding strategy.

Buffalo in the Serengeti

Buffalo (pls. 15, 16) occur over even more of Africa than do impala and topi; until recently, they were almost ubiquitous south of the Sahara. Their biology is described by Sinclair (1977). In their feeding and digestion, they resemble antelopes. Because of their large size (400 to 700 kg), they require a larger quantity of food, although of a lower quality than that of smaller ruminants. They are grazers living in large herds, probably of closed membership, with each herd having a large, discrete home range.

Selection for feeding habitat. In the Serengeti, buffalo occur everywhere except on the short-grass plains. Although they use medium- and long-grass open plains, they concentrate along drainage lines. In the Moru area, their year-round preference for riverine grassland increases in the dry season, when they also avoid hills and dry woodland. The dry-season situation is clear; buffalo concentrate on the vegetation community at the bottom of the catena and avoid all others. In the wet season, although they still prefer riverine grassland, they spread into other habitats, except plains, which are high on the catena.

In the northern Serengeti, the pattern is similar. All vegetation types except *Terminalia* woodland are used in the wet season, with no consistent preference appearing, although open grassland is sometimes heavily favored. In the dry season, they prefer forest and riverine vegetation and avoid open grassland. Bachelor males, living in much smaller herds than the rest of the population, show the same preferences for habitats. However, they also make use of seepage lines, which provide a narrow strip of green vegetation, perhaps too small for large herds to use.

Buffalo have muzzles 12 or 15 cm wide and, hence, prefer habitats with tall grass. They are not well adapted to removing individual leaves from plants such as grass, nor can they easily feed on short grasses. Hence, they prefer habitats with tall grass.

Selection for plant species. Two captive buffalo were able to distinguish between a number of grass species offered as cut samples. Similar ability was shown when a buffalo was offered a choice between grazing natural adjacent stands of *T. triandra* and *P. mezianum.* As in the experiments with fresh-cut grass, *Themeda* was strongly preferred over *Pennisetum* (Sinclair 1977). In all these experiments, the grasses were long, green, and flowering, a growth stage often eaten by buffalo in the wild.

Selection for plant part. By analysing stomach contents, Sinclair and Gwynne (1972) investigated the buffalo's ability to select parts from the available grasses. From the start of the rains, the proportion of grass leaf in the stomach contents rose rapidly, to a peak of 50 to 70 percent in the mid-rains; that of sheath fell to a trough of 20–30 percent, and stem to 10 to 20 percent. Stem rose steadily throughout the dry season, reaching a proportion of 45 percent, while leaf fell to 10 percent or less by the late dry season. The proportion of grass leaf in the buffalo's diet increased with increasing rainfall in the month of sampling, up to about 90 mm monthly rain.

Comparing stomach samples with sward samples clipped where the buffalo were collected (fig. 6.9) shows that buffalo ate grass quite unselectively in the early wet season; thereafter, they consistently avoided stem, although apparently with diminishing success as the dry season progressed (Sinclair 1977)

Causes and effects of selection. The tame buffalo preferred grass species with a high ratio of leaf and sheath to stem (Sinclair 1977). This implied a preference for leaf which is supported by the analysis of stomach contents. This preference was most pronounced at the end of the rains, but it diminished during the dry season, despite an increased preference for riverine habitats, where grass growth persisted longest.

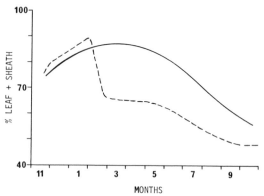

Figure 6.9

Seasonal variation in the proportion of leaf and sheath of grass in the rumina of sampled buffalo (*solid line*), and in the pasture (*broken line*) from the start of the rains to the dry season.

In general, when compared to impala and topi, buffalo are poor at selecting plant parts, and, perhaps, at selecting grass species in natural mixed swards. The buffalos' strategy involves dispersion to many vegetation types in the rains, but retreat in the dry season to the only vegetation communities in which grasses are still long and partly green, where the animals exercise what selection they can by avoiding grasses with tough stems. Buffalo may have to leave some habitats in the dry season because the grass is too short.

Estimates of protein intake increased to a peak in March, then declined steadily until the end of the dry season. In the wet season, intake was up to twice maintenance requirements, but from August to October, it fell to below maintenance. Hence, even the relatively poor selectivity of buffalo would be advantageous at this time.

Protein intake in the dry season differed little between sex and age classes. However, calves and old adults were less able to store fat (as indicated by bone marrow fat content), and these age-groups suffered a higher mortality rate in the dry season.

*Comparison of Impala, Topi,
and Buffalo*

Each of the three species display feeding strategies; for at least part of
the year, each achieves a diet that is better than a random selection from
the surrounding vegetation. Yet their strategies differ, and some of the
differences illustrate important theoretical points about the resource ecology of large herbivores.

The species differ in their selectivity for habitat. Impala were almost
always selective, especially during the seasonal extremes. In the late wet
season, nearly the entire local population occupied just one vegetation
type. By contrast, buffalo showed high selectivity for habitat only in the
dry season, associating randomly with vegetation communities in the wet
season. Topi were intermediate between these styles of habitat selection.

These differences relate to the three species' types of food selection.
Buffalo are so large that their relative metabolic requirements of energy
and protein are low enough to be satisfied by a random cropping of grass
swards during the early wet season. Later they exercised what selection for
grass parts they could as the sward quality declined. However, buffalo have
muzzles over 12 cm broad (compared with 3 or 4 cm in an impala), and
are incapable of distinguishing between plant parts finely enough to
gather a high-quality grass diet in the dry season. Buffalo may be restricted
to their ultimate dry-season habitat by quantity of available fodder as much
as by quality.

Although selective to some extent for green grass leaf in all seasons,
during the rains topi choose a feeding habitat for the quality of its sward
and show little selection for plant parts. As is true with buffalo, topi selectivity increases (sixfold) as grasses increase in size, maturity, and the
proportion of nonleaf components. In the dry season, optimum feeding
conditions are increasingly supplied by those swards offering the greatest
quantities of grass. But, in contrast to buffalo, topi in all seasons seek an
appropriate growth stage of pasture from which to obtain green grass leaf.

Impala, like topi, select plant parts in all seasons, but, unlike topi, they
do not eat only grasses. Impala maintain a high-quality intake by eating
dicot fruits, seeds, seedpods and leaves in the dry season, as the availability
of green grass leaf declines. Because of this, impala appear to be less
restricted by the available plant biomass than are buffalo or topi.

Although all three species can distinguish between grass species, and,
indeed, show similar plant species preferences, they appear to choose

growth stages of grasses according to the plants' physical characteristics rather than their taxonomy. These three species could be characterized as eating bunches of grass unselectively (buffalo); stripping leaves off medium-height grass plants (topi); and picking green leaves from short grass or bushes or seeking fruits and seeds (impala). The spatial distributions of these food items differ at any one time of year. Food plants eaten by buffalo are usually closer together than those eaten by impala (except in the early rains). This applies both in absolute terms and relative to the size of the animal. Such differences are particularly relevant to the herd-forming habits of each species (chap. 8).

Figure 6.10 compares the estimated protein intakes of each species. Buffalo's intake fell steadily from a late rains peak of 10–11 percent crude protein, to a minimum of 2–4 percent at the end of the dry season (Sinclair 1977), which was well below maintenance level, and possibly even below the level at which rumen microorganisms can function. Topi's

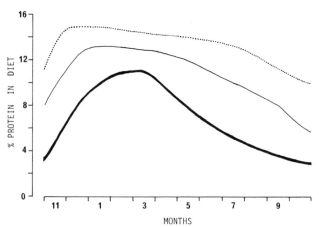

Figure 6.10   Seasonal changes in the dietary protein intake of impala (*dotted line*), topi (*thin line*), and buffalo (*heavy line*), from the beginning of the rains to the late dry season. Data are from Duncan (1975), Sinclair (1977), and Jarman (unpub.).

intake (calculated from data in Duncan 1975) reached a peak of 11–16
percent in the wet season, and remained above 8 percent until the end of
the dry season, when it fell abruptly to 6 percent. Duncan (1975) considers this barely adequate for maintenance. Neither the protein intake of
topi nor that of impala fell as steadily or as rapidly as that of buffalo.
Impala intake was consistently high (12–18 percent crude protein) in
the wet season, and even at the end of the dry season, it was 9–12 percent
protein, largely because of dicot seeds, fruits, and leaves.

From figure 6.10, several points emerge. There is a sequence—impala,
then topi, followed by buffalo—of increase in protein intake after the
start of the rains. Diet quality is most similar in the late rains; and there is
a reverse sequence—buffalo, then topi, then impala— in the rate of decline in dietary quality. That the smallest, most selective, finest-muzzled
species (impala), should benefit soonest by the flush of annuals and new,
short growth produced by the rains is to be expected. Also, with the greatest flexibility in feeding, impala can best reduce the stress of the dry
season. However, absolute factors influence these relative abilities. No
matter how successful an animal is at selecting its food, there is an
absolute limit to the quality of diet it can obtain. At the time of year when
plant parts are least differentiated structurally, during the middle of the
rains, there is least to be gained by selection. And, indeed, at that time,
buffalo, exercising hardly any selectivity, achieved a diet little worse than
that of topi or impala. Relative to their specific protein requirements, buffalo may even have enjoyed the highest-quality diet at this time (over twice
their needs) and impala the lowest (perhaps less than 1.5 times their
needs of roughly 11 percent protein; D. Western, pers. comm.). It may
be that the larger species, which benefit most by the high average quality
of pasture in the wet season, can better afford a strategy of protein storage
in the wet season to carry them through the poverty of the dry season.
Unable to select sufficiently in the dry season, buffalo rely on bulk intake
of grass to maintain their energy intake, and on stored protein for their
nitrogen metabolism. Thus, differences in protein requirements between
herbivores of different sizes leads to the largest becoming the most dominant (in biomass) in the community.

Ecological Separation and
Feeding Strategy

Ecologists point to Africa's species-rich communities of large mammals
to illustrate the axiom that species living together must depend on differ-

ent resources. Evidence for this derives from our knowledge of the feeding strategies of herbivores. Differences between species have usually been sought in either their use of habitats or in their use of plant species, rarely both.

Lamprey's study (1963) remains the most complete. He demonstrated that each species of large herbivore in a community depends upon separate resources because of its different spatial and temporal distribution, choice of plant species, and height at which the animals feed. But that study did not include large populations of the migratory grazers that dominate the Serengeti. Studies of these mixed-species concentrations by Talbot and Talbot (1962) and Stewart (1971; Stewart and Stewart 1970) confirmed extensive overlap in habitats used and grass species eaten. However, the work of M. D. Gwynne (Gwynne and Bell 1968; Sinclair and Gwynne 1972; Jarman and Gwynne, unpublished; Duncan, Gwynne, and Jarman, unpublished) showed that, for a herbivore, grass is a heterogeneous collection of parts as distinct and distinguishable in value as those of a shrub or a tree, and that, at any one time, the parts of one grass plant differ more than the average quality of whole plants of different species. Once this was appreciated, the ecological separation of grazers became apparent. By aiding observation of the animal's patterns of distribution, the development of systematic aerial survey provided further evidence of separation (Grzimek and Grzimek 1960; Watson 1967; chap. 5 above).

Ecological Separation of
Resident Species

Impala, topi, and buffalo are nonmigratory species whose ranges in the Serengeti overlap extensively. They illustrate how species may be ecologically separated through feeding strategies. For a start, separation occurs through habitat preferences, with buffalo and impala avoiding open plains used by large herds of topi. Buffalo breeding herds avoid *Acacia* woodlands favored by impala, which, in turn, avoid long-grass areas used by buffalo. However, separation through habitat preferences is incomplete, and all three species may feed within meters of each other. A degree of temporal separation occurs because of their sequential progress up and down the catena. Overlap is greatest in the late dry season, when all three use vegetation communities low on the catena, and least in the wet season, when they tend to occupy different catena levels.

The types and species of plants they eat are not totally distinct. Impala share little of the browse component of their diet with the other two, but

many of the grasses eaten, such as *D. macroblephora,* are preferred by all
three. However, buffalo also eat some species when mature, such as *T. triandra,* which are avoided by the other two.

In their selection of grass parts, the smaller impala eat a higher proportion of leaf and sheath than do either topi or buffalo, although topi eat
a higher proportion of leaf alone than do the other two. In the dry season,
buffalo were least successful at avoiding grass stem.

Since protein resources in the dry season are limiting, at least for topi
and buffalo (Duncan 1975; Sinclair 1977), overlap indicates that competition is the selection pressure maintaining ecological separation.

Ecological Separation and the Migratory Species

Separation within the migratory species. Gwynne and Bell (1968) describe the food of the three major migratory ungulates—wildebeest, zebra,
and Thomson's gazelle (pls. 9, 17, 19)—and Bell (1969, 1970, 1971)
describes their separation through differences in spatial and temporal use
of habitats and consumption of different grass parts. These species use the
local plains catena sequentially. After the rains cease, zebra are the first
to descend from the upper catena communities, followed by wildebeest,
then Thomson's gazelle. Bell generalized the catena concept to cover the
whole system used by the migratory species, equating the highest catena
levels with the plains, areas of most ephemeral productivity seasonally;
and lower catena levels with the northern Serengeti, regions of extended
productivity. The way in which the three migratory species move within
the whole system is essentially the same as their use of the local catena:
zebra leave the eastern, short-grass plains first, followed by wildebeest,
then Thomson's gazelle.

Bell (1970) explained the grazing succession in terms of feeding
preferences. In the wet season, all three concentrate on the upper catena or
plains, where "the vegetation structure is least robust and the protein
content highest." Since zebra eat higher proportions of stem than do the
other two (table 6.4), they are able to move into longer grass areas
(northern woodlands or lower catena levels) in the dry season earlier
than the others. Bell suggested that zebra open up the tall, coarse grasses
by removing stems, thus exposing the leaves at the base of the plants and
allowing greater access for wildebeest, which need a higher proportion
of leaf in their diet. Wildebeest then remove the majority of the tall grass,
leaving a short sward for gazelle. The latter avoid long grass, perhaps

Table 6.4

The proportions of grass parts in the stomach contents of wildebeest and zebra collected in the western Serengeti in wet and dry seasons. Data are from Gwynne and Bell (1968).

| | Grass Parts | | |
	Leaf %	Sheath %	Stem %
Wet Season			
Wildebeest	61.3	28.9	8.0
Zebra	36.8	43.9	19.3
Dry Season			
Wildebeest	17.2	52.7	30.1
Zebra	0.2	48.7	57.0

because they cannot see predators there, nor run away from them. In the short, grazed areas, they can make use of the few scattered leaves left by the wildebeest and obtain greater access to the high-protein herbs normally hidden in the tall grass. Thomson's gazelle eat large amounts of these herbs (39 percent of rumen contents), whereas wildebeest and zebra are pure grazers. Wildebeest actively select for leaf in all seasons, but zebra select for grass sheath in the dry season, feeding unselectively in the wet season (Owaga 1975).

Two other migratory species await study in the Serengeti: Grant's gazelle (pl. 18) and eland. Elsewhere these species take consistently higher proportions of browse than wildebeest, zebra, or Thomson's gazelle (Talbot and Talbot 1962; Lamprey 1963; Stewart and Stewart 1970). Consequently, they can delay their dry-season departure from the upper catena, or eastern plains, even later than Thomson's gazelle.

Competition between migratory and resident herbivores. In their grazing sequence down the local catena in the dry season, Bell found that buffalo preceded zebra, and topi were intermediate between zebra and wildebeest. Since Gwynne and Bell (1968) have shown that, in the dry season, topi were intermediate between zebra and wildebeest in their selection of grass parts, Bell interpreted these observations as indicating that buffalo were at least as tolerant as zebra of tall, fibrous grass, and that topi preferred a

somewhat taller growth stage of grass than did wildebeest. This appears
to be substantially true; the relatively narrow-muzzled topi can strip leaves
off quite tall grasses, while the broader-muzzled wildebeest feeds more
efficiently on shorter, leafy green swards. Topi in the western Serengeti
were separated in both wet and dry seasons from wildebeest by the
former's preference for vegetation types with longer grass: "the preference of topi for a herb layer that combines a high proportion of vertical
as opposed to horizontal elements produces a distribution radically different from that of wildebeest" (Bell 1969); and Jarman (1976) found
an essentially similar distinction in the Kenya Mara.

The concentration of migratory wildebeest at the height of the dry
season has a substantial impact upon buffalo (Sinclair 1977). Since wildebeest keep to open grassland, and buffalo use riverine vegetation at that
season, there is remarkably little overlap in preference for feeding habitat.
However, wildebeest pass through the grassland parts (but not the thickets
and forest) of the riverine vegetation to reach water, and in the process
they trample and flatten the grass, making it inaccessible for feeding
buffalo. The wildebeest concentrations are so great that, even without
preferring a habitat, the presence of a small proportion of them can
devastate a pasture, and result in competition for habitat. Zebra might also
reduce the remnant dry-season pastures needed by the buffalo. In most
other parts of the buffalo's range in the Serengeti, wildebeest are absent
in the dry season, but the competitive effect will increase as the expanding
wildebeest population increases its range (chap. 5).

Impala and other woodland antelope find wildebeest and zebra passing
through their habitats in the late rains or dry season; and Thomson's
gazelle, some Grant's gazelle, and zebra disperse throughout the central
and western woodlands in the dry season. Transient zebra and wildebeest
reduce the standing crop of grass rapidly. This may benefit impala, since
they otherwise avoid communities carrying tall grass; but the sudden reduction in available grass may force them on to the limited reserves of
browse sooner than if the migratory herds had gone elsewhere. Thomson's
gazelle, although only half the weight of impala and rarely achieving the
same local densities, pose more of a competitive threat, since their dietary
preferences appear so similar to impalas'. They have even finer muzzles,
and so are at least as capable as impala of selecting the high-quality dry-
season dicot foods, such as *Acacia* seeds, *Solanum* fruits, and young leaves
between the thorns of *Acacia* bushes.

Impala are caught between the generalist grazers (zebra and wildebeest) and the selective browser (Thomson's gazelle). The grazers re-

move the grass crop, at least some of which would have been of sufficient quality for impala, forcing impala onto resources of dispersed items of browse, for which they must compete with the smaller and more selective Thomson's gazelle that follow the wildebeest. However, impala are larger than Thomson's gazelle, so they do not need to be as selective, and can continue to eat more grass in the dry season.

Advantages of being migratory or resident. Similar patterns of food requirements and selection characterize the trios of resident and migrant species discussed above. The largest species, buffalo and zebra, descend the catena soonest in the dry season, spending longest on the lower catena, and feeding least selectively on the large quantities of food available. The smallest species considered, impala and Thomson's gazelle, utilize the upper catena longest, obtaining high-quality diets by feeding selectively from the vegetation left by the larger species. But while the residents move annually only within their local land system, the migrants extend their movements to exploit the differences in amount and timing of rainfall within the whole ecosystem (chap. 5). All three migrant species move to the eastern plains for their brief period of productivity in the rains, later moving west and north to regions where higher rainfall and dry-season storms prolong plant growth. As a result, the biomass of migrant populations far outweighs that of residents.

The migrants' basic advantage over the resident is presumably their access to a greater total of plant production each year. Migrants can apply a far greater instantaneous grazing pressure to an area, which they will leave when its pasture is exhausted, than can residents which must survive there year-round. Thus, migrants depress the populations of any residents which share their food and habitat preferences. Only those residents survive that can live on what the migrants leave untouched. Such species include highly selective small antelopes (dik-dik, oribi, grey duiker, bushbuck, steinbuck, klipspringer), many of which live in habitats unused by migrants; specialized grazers (topi, kongoni, waterbuck, reedbuck) partly separated from the migrants by the growth forms of the grasses they eat and the habitats they use; the impala, adapting to the impact of migrants by changing habitats and foods; and the very large buffalo, which survives in habitats and on a quality of food unsuitable for the major migrants.

Migration is likely to arise wherever the amount or timing of productivity of some ubiquitous plant types differs within the region. In the Serengeti, these ubiquitous plants are the major grasses, and, consequently, the migrants are mainly grazers. Migration presumably started because individuals survived and reproduced better if they moved than if they remained in one area. Clearly, an individual wildebeest remaining on the eastern plains would starve in the dry season; one moving away would not. Yet in the west and north, wildebeest could (and some do still) successfully remain year-round, although their numbers would be limited by the local plant productivity, and they would be even more affected than other residents by any other wildebeest that migrated through the area.

If migrating is so advantageous, why do other Serengeti ungulates not migrate? Firstly, they may be too small to move over the distances needed to benefit by productivity differences, at acceptable energy cost (see chap. 7). Secondly, their required foods or habitats may not be sufficiently ubiquitous for them to take full advantage of the regional productivity differences, or for them to find a route without long stretches of inhospitable habitat. The migratory species are remarkably catholic in their use of habitats; no resident ungulate population uses so wide a range of habitats nor extends over so much of the ecosystem. The migrant's needs, not only for food, but also for shade, cover, water, minerals, and their social and antipredator behavior must be so flexible that they can be met over a large and continuous part of the ecosystem. But if that applies, a migrant will have a much greater probability of gaining adequate year-round resources than will a resident of the same species.

Dynamics of Feeding Succession and Plant Community

Vesey-Fitzgerald (1960) proposed the concept of facilitation through a grazing succession whereby large species, through feeding, provided habitats for smaller species. Bell (1970, 1971) suggested that this occurred with zebra, wildebeest, and Thomson's gazelle. As mentioned earlier, the larger species can dominate the herbivore community. Smaller species can coexist because they have two advantages: first, they can survive on the more dispersed remnants of the herb layer left by the larger species; and second, they can make use of the very short new growth at the beginning of the rains earlier than can the larger herbivores. If the small species are sufficiently numerous, as Thomson's gazelle may be, they can maintain the sward in this state and prolong their advantage for a while. But, in general, the dominant ungulate is the one whose feeding strategy is best suited to the phenology of the plant community that the current grazing pressure induces. At present, this appears to be the wildebeest, but might under other circumstances be Thomson's gazelle or zebra.

This interaction between herbivores and vegetation is dynamic at two levels. First, if grazing pressure on the plains were lighter or more selective, grasses would grow longer and zebra would benefit. If zebra numbers increased, wildebeest would later benefit; and their grazing would induce a short grass sward which would eventually lead to invasion by herbs (as now; McNaughten, pers. comm.) and an increase in gazelles. Increasing wildebeest numbers would adversely affect zebra in the early rains by preventing the growth of long grass; similarly, gazelles might affect both wildebeest and zebra. The system might not stabilize because the gazelles would graze too selectively to prevent the development of a sward or coarser, taller grasses, which would benefit zebra and set off the cycle again.

Second, different phases of the cycle may occur simultaneously in different areas. For example, a mosaic of burned and unburned areas in the woodlands would initially favor wildebeest-dominated and zebra-dominated phases of the cycle, respectively. On the burned areas, consistently heavy grazing by wildebeest would later benefit the gazelles, and wildebeest would then move to areas grazed down by zebra, which, in turn, would move to other unburned areas. This cyclic progression would be interrupted by the spatial distributions of fire and local rainstorms, which vary from year to year, affecting the dry-season movements of the migratory herbivores. The relatively low biomass of resident species probably depresses the amplitude or extends the periodicity of the migrants' cycle slightly, and other factors, such as surrounding land use and long-term rainfall, affect the whole system.

Conclusions

Those Serengeti ungulates that have been studied in enough depth all select food at least for part of the year. Selection occurs at a number of levels—for habitat, for plant type or species, or for plant parts. Emphasis given to these levels differs between species. The largest grazers studied, buffalo and zebra, hardly select for plant part or species at all in the early wet season, and buffalo show little selection for habitat in the wet season as a whole. These species have relatively low requirements of dietary protein, so that even without selection, buffalo consume twice the food quality they require for maintenance in the early wet season. By contrast, smaller species, such as impala, select habitat, type of food species, and plant part in all seasons. Despite this selection, their protein intake is relatively less rich in quality than the buffalo's, but does not decline as rapidly, or as far, in the dry season.

Among migratory and nonmigratory species, these strategies lead to a certain degree of resource partitioning and ecological separation of species. The relationships between the species are seen in the grazing succession: smaller herbivores use the upper catena with short grasses first in the rains, and for longest; larger species use them least in the rains, descending to lower levels of long grass first. Larger species have the advantage of tolerating coarser, lower-quality food, and, hence, can obtain the bulk of the available grass. Smaller species can coexist by using the scattered remnants of high-quality food. Larger species may also facilitate smaller species by removing coarse material and exposing the higher-quality items. Thus, differences in food dispersion result in a range of herbivore sizes. But the composition of the community is unlikely to be stable, since external factors (fire, climate, humans) will perturb the herbivore-vegetation system. Management should recognize the potential for fluctuation in species composition and attempt to monitor and model the system in such a way that the outcomes of future perturbations can be predicted.

References

Azavedo, J. C. S., and Agnew, A. D. Q. 1968. Rift valley impala food preferences. *E. Afr. Wildl. J.* 6:145–46.

Bell, R. H. V. 1969. The use of the herb layer by grazing ungulates in the Serengeti National Park, Tanzania. Ph.D. dissertation, Manchester University.

———. 1970. The use of the herb layer by grazing ungulates in the Serengeti. In *Animal populations in relation to their food resources,* ed. A. Watson, pp. 111–23. Oxford: Blackwell.

———. 1971. A grazing ecosystem in the Serengeti. *Sci. Am.* 224(1): 86–93.

Braun, H. M. H. 1973. *Primary production in the Serengeti: purpose, methods, and some results of research.* Ann. Univ. d'Abidjan, ser. E, no. 6, pp. 171–88.

Bredon, R. M., and Wilson, J. 1963. The chemical composition and nutritive value of grasses from semi-arid areas of Karamoja, as related to ecology and types of soil. *E. Afr. Agric. For. J.* 29:134–42.

Duncan, P. 1975. Topi and their food supply. Ph.D. dissertation, University of Nairobi.

Duncan, P.; Gwynne, M. D.; and Jarman, P. J. The status of the Lamu-Garissa topi population, parts 1–4. Unpublished.

Field, C. R. 1972. The food habits of wild ungulates in Uganda by analyses of stomach contents. *E. Afr. Wildl. J.* 10:17–42.

Grzimek, M., and Grzimek, B. 1960. A study of the game of the Serengeti plains. *Z. Saugetier.* 25:1–61.

Gwynne, M. D. 1969. The nutritive values of *Acacia* pods in relation to *Acacia* seeds distribution by ungulates. *E. Afr. Wildl. J.* 7:176–78.

Gwynne, M. D., and Bell, R. H. V. 1968. Selection of grazing components by grazing ungulates in the Serengeti National Park. *Nature* (Lond.) 220:390–93.

Jarman, P. J. 1968. The effect of the creation of Lake Kariba upon the terrestrial ecology of the Middle Zambesi Valley, with particular reference to the large mammals. Ph.D. dissertation, Manchester University.

————. 1971. Diets of large mammals in the woodlands around Lake Kariba, Rhodesia. *Oecologia* (Berlin) 8:157–78.

————. 1972. Seasonal distribution of large mammal populations in the unflooded Middle Zambesi Valley. *J. Appl. Ecol.* 9:283–99.

————. 1973. The free water intake of impala in relation to the water content of their food. *E. Afr. Agric. For. J.* 38:343–51.

————. 1974. The social organization of antelope in relation to their ecology. *Behaviour* 48:215–66.

————. 1976. Behaviour of topi in a shadeless environment. *Zool. Afr.* 12:101–11.

Jarman, P. J., and Gwynne, M. D. Feeding ecology of impala. Unpublished.

Jarman, P. J., and Jarman, M. V. 1973. Social behaviour, population structure, and reproductive potential in impala. *E. Afr. Wildl. J.* 11:329–38.

Lamprey, H. F. 1963. Ecological separation of the large mammal species in the Tarangire Game Reserve, Tanganyika. *E. Afr. Wildl. J.* 1:63–92.

Owaga, M. L. 1975. The feeding ecology of wildebeest and zebra in Athi-Kaputei plains. *E. Afr. Wildl. J.* 13:375–83.

Sinclair, A. R. E. 1972. Long-term monitoring of mammal populations in the Serengeti: census of non-migratory ungulates, 1971. *E. Afr. Wildl. J.* 10:287–97.

————. 1974. The natural regulation of buffalo populations in East
Africa. Part 4. The food supply as a regulating factor, and competition. *E. Afr. Wildl. J.* 12:291–311.

Sinclair, A. R. E. 1977. *The African buffalo.* Chicago: Univ. of Chicago
Press.

Sinclair, A. R. E., and Gwynne, M. D. 1972. Food selection and competition in the East African buffalo (*Syncerus caffer* Sparrman) *E.
Afr. Wildl. J.* 10:77–89.

Stewart, D. R. M. 1971. Food preferences of an impala herd. *J. Wildl.
Manage.* 35:86–93.

Stewart, D. R. M., and Stewart, J. E. 1970. Food preference data by
fecal analysis for African plains ungulates. *Zool. Afr.* 5:115–29.

Talbot, L. M., and Talbot, M. H. 1962. Food preferences of some East
African wild ungulates. *E. Afr. Agric. For. J.* 27:131–38.

Vesey-Fitzgerald, D. F. 1960. Grazing succession amongst East African
game animals. *J. Mammal.* 41:161–70.

Watson, R. M. 1967. The population ecology of the Serengeti wildebeest. Ph.D. dissertation, Cambridge University.

C. J. Pennycuick

Seven Energy Costs of
Locomotion and the Concept
of "Foraging Radius"

The energy used by an animal for locomotion may be represented as one of several items on the debit side of its energy budget. This chapter is concerned with the relationship of this item to the rest of the budget. For this purpose, it will be convenient to consider the debit side of the budget under three headings, as follows:

E_m: The rate of energy expenditure on basal metabolism.
E_f: The rate of energy expenditure on feeding.
E_l: The rate of energy expenditure on locomotion.

The E's are powers or energy rates, that is, they have the dimensions of energy/time. If these rates are not qualified by a further subscript, they are to be understood as mean rates, averaged over one or more complete cycles of feeding and other activities. The additional subscript z indicates an instantaneous rate. For instance E_{lz} is the rate at which energy is used for locomotion while the animal is actually in motion, while E_l is the average rate of consumption for locomotion over a period which includes a mixture of movement and inactivity. The animal's total expenditure, E_x, may now be represented as

$$E_x = E_m + E_f + E_l \tag{1}$$

Its energy income E_i, is the rate at which it extracts energy from its food, and its energy surplus, E_s, may be defined as the excess of income over expenditure:

$$E_s = E_i - E_x \tag{2}$$

E_s is the power supply for growth and reproduction, and it is the quantity which natural selection requires every animal to maximize.

The three components of expenditure distinguished in equation 1 are to some extent interdependent. An animal like a wildebeest must move about to find its food, but cannot eat until it stops. It is the purpose of this paper to consider how the requirements of locomotion combine with other features of the biology of various animals to set limits on the strategies open to them.

The question is of particular interest in connection with the Serengeti fauna, in which the biomass is dominated by species with well-developed migratory habits. Evidently such animals as the wildebeest, zebra, and Thomson's gazelle achieve a greater energy surplus through their migratory behavior than they would if they remained sedentary; otherwise, this behavior would be selected against. The expenditure of fuel incurred in migrating about the ecosystem is quite large, but the resulting increased average rate of food intake more than compensates for this.

Amount of Energy Needed
for Locomotion

To get some idea of the orders of magnitude involved, a very rough estimate can be made of the amount of fuel consumed by a typical wildebeest in propelling itself around its annual migration circuit, for comparison with a similarly rough estimate of the net amount of energy obtained by the same animal from its food in a year. The results reviewed by Taylor et al. (1970) provide the basis for such an estimate. On the basis of measurements of the oxygen consumption of animals walking or running on treadmills, made by a number of different authors, Taylor et al. found that for a particular animal, the instantaneous rate at which metabolic energy is required for locomotion (E_{lz}) is directly proportional to the speed:

$$E_{lz} = kV \qquad (3)$$

where V is the speed. E_{lz} is not the animal's total rate of energy consumption, but the additional rate due to exercise. Basal metabolism continues at the same time. k is a constant with dimensions energy/distance. It is the energy required to propel the animal unit distance. This turns out to be independent of speed, at any rate between a medium walk and quite a fast canter. This remarkable result is not readily explicable from theory, but has been confirmed empirically for a wide range of species of quite diverse morphology.

Taylor et al. go on to show that within a group of broadly similar animals (such as all large cursorial warm-blooded vertebrates), k can be estimated from the body mass alone. The relationship is of the form

$$\frac{k}{k_o} = \left(\frac{m}{m_o}\right)^b \tag{4}$$

where m is the body mass, and the suffix o refers to some datum animal. For large mammals, the relationship may be expressed in the more convenient (if not dimensionally correct) form

$$k = 11 \, m^{0.60} \tag{5}$$

where k is in joules per meter and m is in kilograms. Thus, a typical wildebeest, with a body mass of around 180 kg, requires about 250 J/m for locomotion, irrespective of whether it walks or canters. This result can be used to estimate how much energy is needed for locomotion in the course of a year. From detailed studies of the movements of radio-tagged individuals, D. Kreulen (pers. comm.) estimates that 10 km per day is a fair estimate of the rate of travel of a wildebeest of the Serengeti migratory population, averaged over a whole year. Thus, the energy required in one year is about 9.1×10^8 J. Kreulen further estimates that the net energy income in a year is about 1.2×10^{10} J, so that about 8 percent of this income would be required for locomotion.

Foraging Radius

The energy required for locomotion, though appreciable, is not a very large fraction of even a wildebeest's net energy income, yet its significance is greater than this estimate would suggest. Its influence is indirect, and concerned with the division of the animal's time between different activities. This is sometimes discussed under the heading "time budget," but the latter concept is not directly analogous to, nor indeed separable from, that of an energy budget. It is more satisfactory to consider the division of time in terms of its effect on the energy budget, as defined above.

The notion of "foraging radius" is derived from (but not confined to) situations where an animal has to commute between a foraging area and some fixed point. This is a class of special cases, of which two examples will be considered.

1. The case of a water-dependent herbivore under dry-season conditions (that is, no plant production taking place), in which the animal has to commute outward from a water source, traveling progressively further as the vegetation around the water is grazed down.

2. The case of a female carnivore with immobile young in a den, in which the mother has to travel to forage among a herbivore concentration, and then return to suckle her young.

In each case, the animal is presumed to set off from the foraging area with a full gut, and to travel a distance r to the fixed point, then to travel the same distance back before it can feed again. The distance r is the "foraging radius," so that the total distance out and back is $2r$.

Since the animal is assumed not to eat during the journey, locomotion and basal metabolism are the only debit items that need to be considered in the energy budget. The expenditure on these items must be limited to some amount of energy, e, if the animal is to continue its commuting activity without net loss of body mass. In the case of a carnivore, e may be considered to be the net energy extracted from a full gut load—that is, the animal can set off from the foraging area after feeding to capacity, and arrive back there with a more or less empty gut. Herbivores, both ruminant and otherwise, cannot do this, however, since depletion of their gut contents below a certain point leads to disruption of the gut flora, which then takes some time to reestablish itself. Indirect estimates of e will be obtained below, but it may be noted here that e is much less, in relation to the body mass, in herbivores than in carnivores. In addition, the quality of the diet is more variable in herbivores than in carnivores, and this, too, will affect the current value of e for any particular animal.

Foraging radius in herbivores. Considering a herbivore commuting to and from a water source, its energy expenditure while walking, from equation 1, is

$$E_{xz} = E_m + E_{lz}$$

since E_{fz} will be zero if the animal does not feed while walking. Substituting from equation 3, we obtain

$$E_{xz} = E_m + kV \qquad (6)$$

The time taken to consume e is then,

$$T = \frac{e}{E_{xz}} = \frac{e}{E_m + kV} \qquad (7)$$

Since $2r = TV$, by definition, it follows that

$$r = \frac{eV}{2(E_m + kV)} \qquad (8)$$

Scaling of foraging radius with body mass. It is of interest to see how *r* varies with the body mass of the animal. Since the two terms in the denominator of the right-hand side of equation 8 scale in different ways with the body mass, no simple power law can be deduced directly. However, it is possible to construct empirical relationships between each of the variables in equation 8 and the body mass, and then to deduce the manner in which *r* should vary with the mass. Such a relationship for *k* has already been given in equation 5.

According to Alexander (1976), mechanically similar motion is obtained in different animals if the walking speed varies with the square root of the limb length. In a later paper, Alexander (1977) shows that in a selection of East African herbivores, limb length varies with about the 0.25 power of the body mass, as predicted by McMahon's (1975) scaling law for "elastic similarity." Thus, the cruising speed in any particular gait (for example, the normal walking speed) should increase with the 0.13 power of the mass. This relationship can be calibrated on the wildebeest, whose mass is assumed to be 180 kg, and for which the mean walking speed was found by Pennycuick (1975) to be 0.98 m/s. Thus,

$$V = 0.50 m^{0.13} \tag{9}$$

where V is the walking speed in m/s and m is the mass in kg.

The variation of basal metabolic rate E_m with body mass is well documented. The relationship appropriate to ungulates, according to Rogerson (1968), is

$$E_m = 4.8 m^{0.74} \tag{10}$$

where E_m is in watts.

An estimate for *e* can, in principle, be obtained from a knowledge of the dynamics of the animal's feeding and digestion. This may well be practicable, but is difficult. An alternative, if rather approximate method is to infer *e* indirectly for some animal for which an estimate of foraging radius is available, and then to assume that *e* for other trophically similar animals is proportional to the body mass. Thus, we can begin with Lamprey's (1963) remark that elephants forage out to about 40 km from water in the dry season as an estimate for *r* for this animal, and assume a typical body mass of, say, 2000 kg. Then we have for the elephant:

$$r = 4 \times 10^4 \text{ m} \quad \text{(assumed)}$$
$$V = 1.34 \text{ m/s} \quad \text{(from equation 9)}$$

$$k = 1050 \text{ J/m} \quad \text{(from equation 5)}$$
$$E_m = 1330 \text{ W} \quad \text{(from equation 10)}$$

Substituting these values in equation 8, we get

$$e = 1.63 \times 10^8 \text{ J}$$

for a 2000 kg elephant. Then, assuming that e for other animals is proportional to body mass, we can write

$$e = 82000 \text{ m} \tag{11}$$

where e is in joules.

Figure 7.1. is a double-logarithmic plot of foraging radius against body mass. r has been calculated from equation 8, using values of k, V, E_m

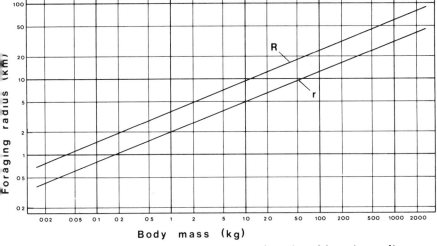

Figure 7.1   Log-log plot of foraging radius versus body mass. The lower line is for herbivores in dry-season conditions, calculated as in the text, while the upper line is the "ultimate foraging radius" obtained by neglecting expenditure on basal metabolism while traveling.

and e calculated from equations 5, 9, 10, and 11, respectively. It can be seen that for practical purposes the curve is a straight line with a slope of 0.40. It can be represented approximately by the formula

$$r = 2000 \ m^{0.40} \tag{12}$$

where r is in meters.

This 0.40 power law describes a very pronounced increase in foraging radius with increasing body mass. The original assumption was that $r = 40$ km for a 2000 kg elephant, although equation 12 actually yields 42 km because of rounding of the coefficients. For smaller animals it yields 16 km for a 180 kg wildebeest, 6.6 km for a 20 kg Thomson's gazelle, and 420 m for a 20 g mouse.

Effect of speed on foraging radius. It has been noted that according to Taylor's Law as represented by equation 3, the direct energy cost of traveling a given distance does not vary, whether the animal walks or runs. The foraging radius, however, is greater if the animal uses a faster gait. For example, the lower line of fig. 7.1 gives a foraging radius of 16 km for a 180 kg wildebeest, on the assumption that the animal walks at a speed of 0.98 m/s (the average walking speed actually observed in the field). If we assume instead that the animal canters at the observed average cantering speed, which was 5.1 m/s, and substitute this speed in equation 8, a foraging radius of 25 km is obtained, 58 percent more than before. This is because the cantering wildebeest takes a shorter time over its journey, so that it uses less of its fixed fuel allowance (e) on basal metabolism, and therefore has more left over to convert into distance.

If the speed could be increased indefinitely, the foraging radius would approach an asymptotic or ultimate value R, where

$$R = \frac{e}{2k} \tag{13}$$

It can be seen by comparing equation 13 with equations 5 and 11 that R must vary with the 0.40 power of the body mass, and, for any given herbivore, it is about 1.94 times the foraging radius for walking, as calculated above. Thus, although changing from a walk to a canter gives a wildebeest a large increase in r, higher speeds would not bring a correspondingly higher advantage.

R has been plotted as the upper line in Figure 7.1. The fact that the lower line, representing r, has the same slope (that is, R/r independent of mass) is not a necessary consequence of the algebra, but results from the particular numerical values used for the constants in equations 5, 9, 10, and 11.

Selection for increased foraging radius in ungulates. The situation depicted above, in which an animal subsisting on marginal food has to travel out and back to a water source at frequent intervals, is probably a fair approximation to the conditions under which most mortality occurs in water-dependent ungulates. As the dry season progresses, with no new plant growth taking place, the animals have to go further and further from the water to find food. There is thus mounting selection pressure for a large foraging radius. From the argument above, it is clear that this selection will favor two attributes: (1) large body mass, and (2) the ability to maintain a fast gait for long periods, that is, without going into oxygen debt.

The use of fast gaits. The term *gait* has been used in various senses. In the usage preferred here, which has been discussed at some length elsewhere (Pennycuick 1975), the distinction between two gaits depends on a *discontinuous* change of coordination required to change from one to the other. Two such transitions could be recognized in the field, from a walk to a trot, and from a trot to a canter. R. McN. Alexander (pers. comm.) recognizes a third transition, from a canter to a gallop, based on a discontinuous change in the phase relationships between the legs. This last transition could not be discerned in the field, however, and so, for the present purpose, only three gaits are recognized: the walk, trot, and canter, with the gallop being regarded as a fast canter. Mechanically equivalent variants are considered to be "stepping patterns" within one gait, and not separate gaits, so that the "trot" includes the pace, and the "canter" includes both transverse and rotatory forms. The trot is not used in animals with sloping backs, such as the topi, kongoni, hyena, giraffe, wildebeest, and so on. Such animals shift directly from a walk to a canter, although some of them, including the wildebeest, are capable of the coordination required to trot. Animals with horizontal backs (gazelle, eland, zebra, lion, buffalo, hunting dog, rhino, and so on) use the trot readily as a medium-speed gait.

A range of speeds is available in each gait, but the extremes are infrequently used. For instance, a migrating wildebeest either walks at about 1 m/s or canters at about 5 m/s. The intervening speeds are not much used, although the canter is more flexible in this respect than the walk (fig. 7.2).

A fast gait yields a large foraging radius, provided that it can be maintained continuously without incurring an oxygen debt. All the Serengeti animals except the elephant are capable of galloping at speeds in

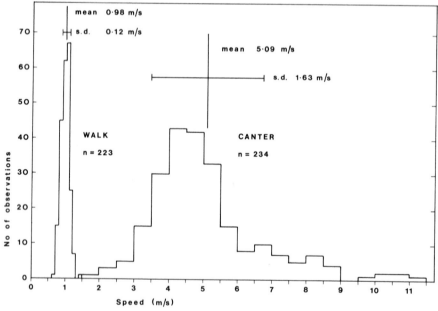

Figure 7.2

Distribution of observed speeds for walking and cantering (including galloping) wildebeest. The methods of observation have been described by Pennycuick (1975). Observations of both sexes are combined in approximately equal numbers, and there are no significant differences between the observations for the two sexes.

excess of 10 m/s, but can only do so for distances too short to be of use for migration. Only the wildebeest and spotted hyena are regularly seen cantering steadily for several kilometers without pausing for breath. Two adaptations are needed for this. First, the mechanics of the animal's locomotion must be such as to allow cantering to be economical in terms of energy consumption. Second, the heart and lungs must have sufficient capacity to supply the demands of the muscles for oxygen and fuel in the fast gait. Locomotor adaptations will be considered further below.

Some of the level-backed animals trot when commuting between water and a grazing area. For example, D. Western (pers. comm.) has observed that at the height of the dry season, zebras alternately trot and walk across the lake bed at Amboseli, Kenya, between water to the southeast of the dry lake, and grazing on the slopes of the Ilaingarunyeni Hills. The distance is about 15 to 20 km, which agrees well with estimates of foraging radius based on the above figures. Zebras walk at about 1.0 m/s and trot at about 2.7 m/s. If we assume m = 200 kg, then the foraging radius works out at 16 km for walking and 23 km for trotting.

Selection for large body mass. Like the dinosaurs, the mammals started their radiation with small forms only, and gradually increased the range of size by evolving medium-sized and large forms, without losing the small ones. Large animals are better able to last out seasonal food shortages than small ones, but they lack resilience in times (like the present) of severe environmental disruption, because of their small equilibrium densities and lower reproductive rates. The need for a large foraging radius may well be a potent source of selection for large size. If two similar species are using the same water source during a prolonged drought, the one with the larger foraging radius will last longer, other things being equal.

Somewhere above the 1000 kg mark, this argument is modified by mechanical considerations, which make trotting and cantering no longer practicable. These gaits depend upon a "pogo-stick" action, in which the animal, in effect, bounces up and down on springs, which are located in the ligaments limiting movement of the distal leg joints. The legs of elephants do not have the necessary elastic structures, and these animals "run" by simply speeding up the walk, without any discontinuous change of gait. They are thus confined to a normal walking speed when migrating, and cannot further increase their range by shifting to a faster gait, as some of the smaller animals can.

Foraging Radius in Carnivores and Scavengers. The preceding argument has been concerned exclusively with the problems of herbivores, but in their role as prey, these animals cause problems for their predators. The studies of Schaller (1972) and Kruuk (1972) have shown that the density of lions and hyenas in the Serengeti is much lower than the potential food supply in the large herbivore populations should theoretically be able to support. The problem for the carnivores is once again one of foraging radius, although in this case the fixed point is not a water source. Although the large carnivores are water-dependent, their food comes to water of its own accord, so there is no need for them to forage far away from it. The problem is that the young of both species are unable to move significant distances for the first few months of their lives, so that nursing mothers have to return to suckle them at regular intervals. Thus, a herd of migratory wildebeest, say, is only available as a food supply as long as it remains within the mother's foraging radius from the den where the young are.

Kruuk (1972) has described how the hyenas place their dens around the western and northern boundaries of the Serengeti plains, where they are between the wet-season and dry-season ranges of the migratory wildebeest. The difficulties are most acute in dry weather, when the wildebeest are in the western corridor. During dry-season conditions, Kruuk records hyena mothers' being away from their dens for three days at a time, and foraging 40 km away.

Hyenas take on a very large amount of food during these expeditions, and return with distended udders, after which they remain with the cubs for as much as two days before setting off again. Greater expenditure, therefore, has to be allowed for in calculating foraging radius in a carnivore, as compared to a herbivore. In the definition of equation 8, it was assumed, in effect, that the herbivore, having filled its gut with food, travels to water and drinks instantaneously, then immediately returns to resume feeding. The carnivore has to spend some time at the den, during which it metabolizes at E_m, and also offloads sufficient energy in the form of milk to supply the cubs' requirements for metabolism and growth for one whole cycle of commuting and nursing. On the basis of Kruuk's data, we can consider a 52 kg hyena with two 10 kg cubs, which spends 40 percent of its time at the den and the other 60 percent traveling and foraging. The distance traveled is $2r$, which includes running about in pursuit of prey, in addition to the double commuting distance to and from the foraging area. The time taken for the whole cycle is T. The mother's metabolic rate is E_{ma}, while the cubs each metabolize at E_{mc}, so that the energy expended on metabolism in one cycle by the three of them is $T(E_{ma} + 2E_{mc})$. Since 60 percent of T is needed to travel a distance $2r$ at a speed V,

$$T = \frac{3.3r}{V} \tag{14}$$

Also, the energy needed to travel $2r$ (regardless of speed) is $2kr$. The sum of the two energy terms is e, the amount of energy that can be expended between feeds, so that

$$e = T(E_{ma} + 2E_{mc}) + 2kr$$
$$= r\left(\frac{3.3}{V}(E_{ma} + 2E_{mc}) + 2k\right)$$

and

$$r = \frac{eV}{3.3(E_{ma} + 2E_{mc}) + 2kV} \tag{8a}$$

This is a modified form of equation 8, with an increased allowance for metabolism of mother and cubs.

The empirical formulas of equations 5 and 11 for k and E_m will provide a first approximation for carnivores, but e must be considerably higher for them, since the variation in food quality which afflicts herbivores should not apply to carnivores to the same extent. They eat meat of similar quality at any season, if they can get it. To judge from the calorific equivalents listed by Cummins and Wuycheck (1971), the substitution of standard-quality meat for low-quality herbage should have the effect of multiplying e by 4 at least. Furthermore, it would seem that the maximum gut load of lions and hyenas is a greater fraction of the body mass than in herbivores, and that a greater fraction (perhaps all) of it can be absorbed before feeding again. If this is represented by a further factor of 2, then e should be around eight times as great as the dry-season herbivore formula of equation 11 would predict. Thus, for carnivores we can suggest that

$$e = (6.6 \times 10^5)m \tag{11a}$$

should give a fair estimate, where e is in joules and m in kilograms, as before.

We now get the following values for substitution in equation 8a:

$$e = 3.4 \times 10^7 \text{ J}$$
$$E_{ma} = 89 \text{ W}$$
$$E_{mc} = 26 \text{ W}$$
$$k = 120 \text{ J/m}$$

If the hyena walks at 0.76 m/s (estimated from equation 9), then $r = 40$ km. If it canters at 3.3 m/s (observed), then $r = 89$ km. Kruuk observed both gaits being used by commuting hyenas, so the practical foraging radius probably lies between these two figures. If the estimates seem on the high side, it has to be remembered that r in this context includes the distance traveled while looking for, and running after, prey, in addition to the transit distance between the den and foraging area. This may well amount to several kilometers.

The commuting stratagem allows hyenas to exploit migratory prey, but only precariously, according to Kruuk. The herbivore migration circuit in the Serengeti is so large that no den site can be chosen that will certainly be within range of the migratory herds throughout the dependent period of the cubs' lives. The favored sites near the edge of the plains maximize the probability of this, but, even so, the herds not infrequently move out of range for a period. When this happens, hyena cubs are liable to starve. Kruuk attributes the low density of hyenas in the Serengeti, relative to prey densities, to juvenile mortality caused by the unreliability of the food supply. Foraging radius may thus be seen as a variable affecting the reliability of the food supply for the young, and hence juvenile survival.

It is interesting to compare hyenas in this respect with the griffon vultures, which also depend on the migratory ungulate herds. As scavengers, these vultures are in direct competition with hyenas, but are not able to dispute with them directly over food, because their body mass is an order of magnitude less. However, their method of cross-country travel, soaring in thermals, is fast and economical, and confers on them a greater foraging radius than can be attained by their mammalian competitors. Taking the Rüppell's griffon vulture as an example, figures given by Pennycuick (1972) indicate that it can soar across country at an average speed of about 12 m/s (a fast gallop for a large mammal), at a total energy cost of about 2.0 J/m. A mammal of similar mass (7.6 kg) would, according to equation 5, need 37 J/m.

Rüppell's griffons rear one chick at a time, and the parents forage alternately, taking turns at guarding the chick (Houston 1976). Each parent thus has to provide for its own metabolism, plus half that of the chick. On a long foraging trip, a parent might spend as much as 8 hours flying, which amounts to one-sixth of its total time, as alternate days are spent on the nest. With this information, a new version of equation 8 can be constructed for the vulture, using the same reasoning as was followed above for the hyena. The result is

$$r = \frac{eV}{12(E_{ma} + \frac{1}{2}E_{mc}) + 2kV} \tag{8b}$$

The actual food requirements of adult and nestling Rüppell's griffons
were determined by Houston (1976), and his figures can be used to give
estimates for the E's in equation 8b. The adult would require about 24 W,
while the nestling's food requirements vary with age, reaching a peak of
around 42 W between 45 and 65 days from hatching. e can be estimated
from the fact that a Rüppell's griffon can lift a crop load of about 1.5 kg
of meat (pers. obs.), with a calorific equivalent of about 5.2×10^6 J/kg,
according to Houston. This yields $e = 7.8 \times 10^6$ J. We already have
$k = 2.0$ J/m and $V = 12$ m/s, as above. Substituting these figures in equation 8b gives $r = 116$ km.

The smaller of the two griffon species, the white-backed vulture, is a
tree-nester, and has a nesting concentration in the Seronera and Mbalageti
areas. This is much the same area as is used by the hyenas for their dens,
and, no doubt, is favored for the same reason, that it is in the middle of
the herbivores' migration circuit. The larger species, the Rüppell's griffon,
being a cliff nester, has no suitable nest sites so conveniently placed. Virtually all of the Rüppell's griffons frequenting the Serengeti nest on the
Gol escarpment, at the eastern edge of the migration range. In the dry
season, when the wildebeest retreat to the western corridor, or into the
Mara Game Reserve to the north, the griffons actually have to commute
100 km or more to find good foraging. As with the mammals, the vultures
have the option of extending their foraging radius for a limited period at
the expense of depleting their fat reserves, and Houston's data suggest
that this does, in fact, occur during lean periods.

Fortunately, dry weather, when this long-range commuting is necessary,
is also good soaring weather, by and large. Another fortuitous feature of
the area is that in dry weather, a thermal "street" often forms from near
Grumechen Hill to the Musabi area, along which the vultures can (and
do) soar at speeds up to twice the "standard" cross-country speed of 12
m/s assumed above. Even without this extra advantage, the griffons' foraging radius is great enough to ensure that a herbivore concentration is
always within commuting range. This must greatly increase the reliability
of their food supply as compared to that of hyenas, and, hence, if Kruuk's
argument is correct, must reduce juvenile mortality. Indeed, Houston
recorded survival rates to fledging approaching 100 percent in both the
griffon species in 1969 and 1970, admittedly in a period when herbivore
populations were increasing.

Mobility without a Fixed Base

Foraging outward from a fixed point is essentially a dry-season problem for the Serengeti herbivores. In the rains, when surface water is more widely available, the animals are able to move around more freely, but marked differences in strategy are seen between different species. Some, such as wildebeest, zebra, Thomson's gazelle, and eland, move much greater distances than others, such as buffalo, topi, impala, and dik-dik. The most migratory species of all, the wildebeest, also has the highest biomass density. Evidently, the energy cost of migration is more than compensated by the additional income so obtained. By constantly moving about, the animals contrive to remain in good feeding conditions for a greater proportion of the year than would otherwise be possible. In other words, migration enables them to increase the quantity $(E_i - E_m - E_t)$ by an amount that more than covers the cost of locomotion.

If we consider a hungry wildebeest falling upon good grazing, it can feed for a maximum time T_f, but then has to stop because its gut is full. By previous definition, it is then able to expend a quantity of energy, e, before it needs to feed again, and some of this can be used for locomotion if the animal so elects. If it decides to keep moving until all of it is consumed, it can travel twice the foraging radius, as defined above—bearing in mind that e, and, hence, also the foraging radius, will be greater in wet-season feeding conditions than in the poor conditions assumed above, because of the better quality of the food.

To develop the argument further, we need a variable about which not very much information is available, E_{az}, the rate at which energy can be extracted from the gut and made available for storage and metabolic use. If E_{az} is equal to or greater than the instantaneous income, then the animal can eat continuously if other considerations allow this. If E_{iz} exceeds E_{az}, then energy accumulates in the gut at a rate $E_{iz} - E_{az}$, so determining the gut filling time as defined above. Starting from the minimum level of gut contents,

$$T_f = \frac{e}{E_{iz} - E_{az}} \tag{15}$$

If the animal now stops eating, the emptying time, T_e, is

$$T_e = \frac{e}{E_{az}} \tag{16}$$

During the emptying process, the animal can engage in some behavior other than eating, for instance locomotion. Of course, the animal need

not wait until e is consumed before it starts to feed again, but, in general,
the *proportion* of its time which has to be devoted to eating is p, where

$$p = \frac{T_f}{T_e + T_f} = \frac{E_{az}}{E_{iz}} \qquad (17)$$

The remaining $1-p$ of the animal's time is uncommitted, and we can now
consider the effect on the energy surplus of putting this time to two
alternative uses. First, the animal may elect to minimize expenditure by
not moving at all, whether it is eating or not. In this case, its energy
surplus is

$$E_s = (E_{iz} - E_{fz})p - E_m \qquad (18)$$

(where the subscript z indicates an instantaneous rate, which applies while
the animal is actually feeding, and its absence indicates a mean rate, as
before). Now let the animal spend the "spare" proportion $1-p$ of its
time migrating at a speed V. The income and expenditure terms in
equation 18 remain the same as before, but an additional expenditure term
$(1-p)kV$ is now added. The energy surplus now becomes

$$E_s = (E_{iz} - E_{fz})p - E_m - (1-p)kV \qquad (19)$$

Migration is worthwhile if it brings the animal to better feeding, such
that the last term in equation 19 is more than canceled by an increase in
the first term, over the value that would have prevailed had the animal
stayed where it was. It is only advantageous to migrate if (1) the vegetation is in poor condition or eaten down, so that $(E_{iz} - E_{fz})$ is low, or if
(2) the animal can be assured of a high probability of finding better
conditions by moving a distance of the order of $2r$ or less.

Since no regular commuting takes place in this case, the term "foraging
radius" loses its original direct significance, but the quantity r still determines how far the animal can go between feeds. The strategy of a migratory animal is to go where the feeding is currently good, rather than to
remain in one area through both good and bad periods. The bigger the
area that is accessible to it in the course of a complete annual circuit, the
greater the likelihood that good feeding can be found somewhere over
an extended part of the year, and this is more true, the more heterogeneous
the area is spatially. In the Serengeti, there is an element of predictable
spatial heterogeneity, in that the plains generally are more productive than
the woodlands during the rains, and vice versa in dry weather. There is
also an important unpredictable element during dry weather, in the form
of patches of good grazing caused by local rainstorms. To exploit these

irregularities, an animal needs a large foraging radius, promoted as before by large body mass and adaptations for fast, economical cruising. It also needs means of deciding where to migrate at any given time. Thus, adaptations for making the food supply less unpredictable have interrelated mechanical and information aspects.

Mechanical Adaptations
of Mammals

The preceding arguments about the ecological significance of locomotion shed some light on the various kinds of morphology and movement seen in different mammals. For example, the contrast between the ungulates and the felids has been illustrated by Hildebrand's (1959) well-known comparison of galloping in the horse and the cheetah. The cheetah achieves a high top speed by dorsoventral flexing of the backbone, but this is expensive in energy. This animal is specialized for a high top sprinting speed, at the expense of stamina. The horse has a rigid-backed style of galloping, with fore-and-aft pitching, which, together with slender limbs of low moment of inertia, is characteristic of ungulates generally. This is comprehensible if one regards their adaptations as being not so much for running away from predators as for allowing economical cruising in the trot or canter, so conferring a large foraging radius.

The considerations of foraging radius developed above would not seem to apply to cats at all. For instance, Schaller's (1972) study shows that lions do not use the commuting strategy of hyenas, as described by Kruuk. Instead, breeding prides confine their activities to a relatively small "pride area," where they depend on the resident ungulates to supply their food requirements. Breeding prides do not use the migratory ungulates as a food supply, except when the latter happen to migrate through the pride area. Other, nomadic lions follow the migrating herds, but these animals are nonbreeders. Lions do not appear to depend on a commuting strategy at any stage of their lives, and considerations of sustained, economical cruising are therefore irrelevant to them.

Cats generally are exponents of the stalk and sudden rush, and their locomotor adaptations reflect this. The lion's forelegs have a dual function, as locomotor organs and as weapons. The thick legs and heavy paws must have a higher moment of inertia than ungulate legs, with a consequent increase in the energy cost of locomotion, especially in the faster gaits.

The "doglike" predators, including the hunting dog and the spotted hyena, are quite different, and more ungulatelike in their locomotion. The limbs are not used as weapons, and are specialized for locomotion only. The heart and lungs are relatively bigger than in cats—for instance, Schaller records the heart as 0.46 percent and 0.57 percent of the body mass (minus stomach contents) in two lions, but 0.95 percent in a hyena. Dogs show some dorsoventral flexing when galloping, but the slow, "lolloping" canter of the spotted hyena is an essentially stiff-backed, ungulatelike motion, which can be maintained at a speed of 3–3.5 m/s for long distances. Kruuk mentions the use of this gait by hyenas during long-distance commuting.

All the large predators gallop when actually running down prey, but hunting dogs and hyenas can keep up a chase longer than lions or other cats. For instance, van Lawick-Goodall and van Lawick-Goodall (1970) say that hunting dogs chase their prey at 13–15 m/s, and can keep this up for 4–5 km, after which they give up if not successful. There is no question of either predator or prey remaining in oxygen balance at this speed, of course. The outcome of the chase is presumably determined by which overheats or runs out of oxygen first, unless one or the other party commits some tactical error.

When migrating, hunting dogs trot at about 3 m/s. This is evidently an economical cruising gait, which they can keep up for long distances. Like hyenas, hunting dogs are confined to a den while raising small pups, and have to commute outward from it. The social arrangements for feeding the young are unusual, and involve regurgitation of food for the mother and pups by other adults. The energetics of this arrangement must be somewhat different from those of hyenas, but the basic requirement of economical cruising locomotion for commuting clearly applies to these animals also.

Information Requirements
for Migration

In each of the cases considered above, the animal makes use of a large foraging radius to increase the reliability of its food supply. In the first case (herbivore commuting from a water source), increased foraging radius confers the ability to last longer under drought conditions, so that the probability that any particular drought will outlast the animal's powers of survival is reduced. No special information is needed to operate this system; the animal merely continues to push further out from the water as the food gets used up. Thus, the characteristics that go with a large foraging radius—large size and adaptations for economical cruising—may be expected to appear in any water-dependent herbivore that is subject to prolonged droughts.

In the other two cases, the hyena commuting from a fixed den and the wildebeest migrating about an extended range, the energy advantages are only obtained if the animal has information about the current and future spatial distribution of its food, on the basis of which it can make appropriate decisions about where to migrate. The animal has to be programmed to make these decisions in response to physical stimuli, such as visible weather signs, and possibly also to internal stimuli, such as a circannual clock.

This migration program has to be acquired before the benefits of a large foraging radius can be realized. It is conceivable that the more predictable features of such programs might be fixed genetically. For instance, in the Serengeti, the wildebeest plan of going south and east when the rains begin, and west and north when they cease, has probably applied for long enough to be genetically determined. However, in many areas, changes in the landscape due to volcanism or human activities must require these programs to be quite flexible, besides being entirely different for different local populations of the same species. Thus, it seems more likely that the information is mainly learned rather than genetic, but it may also be passed from parent to offspring or from animal to animal in a way which leads to the development of a "pool" of information shared by a particular population, so giving an element of conservatism. If this proves to be the case, then the flow of learned information between individuals might be as relevant a criterion for defining a "population" as the flow of genetic information.

A program to exploit short-term, unpredictable variations of food distribution must be based on day-to-day acquisition of ephemeral information. For example, it has been claimed that emus in Australia have a program that consists of walking toward distant thunderheads, and continuing in the same direction (perhaps for days) after the clouds have dispersed, until eventually they arrive in the area where the rain fell. The radio-tracking observations of Leuthold and Sale (1973) suggest that elephants in Tsavo East may do something of this kind.

Both griffon vultures and hyenas living off the migratory wildebeest and zebras commute from aggregated nesting or denning areas, which could well operate as "information centers" in the manner proposed by Ward and Zahavi (1973). The griffon vultures frequent their nesting areas throughout the year, even when they are not breeding and when the migratory animals are far away, which is understandable under the "information center" hypothesis. Such behavior contrasts with that of species which forage in a fixed area and do not require up-to-date information about food distribution. They do not show colonial breeding and do show aggression toward others of their own species. This applies, for instance, to lappet-faced vultures, and to breeding prides of lions.

Conclusion

The strategies considered here fall into two contrasting groups:

1. There are those animals that seek to improve the predictability of their food supply by maximizing the area over which they forage. This calls for a large foraging radius, with consequent selection pressure for large size, and locomotor adaptations conferring economical cruising. It is impracticable to combine this type of strategy with the defense of a feeding area: on the contrary, it seems to call for a type of social organization that tolerates high densities of individuals, and facilitates the transfer of information about food distribution between individuals. Spotted hyenas, griffon vultures, and the migratory ungulates fall in this category.

2. In contrast are those animals that defend a foraging area against others of their own species, so limiting their foraging to the relatively small area that can be successfully defended. The locomotion of lions and lappet-faced vultures, which come in this category, contrasts with that of hyenas and griffon vultures in not being specially adapted to economical cruising. The sedentary ungulates, on the other hand, do not look greatly different from the migratory ones. This may be because a migratory strategy is primitive in ungulates, and a sedentary one secondary, or possibly because even the sedentary ungulates have to commute to water during periods of critical food shortage.

I am most grateful to Dr. D. Kreulen and Prof. R. McN. Alexander for their comments on the first draft of this chapter, in the light of which it was extensively revised, with great benefit. To Dr. Kreulen I am also indebted for providing numerical estimates from his unpublished data for a number of variables needed in the argument.

References

Alexander, R. McN. 1976. Estimates of speeds of dinosaurs. *Nature* (Lond.) 261:129–30.

————. 1977. Allometry of the limbs of antelopes. *J. Zool.* (Lond.) 183:125–46.

Cummins, K. W., and Wuycheck, J. C. 1971. Caloric equivalents for investigations in ecological energetics. *Mitt. Internat. Verein. Limnol.* 18:1–158.

Hildebrand, M. 1959. Motions of the running cheetah and horse. *J. Mammal.* 40:481–95.

Houston, D. C. 1976. Breeding of the white-backed and Rüppell's griffon vultures, *Gyps africanus* and *G. rueppellii. Ibis* 118:14–40.

Kruuk, H. 1972. *The spotted hyena.* Chicago: Univ. of Chicago Press.

Lamprey, H. F. 1963. Ecological separation of the large mammal species in the Tarangire Game Reserve, Tanganyika. *E. Afr. Wildl. J.* 1:63–92.

van Lawick-Goodall, H., and van Lawick-Goodall, J. 1970. *Innocent killers.* London: Collins.

Leuthold, W., and Sale, J. B. 1973. Movements and patterns of habitat utilization of elephants in Tsavo National Park, Kenya. *E. Afr. Wildl. J.* 11:369–84.

McMahon, T. A. 1975. Using body size to understand the structural design of animals: quadrupedal locomotion. *J. Appl. Physiol.* 39: 619–27.

Pennycuick, C. J. 1972. Soaring behaviour and performance of some East African birds, observed from a motor-glider. *Ibis* 114:178–218.

————. 1975. On the running of the gnu (*Connochaetes taurinus*) and other animals. *J. Exp. Biol.* 63:775–99.

Rogerson, A. 1968. Energy utilization by the eland and wildebeest. In *Comparative nutrition of wild animals,* ed. M. A. Crawford. Symp. Zool. Soc. Lond. 21 (London: Academic Press):153–61.

Schaller, G. 1972. *The Serengeti lion: a study in predator-prey relations.* Chicago: Univ. of Chicago Press.

Taylor, C. R.; Schmidt-Nielsen, K.; and Raab, J. L. 1970. Scaling of energetic cost of running to body size in mammals. *Amer. J. Physiol.* 219:1104–7.

Ward, P., and Zahavi, A. 1973. The importance of certain assemblages of birds as "information-centres" for food-finding. *Ibis* 115:517–34

P. J. Jarman
M. V. Jarman

Eight The Dynamics of Ungulate
Social Organization

This chapter concerns the mechanisms and forces producing the various social organizations in ungulate populations of the Serengeti. A population is "organized" if its members occur nonrandomly in relation to each other. This may arise because individuals approach or avoid others. Individuals may also show discriminatory responses: for example, a zebra stallion will drive other stallions, but not necessarily mares, away from his group of mares; an elephant cow will stay with her mother's group rather than join another; and a low-ranking bachelor impala will move out of the path of a high-ranking male, yet remain in the same male herd with him. Each of these actions in some way organizes the population of that species. Although performed by individuals, they are actions typical of those species, and they produce part of the social organization characteristic of zebra, elephant, or impala.

These organizations, and the social behavior that produces them, are as characteristic of each species as that species' reproductive or resource ecology. We assume that they are similarly adaptive, and that individuals whose behavior is grossly uncharacteristic are disadvantaged. Comparative studies of related herbivore species (Crook 1965; Crook and Gartlan 1966; Jarman 1974; Geist 1974) argue for the adaptiveness of species-characteristic social organization in terms of other aspects of the species' ecology; these arguments are brought together by Wilson (1975). In this chapter, we present hypotheses about the ecological causes of social organization, and use studies carried out in the Serengeti to test some of them.

Social Behavior and the
Individual's Ecology

Our arguments relating social behavior to other aspects of the individual's ecology, and, consequently, linking species-specific ecology and social organization, can be laid out as a series of hypothetical points.

1. Among related herbivore species, with comparable feeding and digestive mechanisms, larger species, having lower metabolic requirements per unit of body weight, will tolerate lower-quality food than will smaller species. However, individuals of larger species will require a greater absolute quantity of food than will those of smaller species.

2. In most plant communities, food items of low quality (mature leaves, stems) are much more abundant than those of high quality (new leaves, seeds, fruits, storage organs). High-quality items tend to be scarce in space and time, and must be sought individually. Within selected habitats, low-quality items may be abundant and evenly dispersed (for example, leaves in a grass sward).

3. Among related herbivore species, therefore, individuals of smaller species are likely to have to seek carefully their scarce, scattered food items; individuals of larger species are more likely to find, with little searching, abundant, evenly dispersed items, provided that the animals have selected an appropriate feeding habitat. Individuals of larger species do not, of course, have to differentiate between low- and high-quality food items.

4. A single food item may form a greater proportion of a day's intake for an individual of a smaller, than for one of a larger, species.

5. An individual of a smaller species is more likely than is one of a larger species to be able to manipulate single leaves, fruits, or other discrete plant parts (although the plants' physical defenses may deter individuals of the smaller species more).

6. Food items in a given dispersion will be more widely spaced (scattered) relative to the size and movements of an individual of a small species than to those of one of a larger species.

7. Among related herbivore species, individuals of smaller species will be vulnerable to a greater range of predator species than will those of larger species. Smaller species are less likely than larger to be able to defend themselves against, or to outrun, predators. In smaller species, therefore, all individuals must avoid being detected by predators; in larger species, this may be important only for individuals who are temporarily vulnerable (newborn, sick, in poor condition, and so on). If individuals of larger species are a close match, in speed and strength, for a predator, then each individual should strive not to be outstanding, assuming that predators are likely to concentrate on outstanding targets.

8. Among related herbivore species, individuals of smaller species are, therefore, more likely to depend upon refuges, or upon cryptic coloring,

shape, smell, and behavior, to avoid being detected by a predator than are larger species. Individuals of larger species are more likely than smaller ones to depend upon self-defense, group defense, group alertness, anonymity within a group, and speed, to avoid being killed by a predator.

9. Unless the group remains cohesive and coordinated, the individual risks becoming an outstanding target unwittingly. Thus, at all times, individuals in groups must remain in communication, and their speeds and directions when moving must vary little between individuals.

10. Dense vegetation and broken terrain will disrupt visual communication, and flat, open country will favor it. Individuals of smaller species may be unable to communicate in vegetation that would not affect larger species.

11. Individuals in groups must be able to move at the same speed and in the same direction, especially when seeking and eating food, since these are their commonest mobile activities. Individuals feeding on abundant, evenly dispersed, easily found items are more likely to be able to adjust their speeds and directions of movement to conform with other group members, without much loss of feeding time, than are individuals searching for scarce, scattered items.

Putting these points together, individuals feeding upon abundant, evenly dispersed, easily found items, are likely to be tolerating low-quality food, and are therefore likely to belong to a large species (points 1, 2, and 3). They are more likely than individuals of smaller species to form groups for defense against predators (points 7 and 8). Such groups are more likely to be found where visual communication is favored (point 10), and where individuals can conform to the group's speed and direction of movement because they are feeding upon abundant, evenly dispersed, easily found items (point 11).

Conversely, individuals searching for scarce, scattered items are likely to do so only because they need high-quality items. Therefore, they are likely to be individuals of smaller species (points 1, 2, and 3), which will be more competent to handle discrete items (point 5), and more satisfied by finding a few sought-out items (point 4), than will individuals of larger species. Individuals of smaller species are unlikely to depend upon grouping for defense against predators (points 7 and 8), and would, in any case, form less cohesive and coordinated feeding groups than would individuals of larger species (points 10 and 11), since they have to search for scarce, scattered food items.

These points do not refer to observations on the social organizations of actual herbivore species. Point 1 comprises two standard beliefs of nutritional physiology. Point 2 can be confirmed by looking at plant communities. Point 3 follows logically from points 1 and 2. Points 4, 5, and 6 are generalizations from what we know of herbivores' food requirements, feeding apparatus, and movements. Point 7 depends upon the general observations that smaller predators take smaller prey only, while larger predators may take all sizes of prey, and that smaller herbivores are weaker and slower than larger ones. Point 8 follows from 7, but also rests on a general knowledge of antipredator behavior in mammals. Point 9 is self-evident, if point 7 is accepted, and so are points 10 and 11.

Taken together, the points argue the case for individual herbivores' forming or not forming groups because of their own, and the environment's, characteristics. Some points predict trends which should be found when we compare the behavior of individuals of related species that differ in, for example, size. Such comparisons are made later in this chapter. Some points predict the relative effects of environmental characteristics that are continuous variables: cover, food quality, and density and dispersion of food items, for example. Hypothetically, these might be expected to affect individuals immediately, as well as influencing species evolutionarily. Some of these points are tested by looking for the immediate effects of some variables.

In addition to these socioecological responses to the risk of predation and their feeding environment, discriminatory responses by individuals to the sex, age, rank, or reproductive state of other members of the population produces further organization in the society. Examples are discussed below, and we argue that this organization is usually imposed upon the basic socioecological grouping rather than being a major force causing it. We believe that sociosexual behavior generally determines which individuals will have most offspring (or, more strictly, will promote most genes to the next generation). The validity of this belief is difficult to test because we do not have enough information on the reproductive success of individuals, and individuals that do not conform to the typical sociosexual behavior of their population are rare. However, populations of the same species may differ in their typical sociosexual behavior, and these differences may illustrate the adaptiveness of each style of behavior for individuals in those populations.

Social Organization in Serengeti Ungulates

Because of its diversity of rainfall regimes, soils, and topography, the

Serengeti has an array of vegetation types which differ in the amount, timing, duration, and predictability of their primary productivity (see chap. 2). The vegetation types also differ in their structures, and in the ratios of monocot to dicot biomass, and of ground layer to higher-strata plants. These vegetation types, and the relationships between some ungulates and their resources, have been described earlier (chaps. 4, 5, 6). Just as the herbivores respond to the vegetation with appropriate feeding strategies, so do they with social strategies. The numerous antelope species demonstrate this well.

Social Organization in the
Antelopes and Buffalo

Antelope in the Serengeti range in size from the diminutive dik-dik (4 kg) to the majestic eland (700 kg), and in typical group size from the solitary individuals or pairs of gray duiker, steinbuck, and dik-dik, to herds of thousands of wildebeest, topi, and buffalo. General trends in social organization are clear. The smallest species live singly or in pairs, the largest in huge herds; herds in woodlands tend to be smaller than herds on the plains; cryptic species do not form large groups, and large-herd–formers may practice mutual defense against predators. Each species' society is further organized through behavior that differentiates between sexes, ages, or ranks of individuals.

Dik-dik (pl. 22) inhabit the *Acacia* woodlands, particularly tall *A. tortilis* woodland with *Salvadora persica* or other dense bushes. A male, a female, and their most recent young live and find all their resources within a defended territory. Because they are small and vulnerable, they move and feed cautiously and slowly, freezing when disturbed, and never moving far from cover. They mark territory largely by olfactory signaling, since auditory signaling would reveal them to predators. Visual signaling would be relatively ineffective in their dense habitat, except at short range during territorial challenge or defense. Both male and female dik-dik, and juveniles while still with their parents, mark territory by urinating and defecating on dung heaps, and by depositing on twigs secretion from a preorbital gland (Hendrichs and Hendrichs 1971). Each adult evicts maturing offspring of its own sex.

The best habitat is made up of a mosaic of occupied territories, and evicted, maturing juveniles are forced into less suitable areas, unless they happen to find and acquire a vacant territorial position. Once established, the dik-dik probably remains in the territory for the rest of its life. The rate of turnover of territory-holders is very low in comparison with some other antelope, such as impala (Hendrichs, pers. comm.; and below). Only territory-holders breed successfully.

Dik-dik select carefully their scattered food items of fruit, seeds, young leaves, and buds of a variety of shrub and herb species; grass is rarely eaten. The territory must be big enough, with a diversity of food plants, to produce these scarce, high-quality food items in all seasons; but it must not be too big for the pair to defend. Patrolling and marking the territory, often combined with feeding, take up much of the day, and are usually performed by the partners together. Little time is spent on overt courtship or herding in dik-dik, although these are important components of male-female interactions in many other antelope. The male dik-dik acquires uninterrupted mating rights with a lifelong partner by gaining and maintaining territory. Both sexes maximize their chances of raising young by excluding all other dik-dik from the resources of a territory, we believe. The size of their territory is presumably determined by the area that a pair can defend and by the availability of suitable food at the season of greatest scarcity.

Dik-dik, and other small, solitary or pair-forming, territorial antelope like them, have evolved the strategy of defending a resource-sufficient breeding territory. This behavior organizes their society dichotomously into territory-holders, which are likely to be long-lived, successful breeders, and the rest, among which only a few will find a vacancy and so gain access to the territorial system's security.

While individuals of these small antelope species characteristically remain in one vegetation type in all seasons, most medium and large antelope alter their habitat preferences seasonally, and may even range through several vegetation types daily. They move to different areas at different times of year as plant species grow, mature, and are removed. They are more tolerant of medium- or poor-quality food than are the smallest antelope, so food items are more plentiful and more evenly dispersed for them, allowing them to form cohesive feeding groups as part of their antipredator strategies. These species occupy open woodland, savanna, or plains, all relatively open habitats conducive to intragroup communication.

In contrast to the male dik-dik's possession of only one female, herd formation gives some males the chance to keep others away from a large number of females simultaneously; but mobility of the population may make this difficult. The simplest solution is seen in species like impala and waterbuck, whose populations shift habitat preference seasonally within fairly small individual home ranges, so that there is good local predictability as to where females will occur. Impala (whose flexible social organization will be discussed in more detail later) are medium-sized antelope which feed on a mixed diet of grasses and browse. In the Serengeti, they are typically found in herds of ten to one hundred or more individuals (pl. 23). For much of the year, about one-third of adult males hold contiguous territories covering the area occupied by females at that time. Female home ranges are several times larger than territories, and females move in herds from territory to territory. The territorial male checks each herd as it enters his territory and ousts any young males he finds in it. These young males, together with nonterritorial adult males, form bachelor herds. Impala bachelor herds are allowed into territories, provided that they keep well away from any females in that territory. In the absence of females, the territorial male may mingle with bachelors in his territory (Jarman and Jarman 1973, 1974).

Impala territory is marked by roaring, an auditory signal; by special standing postures (pl. 26), or just by the obvious presence of the territorial male, which are visual signals; and by olfactory marking through urination, defecation, and the wiping of secretion from the forehead onto plants around the territory (pl. 24). Territory is directly defended by display or fighting between the holder and a challenger, with the challenger being a male at or near the top of the hierarchy that exists in the bachelor society. That hierarchy is determined by personal development of the males and by their success relative to other bachelors in displays and sparring. High-ranking bachelors achieve many of the characteristics of territorial males, especially the production of the forehead secretion. The reactions of bachelors to this secretion clearly show that it indicates to them its producer's dominance (pl. 25). The dominance of a high-ranking bachelor is acknowledged by all other males in the local bachelor society; on gaining a territory, the former high-ranking bachelor effectively attaches that personal dominance to an area of land by using the signals of territorial marking. Those signals indicate to all other males that within that territory there is an almost undefeatable male. In this way, the territorial male ensures his undisputed mating rights over any females in his territory (pl. 27).

Holding territory involves strenuous and time-consuming patrolling, marking, displaying, and herding. Territorial males that were initially in good physical condition, lost condition while they held territory; bachelors observed for the same period did not (table 8.1). When in poor condition, a territorial male may be challenged and defeated by a high-ranking bachelor, whereupon he generally returns to the bachelor society, often entering

Table 8.1

Gain or loss of condition of known males while holding territory in the wet, or dry, season, compared with bachelor males observed for the same periods. Four conditions were recognized; values in the table are of numbers of males remaining unchanged, or changing one or two condition categories.

	Condition categories gained or lost				
	+2	+1	0	−1	−2
Wet season					
No. of territorial males	0	4	3	6	0
No. of bachelor males	2	7	3	1	0
Dry season					
No. of territorial males	0	0	0	8	3
No. of bachelor males	0	1	6	4	0

fairly low in the adult hierarchy. After recovering condition and rising in rank, he may again challenge for territory.

A female impala may defend her calf against a small predator, such as a jackal, but impala herds do not attack predators or help an attacked individual. Herds remain well coordinated even under attack, and individuals may benefit from the alertness and communication within the herd (pl. 23); there is an alarm call and an alert, or alarmed, posture.

This social system differs greatly from that of dik-dik. While a dik-dik pair finds all their resources in the territory they jointly defend, the impala territory is not totally resource-sufficient. Female impala wander over many territories each season. Females take no part in territoriality, and an individual male holds a territory for only a few weeks or months at a time, although he may enjoy several bouts of territory-holding in a lifetime. At any one time, only a proportion of the adult males will be holding territory and, hence, be able to gain undisputed matings.

The dual nature of the determination of social organization is illustrated in the impala's system. Group formation is permitted by their habitat selection, feeding style, and food-item abundance and dispersion, and is required by their antipredator behavior. Further organization is

imposed on this grouping by the actions of territorial males, which separate breeding herds of females and juveniles from bachelor herds of
subadult and nonterritorial adult males. The territorial males themselves
are found either alone or temporarily accompanying one or another of the
two herd types. Hierarchical behavior produces detailed organization
within bachelor herds, determining the positions of individuals relative to
each other. Bachelor herds tend not to occupy the area most intensively
used by females, presumably because of the frequent harassment by territorial males that they encounter there.

Detailed organization occurs in breeding herds because calves associate
with their mothers during suckling and herd movement; at other times
they form distinct subgroups, or crèches. In a local population, the
females may group according to similarity of reproductive state. One herd
may contain mainly females with young calves, while another may contain
a large number of pregnant females without dependent young; or this
kind of differential grouping may occur spatially within the herd.

The division of impala society into breeding and bachelor herds by
territorial males imposes secondary, detailed organization upon the grouping—organization primarily determined by other factors. In the Serengeti,
in dry years, territoriality may disappear for a few weeks at the end of the
dry season. When it does, and there are no longer any active territorial
males, all males and females mingle freely; but they still form herds.
These herds are of open membership; that is, they do not contain the
same individuals at all times, since animals may freely leave or join any
herd. Breeding and bachelor herds are also of open membership, which
gives these herds important flexibility of size and composition.

Impala social behavior and resultant organization is representative of
several fairly sedentary species of antelope in the Serengeti. Waterbuck
(Spinage 1969a, b) and kongoni (pl. 21; Gosling 1969, 1974) social
organization contains the same elements—some males holding territory
for varying lengths of time, gaining that rank through the hierarchy in
bachelor society, and open-membership bachelor and breeding herds moving between territories. Similar organization can be seen in the Serengeti
Grant's gazelle (Walther 1965) and Thomson's gazelle (Walther 1964)
populations, even though both species migrate seasonally over a large part
of the Serengeti ecosystem. Male gazelles establish territories, which are
maintained as long as the population occupies a particular region and
abandoned when other gazelles have moved on. Although temporary, the
territories are well defined spatially. They are marked by urination and

defecation, and, in Thomson's gazelle, by deposition of secretion from the preorbital gland. More ephemeral auditory and visual forms of marking are also used.

Most of the Serengeti's wildebeest are even more extensively migratory than the gazelles (chap. 5), but in the western, northern, and eastern extremities of the ecosystem there live small resident populations of wildebeest. Wildebeest are purely grass-eaters and migrate locally or throughout the whole ecosystem in pursuit of green grass with a high leaf-stem ratio and an appropriate leaf table height. When they find such conditions of food availability and dispersion, they can form enormous, rather formless feeding aggregations of many thousands of animals.

On the open plains, undisturbed aggregations are not divided into clearly defined herds (pl. 12). However, within the aggregations, individuals tend to conform with their neighbor's activities, and many hundreds, even thousands, of wildebeest can coordinate rapidly in response to alarm. Despite apparent lack of social structure, the aggregations may serve some antipredator functions through mutual alertness and communication of alarm, anonymity for individuals, reduction of the individual's chance of being the target of an attack, and the creation of a clustered dispersion of prey items (a patchy environment) for the predator. The aggregations fragment into discrete, coordinated herds in the woodlands, where visibility is much less than on the plains.

Unlike the species mentioned above, wildebeest mate during a restricted rut (Estes 1966, 1969). In the sedentary populations, an organization into territorial males, bachelor male herds, female herds, and sometimes separate juvenile herds, is established by a proportion of the adult males setting up a mosaic of spatially defined, marked, and defended territories (pers. obs.). As with dik-dik, impala, kongoni, or the gazelles, possession of a territory is a prerequisite for mating. Wildebeest bachelor herds are largely excluded from the territorial mosaic during the rut, perhaps because of the intensity of territorial defense (Estes 1966, 1969). Territories are fairly small, visibility is good, and much of the defense comes from visual display, although olfactory marking through urination and defecation, and auditory signaling, do occur.

For the migratory wildebeest, the rut coincides in most years with a fast-moving phase of the migration, so that no male can predict precisely where females will be during these three or four weeks. Establishment of a spatially fixed territory prior to the rut could leave a male many kilometers from any female at the critical time. Migratory wildebeest males have solved this problem by abandoning spatially fixed territories. Instead, they accompany the moving females and set up very temporary, even mobile, territories in the midst of the females. These territories are tiny; no more than the area a male can keep free of other males by his immediate presence. The territorial males—for so they are, despite the absence of long-established boundaries to their area of dominance—gallop around and around in a characteristic gait, trying to cut out and encircle a small mob of females, seeking and courting ones in estrus (Estes 1966, 1969; pers. obs.). Their frantic activity drains their resources, and it is probable that during the course of the rut, there is a rapid turnover of males forming the breeding population.

This territorial male behavior does little to organize wildebeest society. Briefly each year, separate male classes are revealed, and females are spatially divided into small, ephemeral groups. But this organization occurs within the framework of the vast aggregations that characterize the mobile phases of the migratory population. Nevertheless, the immediate mating environment that is created is of one male in exclusive control of a temporary group of females.

Buffalo (pl. 15), and perhaps eland, males secure uninterrupted matings within large, highly mobile herds in a different way. These very large species are extremely catholic feeders—eland upon browse and grass, buffalo predominantly upon grass. Eland are nomadic; buffalo live within enormous home ranges, moving seasonally in order to be in the habitat with the most abundant food of sufficient quality. We have observed eland in the Serengeti in large herds of mixed sexes and all ages. There is often suborganization of the herd; calf groups and young male groups can be distinguished within them. These herds may spread over a wide area, so that what first appears to be several scattered groups of eland is, in fact, a group of animals moving with and responding to each other in one large, dispersed herd. Much smaller all-male eland herds can also be seen. In many respects, eland social organization resembles that of the buffalo, which is well understood because of the studies made by Grimsdell (1969) in Uganda and by Sinclair (1974) in the Serengeti.

Except for about 15 percent of the adult males, which occur singly or in small groups, Serengeti buffalo are typically found in herds varying in size from fifty to two thousand animals. Each large herd occupies its own home range, which, at any one season, overlaps only to a small extent with that of neighboring herds. Within each herd, a degree of organization can be seen; subadult males may form recognizable subgroups, and juveniles of both sexes up to two years old are more likely to be near an adult female, presumed to be their mother, than to an individual of any other class (pl. 16). Most importantly, both Grimsdell and Sinclair reported that buffalo herds were of fixed, or closed, membership. Most buffalo females, and probably some males, live and reproduce within the herd in which they were born.

Apart from the fixity of group membership, this type of social organization differs most obviously from that of impala, the gazelles, or wildebeest, in that many males accompany females at all times. Whereas in the other species', males achieved uninterrupted mating rights by keeping away competing males, by defining an inviolable territory containing potentially estrous females—eland and buffalo do not. Instead, males ensure mating rights by exercising their individual dominance in the presence of a detected estrous female. Individual dominance is developed, as in the other antelope, through hierarchical interactions among the males (Grimsdell 1969), but that dominance does not become attached to a defined area any greater or more persistent than the personal zone around a mating pair.

Several factors make this an appropriate system for buffalo or eland. Males have solved the problem of finding highly mobile females in large herds by remaining with them, like migratory wildebeest bulls. However, unlike migratory wildebeest, neither buffalo nor eland mate only during a short rut, and brief defense of a group of females by a male would be unprofitable. Both eland (Kruuk 1972) and buffalo (Sinclair 1974) individuals cooperate in defending an animal from an attacking predator; a wildebeest mother will defend her calf, but other individuals will not help her. Buffalo calves are less fleet-footed at birth than wildebeest calves, and probably depend much more on the close-knit herd for protection. Buffalo herds are far more cohesive than the loose aggregations of wildebeest, and the persistent association of mother and offspring provides internal structure. A male buffalo, trying to detach a group of females in wildebeest fashion, would be obstructed by these cow-calf bonds, and his efforts might expose calves. The evolved organization, which retains male hierarchy as a means of apportioning mating rights, but does not let it lead to territoriality, is responsible for maintaining the integrity of the herd.

*Social Organization
in Other Ungulates*

The hypothetical points outlined earlier, which predicted trends in ecology and grouping among related herbivore species, can be tested only on the antelopes; no other ungulate group is sufficiently represented in the Serengeti. However, some of the nonbovid herbivores differ from antelopes in the organization of their groups. For example, the plains zebra shares the habitat of the wildebeest but eats longer grass with more stem, migrates in a similar pattern (chap. 5), and occurs in loose aggregations of hundreds, often mingled with the wildebeest aggregations. But, unlike wildebeest, zebra mating is not confined to a few-weeks-long rut each year. Klingel (1974) has described their social organization as consisting of "coherent family groups" of one male with one or several females and their young, and separate male groups (pl. 19). There is no territoriality, but the male keeps other males away from his females. Family groups keep slightly apart within the aggregations. The group male and bachelor males will compete to court and mate an adolescent female in estrus, but she will not become a permanent member of any male's group until she is two to two and one-half years old. A male gains females either by taking over another male's group after his death or defeat, or by acquiring unattached adolescent females.

The availability and dispersion of food allow zebra to form large, loose aggregations on the plains. Although alarm can be rapidly communicated through these aggregations, zebra, unlike wildebeest, do not usually close ranks when a predator is detected. Instead, each of the small family groups remains tightly coordinated, and is actively protected by its male, which attacks predators in the group's defense. This differs from the antipredator behavior of any of the antelopes. Zebra females in a group do not defend each other or other females' young, and there is no genetic common ground to justify such mutual defense.

Elephant society is finely organized around a matriarchal system of female grouping. Douglas-Hamilton (1972) described the family unit, consisting of a small number of related cows (for example, a mother and her daughters) and their dependent young, as the basic social unit, with a "kinship group" of two to four family units also evident. Elephant males range independently of the females, and, in the Serengeti, form open-membership, temporary groups (Hendrichs 1971). There is no trace of territoriality, and the ranges of males and female family units overlap extensively.

Serengeti elephants occupy two distinct ranges, in each of which the population follows an annual cycle of movement, dispersion, and aggregation. There are parallels between the social behavior and organization of each elephant population and each buffalo "population" or fixed-membership herd within its nearly exclusive home range. Both species are catholic feeders. In both species, there is a basic structure of females accompanied by offspring; the association is nearly lifelong among elephant, but has not yet been traced beyond a couple of years in buffalo. In both species, females aggregate into herds whose size varies seasonally; some adolescent and adult males separate from the females and are found peripheral to their range; (this is more pronounced in elephants than in buffalo); and males are not territorial, but organize themselves hierarchically. Both species, living in groups of related individuals, practice mutual defense against predators. Despite these similarities, major differences exist: males are usually present with females in buffalo herds, but this is not the case in elephant family units; also, the strength and persistence of the matriarchal structure in elephant society contrasts with buffalo society.

Elephant grouping poses the question What is a herd? All the family units of the population can come together as a spatially close group at times, but are they still in communication to some extent when more dispersed? In elephant, more than in most ungulates, we can see the possibility for learning and tradition affecting their use of resources. Are they similarly important in their social behavior?

Giraffe, like elephant, form groups so widely spread that it is difficult to detect their limits or membership. Groups appear to be of open membership, contain young and adults of both sexes, and do not practice mutual defense. Neither males nor females are territorial, but males may be hierarchically organized. Their sociality is of interest because they have managed to form groups, despite feeding on items (leaves on trees), which, at first sight, seem to occur in scattered clumps. However, giraffe can maintain group cohesion, through visual communication, despite great distances between the individuals. Considering food-item dispersion from the viewpoint of the giraffe, their food trees are no more clumped than are bushes to a browsing impala herd.

Not all the large, nonbovid herbivores form groups. The black rhinoceros is, like the dik-dik, purely a browser. It is seen singly, or in mother-young pairs, or less often in temporary associations between a few adults. There is no overt territoriality, and home ranges may overlap extensively (Mukinya 1973). The prominent dung heaps and the associated urination and defecation behavior could act as social "sign-posting" to space out individuals without necessarily demarcating exclusive territories. It is not yet clear whether mate-acquisition is organized behaviorally among males; it appears that an estrous female will mate with any adult male she happens to meet.

Flexibility in Social Organization

The preceding sections have related the typical grouping of a number of Serengeti ungulate species to aspects of their size, feeding style, food-item abundance and dispersion, and antipredator behavior. We suggested that details of social organization are imposed on the basic grouping by discriminatory behavior relating to reproduction, mainly the mother-young association and the behavior of males in securing mating rights. If these hypothesized relationships are really influential in an immediate fashion, we would expect to find flexibility in the organizations of those species whose supposedly governing environments vary seasonally.

Flexibility of Impala Social Organization

For one species, impala, there are substantial data on variations in social organization that can be used to test the argument that: impala form groups in order to reduce individual risk of predation; impala group size is limited by food item availability and dispersion; impala males impose detailed organization upon the basic grouping through their behavior in securing mating rights.

Impala social organization and the risk of predation. The individual's risk of predation with, and without, practicing antipredator behavior cannot be compared, since all individuals respond to the risk of predation at all times. However, variation in the type of response helps to define the appropriate circumstances for grouping. Several times in the Serengeti, we observed that a sick or injured impala, which was obviously vulnerable to a predator, left the herd and tried to hide in dense cover. As the outstanding target, such an impala would gain no defense from the herd once that group was found by a predator. As an alternative, the vulnerable impala used the dik-diklike strategy of avoiding detection by hiding from the predator.

More importantly, impala females when about to give birth also try to leave the herd and to find cover (M. V. Jarman 1976). If, by chance, a female does not get away from her herd before the final stages of labor, she is very obvious, even to the human observer, whether she is standing in the herd or lying on her side, exposing her white belly. After birth, the mother keeps the baby "lying out" away from the herd. At this stage, the calf is smaller and slower than other herd members, and would be an out-standing target. Its antipredator behavior consists of "freezing" until a predator is nearly on top of it, and then bursting out of hiding; this is, of course, quite different from the antipredator behavior that it will display later when it becomes part of a herd. The mother, during this "lying out" period, usually rejoins the herd for feeding and ruminating, returning to suckle the baby intermittently. After a number of days, she leads the juvenile to a herd. At this stage there is a risk that the juvenile will be the outstanding herd member because of its obvious smallness and slowness. Although our data are few, it appears that the juvenile will be taken into herd society sooner if there are several small juveniles already in the herds. This suggests an assessment by the mother of the anonymity her juvenile is likely to enjoy during its vulnerable first few days in herd society.

The often solitary existence of territorial impala males calls in question the need for grouping as antipredator defense. We have too few data to compare the rates of predation upon the territorial males and upon males in groups. Territorial defense depends in part upon the male being seen, and lone territorial males do not hide in cover. However, lone males remain constantly alert, either for predators or for territorial intruders. They are always harder to approach without disturbance than are bachelor herds or the same individuals when with females. Perhaps vigilance and increased risk of predation are costs of territorial possession, to be balanced against its reproductive benefits.

Impala group size and dispersion of food items. Impala respond to seasonal variations in the quality and availability of their food items by changing their habitat preferences and their diets. In the dry season, or when feeding on dry foods, they need to drink (Jarman 1973), and most home ranges include some riverbank habitat, which is preferentially used at that season. At the start of the wet season, they move away from the rivers to occupy the woodlands on the local slopes or ridgetops; in the central and western regions of the Serengeti, these are *Acacia senegal, A. hockii, A. clavigera,* or *A. nilotica* woodlands. At that season they feed predominantly on new green grass leaves, with some herbs; many of these plants may be annuals. As grasses mature and the rains cease, their diet starts to include progressively more browse, and they move into plant communities such as *A. tortilis* woodlands, where browse is more available. These communities also tend to be on lower catena, alluvial soils, where productivity is more prolonged than on the slopes or ridgetops. Fruits, seeds and seedpods, and leaves of many bushes and trees feature

strongly in their diet in the dry season. There is a close relationship
between diet and past rainfall, and the proportion of browse in the diet
can be predicted, based on the amount of rainfall over the past thirty days;
there is a similar relationship between habitat preference and past rainfall.

When feeding on browse, impala are taking more scattered items that
require more individual seeking and selecting. According to the theory
with which this chapter began, this should mean that individuals must
spread out if they wish to avoid interference when feeding. This is borne
out by our observations. By measuring the approximate area covered by a
feeding herd, and dividing it by the number of animals in that herd, we
can get some measure of the separation of individuals in feeding groups.
Figure 8.1 shows how these values varied seasonally. Despite high dry-

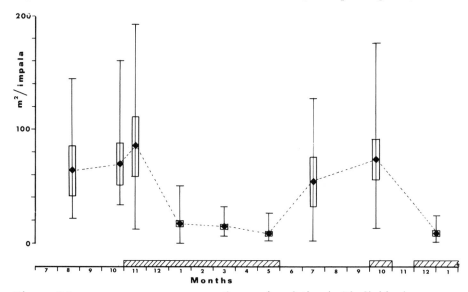

Figure 8.1   Seasonal variation in "individual
                                    space" of impala within female
                                    herds. "Individual space" is calcu-
                                    lated from the area covered by a
                                    feeding herd divided by the num-
                                    ber of animals in the herd. Mean,
                                    ± one standard error, and range
                                    are shown. The shaded bars indi-
                                    cate periods or rain sufficient for
                                    plant growth.

season variance, which possibly reflects the patchy nature of food availability at that season, females were able to group consistently more closely in the wet season, when feeding only on abundant, evenly dispersed grasses, than when including scattered browse in the diet. Bachelor males, even in the wet season, maintained relatively larger distances between each other, reflecting the permanent agonistic tensions of hierarchical male society.

There were also seasonal changes in the amount of time impala spent feeding and digesting. Table 8.2 shows the proportions of time devoted to feeding and ruminating by impala females in various seasons. Although the total amount of time devoted to these activities varied little, the proportion of that time given to feeding rose significantly, at the expense of ruminating time, as the dry season progressed. This suggests that impala switched from thoroughly digesting slowly passaged food of good quality, to rapidly passaging lightly digested food of lower quality.

Decreasing quality of available food is reflected in an increase in the ratio of stem to leaf in the grass fraction of the diet, as well as in the increase in the amount of browse eaten. Impala walk farther to find their day's food as quality falls (fig. 8.2): the threefold increase in the distance moved daily was accomplished with a one-third increase in time spent

Table 8.2		Seasonal variations in the amount of time spent feeding and ruminating by impala females.

Full 24 hours.			
Month	Season	Feeding (h)	Ruminating (h)
January	Early wet	7.95	12.05
May	Late wet	10.08	10.07
July	Mid dry	10.68	7.82

Ratios of feeding: ruminating during 12 daylight hours only		
Month	Season	Ratio of feeding: ruminating time
January	Early wet	0.97 : 1
April	Mid wet	1.83 : 1
May	Late wet	2.14 : 1
July	Mid dry	3.67 : 1
August	Late dry	4.00 : 1

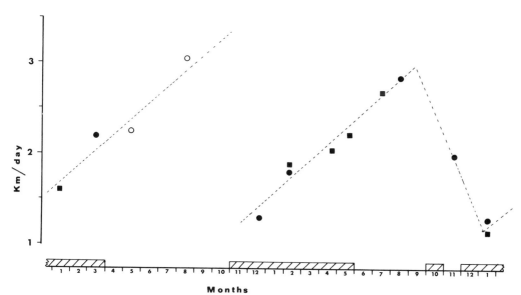

Figure 8.2   The distance moved during the
twelve daylight hours by impala
females in different months. *Open
circles* indicate measurements from
a single known individual female;
solid circles, the average measurement from several such females;
and *solid squares,* measurements of
whole herds. Periods of significant
rain are shown by the shaded bars.

feeding and moving. The animals not only moved further each day, then,
but also moved faster as food quality declined.

The theoretical consequence of these changes predicts a smaller group
size as conditions become more difficult for the faster moving, more dispersed groups to maintain internal cohesion; and, in fact, that was what
we found. Although individuals became quite widely dispersed, group
limits were usually definable. Figure 8.3 compares the size of group, in
which the average female occurred, with the rainfall. Group size was
lowest during the dry season and rose in the rains. A high group size
was maintained until the end of the rains (there was no late-rains sample

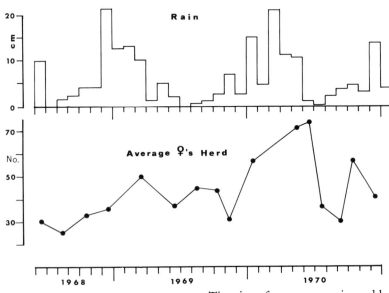

Figure 8.3

The size of group experienced by the average impala female in the Serengeti woodlands, and the Banagi rainfall figures for two-and-a-half years.

in 1969), even though food quality fell steadily from early in the wet season as plants grew and matured. Individual separation within herds remained low throughout the rains, rising rapidly only in the dry season. Only diet, habitat selection, and daily distance moved changed from the start of the rains, and it looks as though these are the aspects of response to food availability which the impala alters in order to sustain close-knit, large herds for as long as possible.

Once diet and distance-moved pass some (undefinable) threshold, feeding herds begin to diminish in size. This conforms to theoretical expectation. We cannot suppose that the individual impala's perception of the threat of predation is any less in the dry season, allowing the animal to feel "safe" in smaller herds. Impala in artificially predator-free environments still form herds; herds do not suddenly become smaller when migratory ungulates move into an area, diluting the predation pressure on resident impala. Predation must be treated by the individual as a constant risk, so that group size consequently remains as high as possible at all times. Hence, reduced group size is due to changes in food rather than to changes in predation risk.

Regional variation in group size. The effects of habitat and dispersion of food items upon impala can be seen in the Serengeti as regional variation in group size within any one season. Mean wet-season group size is greatest in the western corridor and north from Banagi to the Mara. There the woodlands can carry even swards of green grass, produced by high rainfall and maintained by fire and heavy grazing by other ungulates; and on these pastures impala form large, coherent feeding herds. Group size is smaller on the edges of the plains around and west of Seronera. In the dry bush country in the low rainfall region of the south and southwest, grass is relatively sparse and less likely to form even swards. Impala feed more upon browse, and the bushier habitat disrupts visibility. Consequently, group size, even in the wet season, is low.

Impala social organization and the behavior of territorial males. The detailed social organization, imposed on this flexible basic grouping by the territorial males, has to be similarly flexible in some ways. There is, for instance, the problem of females changing their habitat preferences seasonally. The territorial males' response to this is to attempt to move the boundaries of their territories, to include habitat currently favored by the females, by exerting pressure on neighbors in the territorial mosaic (fig. 8.4). Examples of these seasonal boundary shifts are shown in fig. 8.5. This is one aspect of the dynamic nature of the territorial mosaic. Despite the use of olfactory territorial marking, the boundaries are not rigidly fixed, but are constantly being adjusted according to the pressures exerted by neighboring males. In the study area where these measurements were made, there was less of the preferred wet-season habitat (upper catena woodlands) than of the dry-season habitats. Consequently, the mean size of territory declined in the wet season as all territory-holders strove to acquire limited upper-catena habitat (table 8.3). Territory size, then, is another flexible feature of the system, responding, to some extent, to immediate demand for territories in that habitat.

Territory size also varied with vegetation complex. In areas with open woodland or with grassland interspersed with woodland, territories tended to be much larger than in closed, continuous woodland. This reflects the impala's general avoidance of grassland, and territorial boundaries crossing grassland areas were ill-defined and poorly defended.

The territorial males' responses to the females' seasonal shifts in habi-

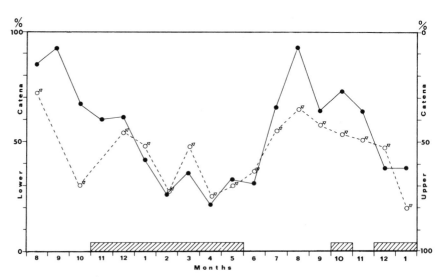

Figure 8.4   The proportion of impala females (*solid circles*) recorded in the upper catena (slopes and ridges) and lower catena (alluvial flats) habitats of the Banagi study area, compared with the proportions of sightings of territorial males (*open symbols*) in those two habitat classes. Both sexes use upper catena habitats preferentially in the wet season. In the dry season, females use predominantly lower catena habitats, but males maintain a presence in the upper catena habitats.

tat preference can be considered as strategies to increase the individual's opportunities to mate. One simple strategy is for a male to hold a territory containing a diversity of habitats, so that no matter how the seasons change, there will always be some of the currently preferred habitat to attract females into his territory. Another strategy depends upon the uneven distribution of impala births and estrus periods through the year; two-

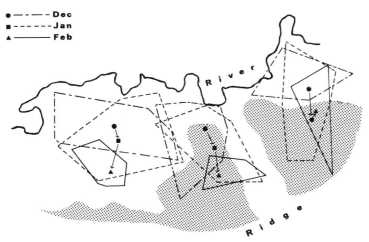

Figure 8.5   Seasonal movements of territorial boundaries (*lines*) and males' centers of activity (*symbols*) by three impala males near Banagi. Territories move from the river toward upper catena habitat (*shaded area*) on the ridge as the wet season (December to February) progresses and territory size decreases.

Table 8.3   Seasonal changes in mean territory size of impala males in the Banagi study area. In 1969, the dry season was more prolonged and severe than in 1970; the few remaining territories were very large.

	Mean area of territories (hectares)	
	1969	1970
Dry season	46	26
Wet season	21	19

fifths of births result from conceptions in April and May. Males holding territory at that time stand an above-average chance of securing matings. In the Banagi study area, there was a particularly marked turnover of territory-holders in March, giving a strong impression of a new cohort of males establishing themselves in time for this period of peak estrus frequency. These males were in their prime, having had previous territorial experience, but not yet showing signs of old age. Of course, among these, the most successful were those holding territory not only at the right time of the year, but also in the right habitat, that is, that currently preferred by the females. Although this seems a complex strategy, our data suggest that males did compete for these ideal times and places in which to hold territory.

Success had its costs, however. Males whose territories were visited by many female herds may have suffered, relative to the unsuccessful males with few female visitors, in one important respect. Loss of territory was linked to decline in condition of the male; males in poor or bad condition were more easily defeated by fit challengers. The depleting effect of holding territory did not come only from the energy expended on courting and mating females; a male which got no matings might still lose condition. This was probably due to the amount of time and energy spent being alert, patrolling the territory, and driving bachelors away from females. In particular, males lost feeding and ruminating time when with a female herd, since they then spent much of their time trying to coerce females into remaining within the territory. For instance, the activities of two territorial males during the 12 daylight hours were recorded within two weeks of each other in March, 1969 (Jarman and Jarman 1973). One male was accompanied by females for only 20 minutes that day, the other had females in his territory for 8 hours; the first male fed for a total of 6.33 hours, the second for 4.40 hours.

Figure 8.6 shows the relationship between the amount of time a territorial male spent feeding and ruminating and the proportion of the twelve daylight hours for which he had females in his territory. The effect of this decreased feeding time will presumably be least severe when food is abundant and of good quality, and will become more severe as the dry season progresses. A male's time is most taken up by females when they are near his territory's boundaries. When they first enter, he attempts to round them up and herd them deeper into the territory. Whenever they stray toward the boundary, he attempts to turn them back. When they finally leave, he does all he can to stop them, short of attacking them. He

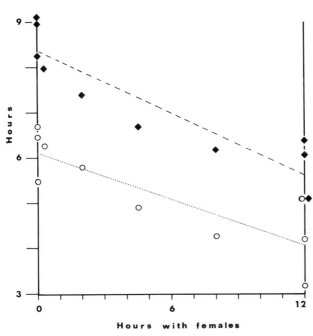

Figure 8.6   The amount of time spent feeding
                       (*open circles*) and feeding-plus-
                       ruminating (*diamonds*) by territo-
                       rial males during the twelve
                       daylight hours, plotted against the
                       number of these hours for which
                       females were within the territory.

will only settle to feed when they are calmly feeding or resting well within
the territory. It follows that two herds passing through a territory, each
spending one hour there, will cost the male more time and energy than one
herd spending two hours there.

   As the dry season progresses, females need to obtain more food from
more scattered, selected food items. As described above, this makes them
move farther each day, move faster, and break into smaller groups, thereby
compounding the territorial male's problems geometrically. He is confronted by a greater number of faster moving herds passing through his

territory, and, within each herd, the individuals are more widely separated and are spending more time moving and feeding and less time resting or ruminating. Consequently, although smaller in numbers, the herds may be more difficult to control.

The total effect of these changes is that territory becomes increasingly costly to maintain, at a time when the number of females coming into estrus is, in any case, low. As a result, in dry years, most males cease being territorial at the height of the dry season. When that happens, we see a population stripped of the organization imposed by the territorial males. Although some low levels of hierarchical behavior among males remain, all sex and age classes mingle freely in herds. The most important point is that the society still forms herds, definitive proof that grouping is not solely the result of the territorial males' herding activities.

This collapse of territoriality does not happen in the Serengeti in all years or in all regions, nor is it an abrupt event. Males gradually decreased the effort put into certain aspects of territoriality. For instance, at the height of wet-season territoriality, a male will evict from the female herd juvenile males as young as four months old, but during the dry season, juvenile males of increasing age up to eighteen months are allowed to remain unmolested in the female herds (Jarman and Jarman 1973, fig. 2). At the same time, the territorial male segregated himself most completely from bachelor males at the height of wet season territoriality (in April), and spent increasing proportions of his time with other males as territoriality declined in the dry season (table 8.4). Other behavior characteristic of territoriality, such as scent-marking, frequency of roaring (an auditory expression of dominance), and even the intensity with which females were herded, also waned in the dry season. The last thing to disappear was the male's attachment to one potentially territorial area. Even that, in certain circumstances could be compromised without being totally surrendered. During one dry season, better pasture conditions, resulting from a localized rainstorm, occurred next to the study area. All the impala left the study area and moved to this green area, except for one or two territorial males, which spent their time moving back and forth between their territories and the green area, maintaining a presence in the former but feeding in the latter.

The impala's social organization is clearly responsive to the environment containing the population's resources. The responses, which are primarily seen in group formation, are flexible. The territorial males' behavior responds directly to the environment in some ways (for example,

Table 8.4			The proportion of sightings of all territorial males with bachelor males in the Banagi study area. The dry season in 1969 was more severe than that in 1970.
Year	Month	Season	Proportion (as %-age) of sightings of territorial males with bachelors.
1969	March	Wet	16
1969	May	Wet (late)	20
1969	August	Dry	48
1969	December	Wet	10
1970	February	Wet	7
1970	April	Wet	3
1970	July	Dry	22
1970	October	Dry	23
1971	January	Wet	5

choice of place and time for territoriality), but is also affected indirectly through the females' responses to environmental changes (for example, variations in group size and rate of movement). The male's use of territoriality to secure mating rights organizes the impala's local society in considerable detail. Yet the details vary—for instance, the age at which juvenile males are excluded from female herds—as territoriality waxes and wanes. Indeed, territoriality, and with it the sociosexual organization, can cease completely.

Such flexibility in territorality makes sense in a population in which mating opportunities occur neither constantly nor fully predictably. We suggest that males, in effect, balance the cost of holding territory against its potential reproductive benefits. When immediate benefits are high, as in the habitats favored by females at the April estrus peak, demands for territory are high, leading to high immediate costs, in terms of the level of aggression that must be maintained to gain and hold the best territories. However, when the estrus frequency falls, maximization of mating opportunities may involve holding larger, more diverse territory and gradually reducing the expenditure on aggression in order to hold the territory as long as possible; but the level of aggression should obviously not fall below that of any other male's challenging for that territory.

*Flexibility of Topi Social
Organization*

Topi are larger than impala (they weigh about 100 to 140 kg), and are less sexually dimorphic (pl. 20). Within the Serengeti, we can compare populations of topi differing greatly in the local extent and distribution of the habitats they use seasonally. These differences appear to affect the ranges of individuals and the sizes of groups they form. These, in turn, have produced the males' differing styles of territorial use to secure matings.

Topi are absent from short-grass plains and dry bush country in the south and east of the Serengeti. They occur in low densities in the central woodlands and reach high densities on the plains in the high rainfall areas of the west and north. Duncan (1975) studied their feeding, movements, and social organization in the center and west of the park. Unless otherwise acknowledged, information in this section comes from his study.

Topi feed purely upon grass, selecting consistently for green leaf. In the wet season, they obtain this diet most readily from short to medium leafy swards; in the dry season, they select what green leaf they can find in grasslands with a high standing crop of tall grasses. Their pursuit of these favored feeding conditions is evinced by the seasonal variation in habitat preference. In the woodlands, these favored habitats occur as interdigitating patches, a few hectares in area, at different levels of the local catena—during the dry season, they consist of long-grass habitats in the alluvial depressions; during the wet season, medium-length grass on the eluvial ridges or colluvial slopes. On the plains, by contrast, the equivalent habitats cover wide areas and are not interdigitated, since the catena is much larger, with low relief and little incised drainage. In those circumstances, movement between habitats means traveling many kilometers.

The various social organizations of Serengeti topi populations reflect their differing environments. In the woodlands, where each season's favored habitat occurs as small patches, topi females live in small herds of about two to eight animals. Duncan suggests that these groups have closed membership. The occupied area is completely covered by a mosaic of territories, each held by one male, and each containing examples of all habitats needed during the year. The sizes of female home ranges and of male territories are similar—about 1 to 3 sq. km—and female herds tend to stay within one male's territory. The male maintains the boundaries of the territory against other males by scent-marking and display, and he

spends a disproportionate amount of his time near the periphery of the
territory. Females use all parts of the territory equally (with due allowance
for seasonal habitat preference), and are aggressive toward strange topi.
However, this aggression does not seem to amount to territorial defense
in the usual sense (our interpretation).

Like impala, a topi male will defend a territory even if it contains no
female. In Duncan's woodland study area, the size of female herd in each
territory was related to the extent of the dry-season habitat the territory
contained. Males could retain territories, and individual females in them,
for several years, and only territorial males were seen to mate. These
woodland territories appear to contain the year-round requirements for a
group of females; the male is not defending a rutting territory only.

Topi on the western plains were organized in a distinctly different
fashion. Male territories were restricted to those that would be in use
during the rut, in the wet season. Territories were small, and Duncan
reported in one area a cluster of very small territories, reminiscent of the
"leks" of Uganda kob (Buechner 1961); we have seen a similar cluster
by the Talek River, Kenya, in the extreme north of the Serengeti-Mara
ecosystem. On the western plains, topi herds can be very large (table
8.5), and are of open membership for both sexes. Home range sizes of
females are fifty times greater than in the woodlands.

In this organization, territorial males usually were found on their terri-

Table 8.5   Topi female grouping and home-range size, and male territory size, in the central woodlands and on the western plains of theSerengeti. Figures are taken from, or calculated from, data in Duncan (1975).

	Region	
	Woodlands	Western Plains
Size of group containing the average female	4.1	1098.0
Area of female home range (km²)	1.5	183.5
Area of male territory (km²)	1.2	0.17
		"Lek" = 0.006

tories in the wet season, but visited them only occasionally to mark them in the dry season, spending the rest of their time with the large, mobile herds on the preferred dry-season habitat. In the wet season, the herds, containing hundreds of females and a number of nonterritorial males, might overlap several territories simultaneously. Each male would try to delay the moving females by herding them within his territory. Nonterritorial males avoided territory-holders, and tended to become concentrated on the periphery of the herds. Even in the wet season, territory-holders might, for short periods, forsake their territories and move with the herds, during which time they behaved within the herd as though they were nonterritorial.

The behavior of a male intermittently traveling 10 to 15 km from preferred dry-season habitat to maintain olfactory marks on his territory—which would not become reproductively functional until the wet season—is reminiscent of the few commuting impala males mentioned earlier. Topi territories change owners much less often than did those of impala (one owner for many years in woodland topi, compared with several owners per year in impala), and provide reproductive rewards only during a short rut. Presumably, their full-time (woodlands) or part-time (western plains) maintenance during the rest of the year guarantees the males their possession during the rut.

The groupings seen in Serengeti topi comply with the hypotheses that large herds occur in the open-plains habitats, where food items are abundant and evenly dispersed, while smaller groups occur in the woodland, where visibility is restricted and food items are less abundant or evenly dispersed. Duncan (1975) has related individual movement, grouping, and type of group membership (open or closed) to the extent and dispersion of habitats, rather than of food items alone. He asserts that small, closed groups of topi females defend food resources where the limiting dry-season habitat occurs in small, discrete patches. Males gain the right to associate with these female groups by superimposing a territory upon their annual range. The range of females on the western plains, moving tens of kilometers seasonally between habitats, would be impossible to defend either as an exclusive feeding area or as a territory. Similarly, the huge herds are too large for defended possession by a single male. Territory on the western plains is an exclusive area within which a male herds and courts transitory females, and territories are located (like those of some impala males) in anticipation of female presence during the rut. Even in the dry season, claims to a territory are maintained by occasional visits to re-mark boundaries. It seems reasonable to say that these different styles of use of territory are adaptive in their socioecological contexts, set by the grouping behavior and movements of the females.

Duncan's study presents a particularly clear example of two distinctly different social organizations arising in one species because of demonstrable differences in food and habitat dispersion. The sedentary and migratory wildebeest populations in the Serengeti show social organizations that are just as distinct, but, unfortunately, we lack comparable data on food and habitat use, grouping, and individual movements in the two population types.

*Flexibility in the Social
Organization of Other Ungulates*

Sinclair (1974) has shown that buffalo in the Serengeti vary their group size seasonally. Mean herd size falls from about five hundred in the wet season to about one hundred or less at the height of the dry season. At the same time, more males leave the breeding herds, and the size of bachelor herds increases. Buffalo herds split, despite the closed nature of their membership. The herd fractions remain within the common home range, and can unite again. Sinclair argued that this splitting occurred in response to the decreasing size of patches of habitat which, in the dry season, still carried appropriate food. The decrease in the proportion of males with the females appears to be the only change in detailed organization within the herds.

Elephants also vary their group sizes seasonally, each population or subpopulation tending to congregate at the start of the rains. At that time, bulls may be closely associated with the congregation, whereas at other times they may be dispersed tens of kilometers from the nearest females. In neither buffalo nor elephant, then, does the closed nature of society or the long-term association between mother and offspring prevent group size from adjusting itself to seasonal variations in resources.

Our limited data suggest that some other species with open-membership groups show a seasonal fall in group size as they move from wet-season occupance of the eastern plains to dry-season use of woodlands; this applies to kongoni and Thomson's gazelle. Zebra display this at the level of the feeding aggregation, but not at the level of the family group, which is of fixed membership and size. Its internal organization clearly does not vary seasonally, nor does the internal organization of the dik-dik pair, where the male and female maintain their territorial behavior throughout the year.

Conclusion

This chapter began with a set of hypothetical points which suggested why an individual ungulate should or should not join others to form a group. The points referred to tendencies among related species, and, hence, could be tested only on the antelopes, since no other ungulate group is represented in the Serengeti by a range of species. However, if comparisons are made outside the Serengeti, the points predict correctly the relative grouping tendencies of black and square-lipped rhinoceros, common and pigmy hippopotamus, giraffe and okapi, and the savanna and forest subspecies of both elephant and buffalo. The predictions are less successful among the pigs, where the warthog's use of burrows for refuge confuses the trend; and among the zebras, where the larger, but arid-dwelling, Grevy's zebra forms smaller aggregations than does the common zebra, living on wetter, more abundant pastures. Among the Serengeti antelope species, the broad predictions of the hypothetical points hold true.

Seasonal and regional variations in grouping within the Serengeti are reported in several species. For impala, the type of food items being sought affects the distance between neighbors in feeding herds and the rate of movement of herds. The type of food consumed varies seasonally, regionally, and with habitat; the last also affects visibility. These are enough to account for variations in group size, if one assumes that individuals try to form cohesive, coordinated herds for defense against predators. The examples of flexibility in group size in impala and buffalo, and in other, less studied species strongly support the general correctness of the hypotheses.

In all ungulate species, either temporary or persistent detailed organization is seen within the environmentally determined grouping. A persistent male-female association forms a detailed organization in the grouping of dik-dik and zebra. Mother-young association creates temporary detail in the organization of all societies, but lasts several years in buffalo and a lifetime in elephant, producing persistent organization. Zebra, buffalo, and elephant show seasonally varying aggregations which are larger than these persistent associations between individuals. Impala and topi represent species in which individual males strive to secure matings by excluding other males from territories and, in the process, secondarily organize the grouped society. In both species, the type of behavior displayed by the territorial males varies with the grouping, movement, and habitat choice of the rest of the population. This shows that the male-imposed organization is largely secondary to the environmentally determined basic grouping.

Because of the activities of the territorial males, bachelor impala tend to be displaced from the area currently occupied by females (Jarman and Jarman 1973). Similar differential distribution of classes of animals in the population is seen in several other species, although the behavioral origins are not always clear. Within each population, elephant males in the Serengeti are widely distributed beyond the area being used by female groups, especially in the dry season. In buffalo, lone males and small bachelor groups can be found in areas in which the large breeding herds never occur (Banagi, for instance); the same is true of eland in some areas (C. Hillman, pers. com.). Wildebeest, topi, and Thomson's gazelle show at all times a similar partial separation of males from females, perhaps seen only as fluctuations in sex ratio among animals counted along a transect. This spatial organization of a population is one of the most interesting of the ecological consequences of social behavior, since it may influence the resources available to the separate classes of individuals in the population. Buffalo males that withdraw from the herds, or impala bachelors retreating from the territorial mosaic, must balance the risks of segregation and the possibly different resource environment against the reduction of hierarchical stress or harassment by territorial males. At the same time, their withdrawal increases the food available to females. The influence of this may be very slight, but needs to be studied more closely.

Differential distribution of classes of individuals may be an immediate consequence of social behavior. Long-term adaptive consequences can also be inferred, one of which is sexual dimorphism. Species such as dik-dik, in which one male pairs with one female for life, show little visible dimorphism. Species such as impala, in which there is intense competition between males to hold territory and gain matings, often have males which are one and a-half times as heavy as females. Males have large and flamboyant horns, and develop shapes, smells, and sounds quite different from those of females. In fact, the differences between males and females, although so striking to the human observer, are less important than those between the males themselves, since it is those differences which will influence the apportionment of mating rights.

Much of the aesthetic impact that the Serengeti's ungulates make upon visitors and scientists derives from the relationships discussed in this chapter. The sight of enormous herds of buffalo, wildebeest, or topi, the sighting of dik-dik, always in pairs, the jealous herding by one impala male of a group of females—all reflect the sociobiological strategies of the species, and all add to the diversity of the observer's experience of the Serengeti.

The ecological adaptiveness of social organization has been argued in detail for some large carnivores in the Serengeti (Kruuk 1966, 1972; Schaller 1972; Bertram 1973). For groups such as primates, small carnivores, or rodents, the comparative studies have yet to be made. There has been a tendency at times to consider behavioral studies as unimportant luxuries, compared with management-oriented pragmatism of ecological studies. This chapter has tried to show that behavior and ecology are inextricably interrelated, and full understanding of the dynamics of the Serengeti's vertebrate ecosystem will require studies of the sociobiology of many more of the component species and their resources.

References

Bertram, B. C. R. 1973. Lion population regulation. *E. Afr. Wildl. J.* 11:215–25.

Buechner, H. K. 1961. Territorial behaviour in Uganda kob. *Science* 133: 698–99.

Crook, J. H. 1965. The adaptive significance of avian social organisation. *Symp. Zool. Soc. Lond.* 14:181–218.

Crook, J. H., and Gartlan, J. S. 1966. Evolution of primate societies. *Nature* (Lond.) 210:1200–1203.

Douglas-Hamilton, I. 1972. On the ecology and behaviour of the African elephant. D.Phil. thesis, Oxford University.

Duncan, P. 1975. Topi and their food supply. Ph.D. dissertation, University of Nairobi.

Estes, R. D. 1966. Behaviour and life history of the wildebeest. *Nature, Lond.* 212:999–1000.

Estes, R. D. 1969. Territorial behaviour of the wildebeest (*Connochaetes taurinus* Burchell, 1823). *Z. Tierpsychol.* 26:284–370.

Geist, V. 1974. On the relationship of social evolution and ecology in ungulates. *Amer. Zool.* 14:205–20.

Gosling, M. 1969. Parturition and related behaviour in Coke's hartebeest, *Alcelaphus buselaphus cokei* Gunther. *J. Reprod. Fert.*, supp. 6, pp. 265–86.

Gosling, M. 1974. The social behaviour of Coke's hartebeest (*Alcelaphus buselaphus cokei*). In *The behaviour of ungulates and its relation to management,* ed. V. Geist and F. Walther, n.s. no. 24, pp. 488–511. Morges, Switzerland: I.U.C.N.

Grimsdell, J. J. R. 1969. The ecology of the buffalo, *Syncerus caffer,* in western Uganda. Ph.D. dissertation, Cambridge University.

Hamilton, W. D. 1971. Geometry for the selfish herd. *J. Theor. Biol.* 31: 295–311.

Hendrichs, H. 1971. Freilandbeobachtungen zum Sozialsystem des Afrikanischen Elefanten, *Loxodonta africana* (Blumenbach, 1797). *Ethol. Stud.* 1:77–173.

Hendrichs, H., and Henrichs, U. 1971. Freilanduntersuchungen zur Ökologie und Ethologie der Zwerg-Antilope *Madoqua* (*Rhynchotragus*) *kirki* Gunther, 1880. *Ethol. Stud.* 1:9–75.

Jarman, M. V. 1976. Impala social behaviour: birth behaviour. *E. Afr. Wildl. J.* 14:153–67.

Jarman, M. V., and Jarman, P. J. 1973. Daily activity of impala. *E. Afr. Wildl. J.* 11:75–92.

Jarman, P. J. 1973. The free water intake of impala in relation to the water content of their food. *E. Afr. Agr. For. J.* 38:343–51.

Jarman, P. J. 1974. The social organisation of antelope in relation to their ecology. *Behaviour* 48:215–67.

Jarman, P. J., and Jarman, M. V. 1974. Impala behaviour and its relevance to management. In *The behaviour of ungulates and its relevance to management,* ed. V. Geist and F. Walther, n.s. no. 24, pp. 871–81. Morges, Switzerland: I.U.C.N.

Klingel, H. 1974. A comparison of the social behaviour of the Equidae. In *The behaviour of ungulates and its relation to management,* ed. V. Geist and F. Walther, n.s. no. 24, pp. 124–32. Morges, Switzerland: I.U.C.N.

Kruuk, H. 1966. Clan-system and feeding habits of spotted hyaenas. *Nature* (Lond.) 209:1257–58.

————. 1972. *The spotted hyena.* Chicago: Univ. of Chicago Press.

Mukinya, J. G. 1973. Density, distribution, population structure and social organisation of the black rhinoceros in Masai Mara Game Reserve. *E. Afr. Wildl. J.* 11:385–400.

Schaller, G. B. 1972. *The Serengeti lion.* Chicago: Univ. of Chicago Press.

Sinclair, A. R. E. 1974. The social organisation of the East African buffalo (*Syncerus caffer* Sparrman). In *The behaviour of ungulates and its relation to management,* ed. V. Geist and F. Walther, n.s. no. 24, pp. 676–89. Morges, Switzerland. I.U.C.N.

Spinage, C. A. 1969a. Naturalistic observation on the reproductive and maternal behaviour of the Uganda defassa waterbuck, *Kobus defassa ugandae* Neumann. *Z. Tierpsychol.* 26:39–47.

Spinage, C. A. 1969b. Territoriality and social organisation of the Uganda defassa waterbuck, *Kobus defassa ugandae* Neumann. *J. Zool.* (Lond.) 159:329–61.

Walther, F. 1964. Einige Verhaltens beobachtungen an Thomsongazellen (*Gazella thomsonii* Gunther, 1884) im Ngorongoro Krater. *Z. Tierpsychol.* 21:871–90.

Walther, F. 1965. Verhaltensstudien an der Grantgazelle (*Gazella granti* Brooke, 1872) im Ngorongoro Krater. *Z. Tierpsychol.* 22:167–208.

Wilson, E. O. 1975. Sociobiology—the new synthesis. Cambridge: Harvard Univ. Press.

Brian C. R. Bertram

Nine Serengeti Predators and Their
Social Systems

Five important carnivore species inhabit the Serengeti, and form the
subject of this chapter. They are the lion (*Panthera leo*), leopard (*Panthera pardus*), cheetah (*Acinonyx jubatus*), wild dog or hunting dog
(*Lycaon pictus*), and spotted hyena (*Crocuta crocuta*). All five feed
mainly on the flesh of newly killed large grazing mammals, and, therefore, one might suppose that they occupy the same or a similar ecological
niche. On closer examination, however, it is evident that these predators
make use of different sections of the whole biomass of potential prey.
They do so for a variety of reasons, some of which are related to habitat,
some to hunting method, some to size, and some to social organization.

My aims in this chapter are twofold. First, I want to summarize the
available information on the hunting methods and prey of these five
species of predator, and to correlate and discuss briefly some of the interspecific differences. Second, I want to summarize information on the
diverse social organizations of these predator species and to consider some
of the evolutionary reasons for the observed differences.

I am considering the "dynamics" of predation and of predator social
systems in a broad sense and mainly in an evolutionary sense. I am interested in the evolutionary effects of the predator on its prey, and vice
versa; in the ways in which predator social systems operate; and in the
factors influencing the way these systems have evolved and the way in
which they are continuing to evolve.

Predation

Table 9.1 summarizes data on predation by the five predator species
under consideration and lists the original sources from which this information has been obtained. It should be noted that these sources are, in

Table 9.1   Predation by Serengeti predators.

Species	Cheetah	Leopard	Lion	Hyena	Wild dog
Approximate weight of adult (kg)	40–60 kg[Y]	35–60 kg[Y, E]	100–200 kg[C]	45–60 kg[R]	17–20 kg[Y]
Number in Serengeti ecosystem	220[Y]–500[M]	800–1200[E, Y]	2000–2400[P, Y]	3000[R]–4500[N]	150[L]–300[Y]
Habitat in Serengeti	Especially plains, but woodlands too[M, E]	Woodlands only[E, Y]	Mainly woodlands, but plains too[E, Y]	Mainly plains[R]	Plains and woodlands[L, Y]
Hunting—time of day	Entirely by day[Y, M, X]	Mainly at night[A]	Mainly at night[E, Y]	Night and dawn[R]	Mainly by day[J, T, Y]
Number of animals hunting together	1[Y, M]	1[A]	1–5[E, Y]	1–3 (for wildebeest & gazelles), 4–20 (for zebras)[R]	Whole pack[J, R, T, Y] i.e. 2–19
Method of hunting	Stalk, then long fast sprint[Y, M, X]	Stalk to close range, then short sprint[E, A]	Spread out: stalk then short sprint[E, Y]	Long distance pursuit; others join in[R]	Long-distance pursuit[J, T, Y]
Distance from prey when chase starts	10–50 m[M] 50–70 m[Y]	5–20 m[E]	10–50 m[Y]	20–100 m[R]	50–200 m[Y]
Speed of pursuit	Up to 95 km/hr[Y]	Up to 60 km/hr[Y]	50–60 km/hr[I, Y]	Up to 65 km/hr[R]	Up to 70 km/hr[Y]
Distance of pursuit	Up to 350 m[U, Y]	Up to 50 m[E]	Up to 200 m[Y]	0.2–3.0 km[R]	0.5–2.5 km[Y]
Measured success rate	37[T]–70%[Y]	5%[A]	15–30%[I, Y]	35%[R]	50[T]–70%[Y]
Commonest species taken (in order of frequency in diet)	Mainly Thomson's gazelle[Y, M], also hare[M], Grant's gazelle[Y, M] impala[E, U]	Impala, Thomson's gazelle, dik dik, reedbuck, many others[A, E]	Zebra, wildebeest, buffalo, Thomson's gazelle, wart-hog, others[E, Y]	Wildebeest, Thomson's gazelle, zebra[R]	Thomson's gazelle, wildebeest, zebra, others J, R, T, Y
Health of prey	Healthy[Y]	Healthy[E]	Healthy[E, Y]	Sick and healthy[R]	Sick and healthy[R, Y]

Species	Cheetah	Leopard	Lion	Hyena	Wild dog
Age and sex of prey	Especially small fawns[Y]	All ages: only the young of topi, wildebeest, zebra[E]	All ages, but disproportionately more young and old[E, Y]	Especially males and young of wildebeest and gazelles; female zebra[R]	Esp gazelle adult males; wildebeest calves; zebra adult females[J, R, T, Y]
% of kills which are partially or wholly lost to other carnivores	10–12%[M, Y]	5–10%[E, F]	Almost none[E, Y]	5% (20% in Ngorongoro)[R]	50%[R]
% of diet obtained by scavenging	None[M, Y]	5–10%[E, F]	10–15%[E, Y]	33%[R]	3%[Y]
Important interference competitors	Hyenas; possibly lions[M, R, Y]	Possibly lions[E]	None[E, Y]	Possibly lions[R, Y]	Hyenas[R, Y]

A   Bertram 1974
B   Bertram 1975a
C   Bertram 1975b
D   Bertram 1976
E   Bertram, unpub. obs.
F   Bertram, pers. obs.
G   Eaton 1970
H   Eisenberg and Lockhart 1970
I   Elliott, Cowan, and Holling 1978
J   Estes and Goddard 1967
K   Frame and Frame 1976a
L   Frame and Frame 1976b
M   Frame and Frame 1977a

N   Frame and Frame 1977b
O   Hamilton 1976
P   Hanby and Bygott, chap. 10, below
Q   Herdman 1973
R   Kruuk 1972
S   Makacha and Schaller 1969
T   Malcolm and Van Lawick 1975
U   McLaughlin 1970
V   Muckenhirn and Eisenberg 1973
W   Sadleir 1966
X   Schaller 1968
Y   Schaller 1972

many cases, contradictory. Often this is because of differences between habitats, caused by time or distance. I suspect that it may sometimes be due to differences between the individual predators or groups of predators observed, and sometimes to differences in observation methods. Therefore, for the purposes of this chapter, table 9.1 should not be used as a source of data, but as an outline.

Hunting Methods

As can be seen in table 9.1, the five species differ widely in their hunting methods. These differences can be related to the social organization of both predator and prey, and help to explain the coexistence of various carnivore species.   Cheetahs (pl. 23) hunt by overtaking their prey at the end of a high-speed chase. Unless it has caught its victim within about 300 m, the cheetah stops, probably because it cannot tolerate any further rise in body temperature (Taylor and Rowntree 1973). Such high-speed chases probably require good visibility and freedom from obstructions, which may explain why cheetahs hunt entirely by day and usually in open country. They stalk to within about 40 m of their prey before starting to sprint—except in the case of small gazelle fawns, which they pursue more slowly from a much longer distance.

Both wild dogs and hyenas hunt in groups (pls. 32, 35, 36) and, in long-distance chases of up to 3 km, run down their prey until it is exhausted. Both do their hunting mainly in open country. Obstructions are likely to be less of a problem for a chaser than for a sprinter, but, nonetheless, they reduce the ability of the hunters to observe their intended prey, to gain on it by cutting corners, and to see where their companions are. It is likely that occasional obstructions generally hinder the pursuers more than the pursued animal. Hyenas hunt mainly at night, when their visual ability may give them an advantage over their prey.

Leopards (pl. 34) hunt alone, by stalking close to their prey and pouncing or sprinting from very short range. Such a method of hunting requires a considerable amount of cover. The vegetation of the Serengeti woodlands provides cover, as does low night visibility. Leopards are not found on the plains, and they do most of their hunting at night.

Lions use hunting methods similar to those of leopards, but usually hunt in groups (pl. 30). Individuals spread out and take different routes in stalking the prey, with the result that they sometimes surround it. When a prey animal detects a lion, which may or may not have already started to rush, its escape route may well pass within a range of another lion, which can catch it. Group hunting helps lions to hunt successfully in country with less cover than they would otherwise need, and it is mostly done at night.

All five predators are opportunists, quickly taking advantage of prey which is vulnerable because of injury or location making them easy to catch. Cheetahs eat only food that they have killed themselves. In contrast, wild dogs, leopards, lions, and hyenas (in increasing order) scavenge a proportion of their food supply from the carcasses of animals which have died of other causes. The ability to get a large proportion of food intake by scavenging is favored by differing adaptations. One of these adaptations is the ability to travel long distances with low expenditure of energy. Vultures can do this superlatively (chap. 11); and hyenas, because of their shape and gait, can probably do so more efficiently than the cat species (chap. 7). Because of their digestive abilities, hyenas can utilize carcasses more efficiently than can other predators (Kruuk 1972). Lions can usually supplant other predators from carcasses (Schaller 1972) because they are larger than other predators and are often in groups. Similarly, being in a group often helps hyenas to displace single lions from carcasses (Kruuk 1972). Thus, how much scavenging each species is able to do probably depends to a great extent on the local circumstances, and particularly on the abundance of other would-be scavengers.

Three of the five species usually hunt in groups, and two solitarily. It is worth examining the reasons for the difference, for if social (or solitary) hunting benefits one species, one might expect that it would benefit all. The hunting methods of hyenas and wild dogs involve a long and tiring pursuit, at which several pursuers are probably better than one, especially if the prey does not run straight. In addition, the large prey animals that are usually hunted would often be able to defend themselves against a single dog, hyena, or even lion. From observations of hyenas hunting wildebeest calves, Kruuk (1972) reported a success rate of 15 percent for a single hyena, 23 percent for two hyenas, and 31 percent for three or more hyenas. The increase in success rate was mainly due to the wildebeest mothers' inability to attack two hunters simultaneously.

Schaller (1972) observed that lionesses had a success rate of about 15 percent when hunting alone and that this was approximately doubled when two or more lionesses hunted together; in this case, the improvement was due to the animals' surrounding their prey, not to any form of defense by the prey.

Although it seems clear why these three predators usually hunt in groups, it is less obvious why leopards and cheetahs would not also benefit by doing so. However, for cheetahs, the necessarily small size of the prey that they can catch, the high success rate normally achieved, and the straight, long-distance chase in open country all mean that a cheetah would probably get less food per unit amount of energy expended if it were to hunt with another than if it were to hunt alone. For leopards, the problem is one of remaining undetected while getting in extremely close range of the prey. Small prey animals in thick cover can escape rapidly into the third dimension—up trees, through obstructions, or down holes, where two leopards are no better at following than one. A leopard essentially has to catch its prey *before* it can flee; unlike the lion, which catches its prey as the prey is fleeing. Two hunters are more likely to be detected during the stalking approach phase. For leopards, hunting the small prey animals which are commonest in thick cover, the success rate would need to be more than doubled for it to be worthwhile to hunt in pairs versus alone.

Table 9.1 shows a wide range of hunting success rates, both within and between species. However, there are considerable problems in measuring the success rates of predator hunts and in comparing them meaningfully between species. First, it is difficult to define what a hunt is. Lions and leopards may move a few yards toward prey and then wait to see if an opportunity arises—usually it does not. Wild dogs and hyenas may "test" herds of prey animals by running at them briefly to break them up or to look for vulnerable individuals. Whether or not such testing or waiting is scored as an "attempt" enormously influences the measured success rate. Second, success rates should perhaps be considered partly as optimism rates. Whether or not a stalking predator succeeds in a hunt is determined largely by chance factors, such as whether a prey animal raises its head at a particular moment, which way it runs, and whether it collides with another animal as it runs away. The lower the predictability of the outcome, the more necessary it is for the predator to embark on apparently unpromising hunts, and, therefore, the lower the resultant success rate will be. It can be seen in table 9.1 that cheetahs and wild dogs both have high hunting success rates, which is probably a reflection both of the high predictability of the outcome of the hunt, and of the very high energy expenditure in chases, penalizing unsuccessful attempts. Thus, I suggest that lions and leopards have to be "overoptimistic," but that cheetahs and wild dogs need not be and cannot afford to be.

*Prey Taken, and Competition
between Predator Species*

The differing sizes of the various predator species, their distributions over the various habitat types within the ecosystem, and the variety of their hunting methods all result in a relatively low degree of overlap in their food species and in the categories of those food species eaten. The extent of exploitation competition (as opposed to interference competition; see chap. 10) between them is therefore probably slight.

Although lions feed on a number of the very large resident herbivores, such as buffalo and giraffe, which no other predator can take, they feed mainly on zebra and wildebeest when these migratory species are within their pride's areas. A particular area is likely to contain wildebeest for only a small proportion of the year, while zebra, with their less aggregated distribution, are likely to be present for a large part of the year (chap. 5). Zebra, therefore, constitute a greater proportion of a lion's diet, despite their being less abundant, and despite the fact that they are harder to catch than wildebeest. Gazelles, despite their abundance, are rarely taken by most prides, for there are usually other prey species present when the gazelles are. Lions, too, take a toll of the main resident prey species—topi, kongoni, impala, and warthog—especially when the migratory species are lacking.

Precise figures cannot be given here because there are considerable methodological problems in determining the relative numbers of different species taken by any predator, especially when kill data are gathered by searching for predators on kills. For example, lions were observed to remain for an average of eighteen hours at wildebeest kills, as opposed to only five hours at impala kills (unpub. obs.). For this reason alone, an observer is some three and a-half times more likely to find a wildebeest kill than an impala kill, and his prey data are distorted accordingly. Similarly, the skin and skeleton of a gazelle eaten by leopards remains in the branches of a tree for weeks, and so is far more likely to be detected than, for example, a steerling, which takes some ten minutes to consume (unpub. obs.). Further biases in the collection of kill data can arise through the observer's utilization of searching time; thus, small animals killed at night are much less likely to be discovered than those killed during daylight. An attempt can be made to correct for some of these biases in one's own data, but not in the data reported by independent observers.

Leopards' prey is made up mainly of small animals and covers a wide variety of species occurring in the woodlands (Bertram 1974). Over the whole of the Serengeti, impala are probably the most often taken, followed by Thomson's gazelles, reedbuck, dik-dik, very young wildebeest, topi, kongoni and zebra, hares, large and small birds, and several small carnivores. A list of one hundred fifty leopard kills contains over thirty species (unpub. obs.), whereas a comparably sized sample of lion kills would contain about twelve.

Cheetahs feed mainly on Thomson's gazelles, although they include a few hares, Grant's gazelles, and impala as well. They have to follow the migratory movements of the gazelles, making large ranges essential (see table 9.1).

Similarly, wild dogs move extensively to follow the movements of their migratory prey species, consisting mainly of gazelles, zebras, and wildebeest. Larger packs can take larger prey animals (Malcolm and Van Lawick 1975). Neither wild dogs nor cheetahs are abundant enough to offer any detectable ecological (exploitation) competition to any of the other predator species.

Hyenas feed mainly on wildebeest calves, adult wildebeest, and gazelles. Because they are the most abundant predator, they are the most likely to be a significant competitor of any of the other predator species. Their competitive effect on wild dog and cheetah by exploiting the same food supply is reduced because hyena scavenge an appreciable proportion of their food, whereas the other two species do not.

Ecological competition between predator species is reduced by their taking different sections of the prey population. The predator species that hunt by pursuit (hyenas and wild dogs) tend to take the slower individuals of their prey species—small young, very old animals, and sick or injured ones (Kruuk 1972; Estes and Goddard 1967; Malcolm and Van Lawick 1975). Lions, which rely on ambushing their prey, take a more random selection of the population (Schaller 1972; unpub. obs.), probably because chance plays a large part in determining which individual from a group falls victim. Nonetheless, a greater proportion of lions' victims are in some way less fit than their companions than is the case in the live population (Schaller 1972).

Prey is vulnerable to different predator species in different ways. Wild dogs, for example, kill mainly adult female zebras, which tire sooner than stallions; lions, on the other hand, catch the two sexes about equally. The males of many antelope species are solitary within their territories. This makes males much more dispersed than females, and they are therefore more likely to be encountered by a hunting predator; being solitary probably also makes a male antelope more vulnerable to predators, since the animal has only its own sense organs to rely on to detect the predators' approach. As a result, the adult males of most ungulates are preyed upon more than females (Schaller 1972; Rudnai 1974).

Hyenas catch a large number of wildebeest calves at the animals' vulnerable early stage; at this time, the wildebeest herds are on the plains and out of reach of most lion prides. By the time the wildebeest have migrated into the woodlands, their calves are less vulnerable, and the lions catch mainly adults. Subadult male impalas tend to be at the edges of herds, driven from the center by dominant animals; in this position, they are vulnerable to approaching predators, and both lions and leopards take a higher proportion of young male than young female impalas (Jarman, pers. comm.; unpub. obs.). Holes provide warthogs with a haven against wild dogs (and usually against lions), but they also become a trap when the ground is so soft that the dogs are able to dig them out.

Overall, the extent of exploitation competition between the different predators is probably minute, in the sense that none appreciably reduces the uncaptured food supply of any of the others. On the other hand, interference competition between some of them may be of more significance. The work of G. W. and L. H. Frame indicates that hyenas drive off wild dogs from a considerable proportion of their kills (pl. 37) and suggests that this may be a factor in the recent decline of wild dogs (chap. 10). The effect of hyenas on cheetahs may be similar, although the short hunt of the latter is less likely to be detected by hyena.

Effects on prey species. The lack of exploitation competition between predator species in the Serengeti is partly due to the fact that predation has little effect on the size of the ungulate populations. In almost all cases, regulation of prey populations appears to take place through their food supply, with predation playing only a minor role (Schaller 1972; Sinclair 1977). The extensive mortality of wildebeest in bad dry seasons, and their dramatic increase in recent years, both indicate that predation does not control the numbers of this species; nor has it been shown to regulate the numbers of any of the other major herbivore species.

One reason for the relatively small impact of predation on prey population sizes is that much of the prey taken is "expendable." The removal by predators of old animals which are past the age of successful reproduction obviously causes no reduction in the reproduction rate; indeed, in some circumstances, it could even cause a slight increase, by reducing the number of competing conspecifics that are relatively inefficient because of age. Similarly, the removal of any animals which are so injured or diseased that they are unlikely to reproduce successfully will obviously not reduce the reproductive rate. The same applies to the males of almost all prey species; by analogy with domestic stock, it would be unlikely that the removal of even 90 percent of the male wildebeest would have any detectable effect on the number of young born, and, by reducing competition, it might even aid their survival. Data on these points is lacking.

A second reason for the lack of importance of Serengeti predators in regulating the population sizes of their prey is that some 80 percent of all potential prey is migratory and tends to move to the same places at the same times. Predators are generally less mobile than ungulates because they have heavy legs and helpless young (pl. 29). They tend to remain in one area, rather than attempting to cover the distances the migrants do. Schaller (1972) showed that the nomadic lions which followed the migratory prey concentrations fared much worse than resident ones. Kruuk (1972) suggested that the long distances which had to be covered by hyenas commuting between their dens and their migrating food supply limited their population size in the Serengeti; hyenas do not attain nearly as high a population density there as in Ngorongoro, which has a higher resident-prey density.

The resident predator population in an area obviously cannot be permanently above the level which can be supported by the prey species there. Bertram (1973) showed that lion population regulation was achieved by social mechanisms influencing whether subadult females were recruited or expelled by the local population. Schaller (1972) suggested that the number of lions in an area was determined by the number of resident prey within that area in bad seasons, when the migrants were elsewhere. In the Serengeti, the ratio of predator biomass to prey biomass is low, compared with other African national parks (Schaller 1972): lions, for example, are almost certainly less abundant there than they would be if the distribution of prey was more uniform over time. The huge moving aggregations of wildebeest mean that most wildebeest are out of the ranges of all but a small proportion of the lion prides in the Serengeti for most of the time, and are at times even beyond the reach of most hyenas.

When the migrants move into an area—for example, when they move into the woodlands of the Serengeti in the dry season—they provide a buffer against predation for the resident prey. During this time, most lion kills are of zebra and wildebeest, and the pressure on the resident impala, warthogs, buffalo, giraffes, topi, and kongoni is somewhat reduced (unpub. obs.). For the resident grazers this is a blessing in disguise, because the migrant population could denude the area of food. Sinclair (1977) calculated that food competition from wildebeest depressed the buffalo population by 18 percent, which far more than counters the effects of a temporary easing of the predation pressure on buffalo.

If predators do not regulate the population sizes of their prey species, what other effects do they have? A form of "purifying" of the prey population has often been suggested; this argument is based on group selection and implies that, in the long term, the prey species "benefits" through having diseased or defective individuals removed by predators. On the other hand, if population regulation takes place via food competition within the species, those diseased or defective individuals will compete less well and will be removed by natural selection, even in the absence of predators.

On an evolutionary time scale, it seems probable that predation is at least partly responsible for many aspects of the form and behavior of ungulates—their alertness, their sprinting ability, their coloring, their highly developed sense of smell, and their fear of thickets. Over the generations, one particular constituent of natural selection—predation— has acted in a directional way by removing animals that were less alert, more conspicuous or more incautious than their companions. As well as this directional selection, it is possible that predation also provides a form of stabilizing selection for prey coat colors. For example, if a lioness fixates on a particular individual when attacking a group of zebras, she may do better to choose one that appears different from the others; a mutant with a coat color different from that of its fellows would therefore be more vulnerable because it would be more likely to be selected. Kruuk (1972, p. 154) cited examples of such selection, but data are scarce because of the problems of observing the selection process itself.

The behavior of wildebeest when traveling in lines and when crossing rivers is probably a functional response to predation. A wildebeest following in the tracks of the one ahead is, I presume, less likely to encounter a lion than is a wildebeest forging its own new route, so grouping and following tend to be selected for. Wildebeest are particularly vulnerable when crossing rivers because they have to slow down and because riverine vegetation provides cover from which lions can ambush them. Each individual, therefore, tries to cross as soon as possible, sometimes plunging in too soon and landing on its predecessor. The risk of being trampled and drowned by following wildebeest is an extra selective pressure which tends to make the wildebeest plunge in sooner still, and which, consequently, makes it even more likely that its predecessor will suffer. Thus, I suggest that the risk of predation has been responsible for a form of behavior with a chain of consequences which today may be responsible for as many wildebeest deaths at rivers as lions cause (see chap. 4).

Predation is, of course, only one aspect of natural selection. What is continuously being selected for is not only the ability to survive against predators, but to reproduce more effectively than conspecifics and to pass on more genes to future generations. As animal's behavior obviously has to be a compromise between somewhat incompatible needs—the need to keep watch for predators, to ingest enough food, and to reproduce. Time spent in one activity reduces the time available for others. By remaining with companions who may help detect predators and may also become substitute victims in a predator attack (Bertram 1978), animals rely on grouping to lessen their individual risks.

One of the problems a prey animal faces is that it may be preyed upon by more than one species of predator. The prey cannot predict by what species or number of predators it may be attacked, nor when, nor by what methods. Therefore, it is difficult for it to evolve defense methods which are effective in all circumstances. Thomson's gazelles have to be both good sprinters to escape from cheetahs and good long-distance runners to get away from wild dogs and hyenas. Yet the structural and physiological requirements necessary for these two kinds of adaptation are incompatible; the gazelle would need different musculature and cooling methods, according to the type of predator that was pursuing it (Taylor and Lyman 1972; Taylor and Rowntree 1973; Taylor et al. 1971). Thus, in an evolutionary sense, cheetahs and wild dogs may be said to benefit one another in preventing their prime common prey species from evolving escape behavior and adaptations completely effective against either of them.

Social Organization

Table 9.2 summarizes the information available on the social organization of Serengeti predators and lists the source of this information.

It can be seen from table 9.2 that the two predator species which hunt solitarily—the cheetah and the leopard—also live in social groups of only one adult animal. By contrast, lions, hyenas, and wild dogs all live in long-lasting social groups. However, the organization and the dynamics of these social groups differ among the three species. In describing and examining some of the evolutionary causes and consequences of these differing social organizations, we will consider: (1) relatedness among group members; (2) avoidance of incest and the transfer of individuals between social groups; (3) sexual dimorphism; (4) dominance relationships; and (5) reproduction within the group.

While considering the differences, it is worth bearing in mind a little of their phylogenetic history. Kleiman and Eisenberg (1973) pointed out that the wild dog social group has apparently developed from the closely bonded pair which is characteristic of many of the canidae, whereas the lion social group is based on the permanent association of daughters with their mother. I suggest that the hyena clan has come about through neither of these routes but that it is a result of the loose association of many largely solitary individuals.

Degrees of Relatedness among Group Members

Within the social group of each predator species, at least some of the members are genetically related to one another—in some cases closely, and in others distantly or not at all. I shall interpret many of the differences between species, and many of the interactions between conspecifics, in the light of our understanding of the operation of kin selection as first outlined by Hamilton (1964). Briefly, what a vertebrate passes on to its offspring is an assortment of 50 percent of its own genes, and, through parental care, it improves the chances that those genes will survive and replicate themselves in their turn. However, replicas of an animal's genes are found not only in its offspring but also in its other close relatives; this is because a gene in an individual, *A,* must have come from one of *A*'s parents, who might well have passed on another replica of the same gene to *A*'s sibling. The same genes can survive and replicate themselves not only through the individual's own offspring, but also through other close relatives. More distant relatives share by common descent a smaller proportion of their genes. Thus, if the process of natural selection is one of competition between genes for survival and proliferation, those genes which dispose an animal to assist close relatives will be selected for in the same way as are genes which dispose an animal to exhibit good parental care.

However, there are practical limits to how many relatives can be assisted. Less close relatives, because they are less likely to contain replicas of an animal's genes, are less genetically "valuable" to that animal. Thus, the degree of relatedness is likely to be an important factor in influencing

Table 9.2   Social organization of Serengeti predators

Species	Cheetah	Leopard	Lion	Hyena	Wild dog
Name of social group	—	—	Pride	Clan	Pack
No. of adults in social group: average (and range)	Usually 1[M, Y]	Usually 1[E, Y]	9 (4–15)[D, E, Y]	55 (35—80) in Ngorongoro; more flexible in Serengeti[R]	7 (2—19)[K, Y]
Usual no. of adults in temporary group	1[M, Y]	1[E, Y]	1–8[E, Y]	1–4[R]	All[K, Y]
Range size of female	1300 km²[M]	10–25 km²[E, H, V] Male range larger[O]	20—150 km²[E, S, Y] Males may cover 2 or 3 pride ranges[E, S, Y]	20–40 km²[R] Male range same as female[R]	1500 km²[K] Male range same as female[K]
Advertising of territory: by voice	None[Y]	Rasping[H]	Roaring[E, Y]	None[R]	None
by scent	Urine-spraying (male) Feces[G, Y]	Urine-spraying[E, H] (both)	Urine-spraying[E, Y] (male) Scraping[E, Y]	Anal gland pasting[R] Pawing[R]	Urine-marking[K, L]
Marked sexually dimorphic characters	None	Size	Size, mane	None	None
Male wt. as % of female wt.	120%[Q]	150%[E, Y]	150%[C]	88%[R]	115%[Y]
Average number of: adult males	0	0	2–3[D, Y]	15—30[R]	2–6[K, Y]
adult females	1[M, Y]	1[A, E, Y]	3–9[D, E, Y]	15—30[R]	2–3[K, Y]
No. in breeding group which reproduce: females	—	—	All[B, D, Y]	Probably all[R]	1[K]
males	1[M, Y]	1[E, Y]	All[B, D, Y]	All[R]	Usually only 1[K, Y]

Species	Cheetah	Leopard	Lion	Hyena	Wild dog
Gestation period (days)	90–95[Y]	99–100[W]	111–119[W]	110[R]	69–73[Y]
Average litter size	3–5[M, Y]	1–3[Y]	2–4[B, E, Y]	2[R]	9–10[L, Y]
Suckling period	2–3 months[Y]	?	6–8 mo.[E, Y]	12 months[R]	1¼ months[Y]
Dependent period	14–18 months[M, Y]	18 mo.[A]	24–30 mo.[E, Y]	18 months[R]	12–18 months[Y]
Communal suckling?	No	No	Yes[B, D, Y]	No	No
Other cooperative feeding	None	None	None; but male tolerance toward cubs[D, Y]	None	Regurgitation by all pack members[J, L, R, Y]
Which sex transfers from natal group	—	Males probable leave region; females remain[E]	All males[D, E, Y]	Neither much, but males more[R]	All females[K]
Pair bonds	Sometimes between two males[E, M]	None[A, E]	None; but sometimes same-sex companionships[E, Y]	Occasional, loose, and temporary, between male and female[R]	Firm, between dominant male and female[K, L]

A Bertram 1974
B Bertram 1975a
C Bertram 1975b
D Bertram 1976
E Bertram, unpub. obs.
F Bertram, pers. obs.
G Eaton 1970
H Eisenberg and Lockhart 1970
I Elliott, Cowan, and Holling 1978

J Estes and Goddard 1967
K Frame and Frame 1976a
L Frame and Frame 1976b
M Frame and Frame 1977a
N Frame and Frame 1977b
O Hamilton 1976
P Hanby and Bygott, chap. 10, below
Q Herdman 1973
R Kruuk 1972

S Makacha and Schaller 1969
T Malcolm and Van Lawick 1975
U McLaughlin 1970
V Muckenhirn and Eisenberg 1973
W Sadleir 1966
X Schaller 1968
Y Schaller 1972

how one animal behaves toward another. For this reason, I shall discuss the degrees of relatedness found among group members within each species before considering other aspects of their social organization.

The average degree of relatedness among the members of lion prides has been calculated by Bertram (1976) from certain assumptions (based on the data of Schaller 1972, and of Bertram 1975a, and on unpublished observations) about the reproduction of the "typical" pride shown in figure 9.1. Here I shall summarize the argument only very briefly. Lion prides are assumed to last for generations, and strange females are not permitted to join: all the adult females in the pride were born in the pride, and, therefore, they are all genetically related to one another. Their cubs, too, are related to one another. Any male cubs which survive to adulthood

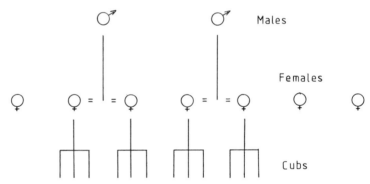

Figure 9.1   A representation of the reproduction of a typical pride. Two related companion males mate equally with four of the seven related lionesses, whose reproduction is synchronized. The resulting cubs are reared communally. Enough surviving female cubs are recruited into the pride to replace adults which die. Surviving male cubs leave the pride together and subsequently take over another pride together, as the breeding companion males. See Bertram (1976) for details.

leave the pride together, remain together, and, as a group, become the breeding adult males in a different pride for a few years only. The adult males in a pride are interrelated, then, and so are the adult females, but the males are not related to the females. The pride lionesses are, on average, related by 0.15, or, in other words, are roughly as close as first cousins. Cubs, because they are fathered by fewer and more closely related males, are, on average, related by 0.22, or are roughly as close as half-brothers; a similar degree of relatedness prevails among the adult males in possession of a pride, because they are the surviving males of a peer group of cubs from another pride. Details of the method of calculation are given elsewhere (Bertram 1976), where it is also shown that, even if the "typical" pride is not as I have assumed, the resultant degrees of relatedness are nonetheless little affected.

In wild dog packs, usually only one dominant pair reproduces (Frame and Frame 1976a). The adult males in the pack are usually the dominant male, his full brothers, and his sons; it can be calculated from the data given by Frame and Frame (1976a) that they are, on average, related to one another by about 0.38. On the other hand, the adult females, of whom there are fewer, are the dominant female, and either her full sisters or her daughters, so that they are all related to one another by 0.5. Thus, the average degree of relatedness among animals of the same sex in a wild dog pack is considerably higher than that in a lion pride, and, therefore, so is the degree of relatedness between any adult and the puppies in a pack. In wild dogs, the degree of relatedness between any two known individuals can be calculated almost exactly, because (unlike in lions) there is little uncertainty about parentage.

From the data given by Kruuk (1972), it can be calculated that the average relatedness values among members of hyena clans are very low (probably less than 0.03), partly because the clan is a much larger and looser social group than either a lion pride or a wild dog pack and partly because reproduction is not confined to a small proportion of the adult population.

Incest, and Transfer between Groups

Incest—reproduction between close relatives—is known to have generally deleterious consequences (see Greenwood, Harvey, and Perrins 1978; and Wilson 1975). Brother-sister matings, or parent-child matings, are likely to result in fewer offspring, which, in turn, tend to be less fit. A social carnivore, living in small groups of relatives, would be expected to be particularly likely to mate with a close relative. In fact, incest rarely occurs. In humans, it is avoided through specific "mechanisms," such as cultural taboos and sexual inhibitions developed in childhood through close familiarity among siblings (Bischoff 1975). By contrast, in the three social carnivores we are considering, incest avoidance appears to result largely from the geographical or statistical separation of potential mates who are relatives.

In hyenas, Kruuk's (1972) data show that clans are large, and most clan members are not closely related to one another. There are no data on the choice of mating partners, but even if hyenas mated randomly with other clan members, incest would be rare. The chances are further reduced by the fact that, in the Serengeti at least, the clans' composition changes with time, as some individuals, particularly males, leave one clan and join another.

In lions, subadult males at about three years old almost always leave their natal pride (Schaller 1972; unpub. obs.), largely, I think, because they are driven out by the adult males. When, a couple of years later, they take over a pride as the breeding adult males, it is very unlikely to be the pride in which they were born. Although there are some indications that they are likely to take over a pride near the area of their birth, they may take over one far away. Thus, the males are only distantly related, if at all, to the lionesses they mate with. After they have been in a pride for three or four years, they may have sexually mature daughters. However, because of intense competition from rival groups, few groups of males retain possession for as long as that; and those which do tend to be the larger groups, where a male mating with a young lioness is less likely to be her father than would be the case in a group with only a pair of males.

In wild dogs, subordinate females leave their pack together and join another pack, often a neighboring one, which is without a breeding female (Frame and Frame 1976a). Because the wild dog breeding population is small, the pack which transferring females join may consist partly of relatives, but not to a great extent; certainly these females are not as close relatives as are the other animals in their own pack.

In lions and wild dogs, the animals which emigrate do so in their own "interests," in order to increase their own reproductive output. Young male lions are prevented by the adult males from mating in their natal pride, and so they do better to seek another pride either without males or with males which can be ousted. Subordinate female wild dogs cannot reproduce, but when they transfer to a pack which lacks a dominant female, one of them quickly becomes dominant and reproductive (Frame and Frame 1976a). I suggest that this system is favored in wild dogs by the distorted sex ratio, there being consistently fewer females than males among adults. There are three contributory causes of this distortion: the unequal sex ratio at birth (Schaller 1972; Kleiman and Eisenberg 1973); a probable raised mortality rate among lone emigrating females (Frame and Frame 1976a); and, possibly, a higher mortality among adult females than males within packs (from the data given by Frame and Frame 1976b). The result is that a pack is more likely to be without a female than without a male, and, therefore, transferring females are more likely to find a reproductive place than males would be. There is a positive feedback here, for if subordinate females, rather than males, tend to leave, then a pack is even less likely to contain a replacement female when the dominant one dies. In lions, the mortality rate among adult males is higher than that among females (unpub. obs.), which favors the reverse.

Sexual Dimorphism

The three social species differ in the degree and the direction of the sexual dimorphism they exhibit. In wild dogs, the two sexes are much the same size (Schaller 1972); female hyenas are about 14 percent heavier than males (Kruuk 1972); and in lions, the males are some 50 percent heavier than the lionesses (Bertram 1975b), and have manes. It is worth examining the evolutionary reasons for this variety, and the relationship between sexual dimorphism, social organization, and hunting methods in these three social species.

Although it is characteristic of all the cat species that the males are somewhat larger than the females, lions are among the most sexually dimorphic with respect to size, and they are the only species with a striking difference in appearance between the sexes. I suggest that the marked sexual dimorphism is related to their social way of life, and that both are unique among cats. A pride of lionesses provides an abundance of reproductive opportunities: a pride male spends 15–20 percent of his time mating (unpub. obs.). There is intense competition between rival male groups for possession of a pride and its territory, as is evidenced by severe and often fatal fights, by the short period of tenure of a pride, and by the strong effect of the number of male companions in determining the length of this period (Bertram 1975a). The males' large size, together with the mane that makes them look even larger and may also protect the neck (Schaller 1972), probably makes them more effective competitors. It is likely that, as in many other species, large size generally helps in fights. On the other hand, a male, with his huge size and mane, must presumably find stalking and concealing himself more difficult. There are no data on the hunting success rates of adult males, because they do so little hunting —much less than the lionesses in their pride (Schaller 1972; Elliott, Cowan, and Holling 1977). This may be partly because they are less good at it, but it is also because they have no need to hunt: they can follow the females and, with their greater strength, can appropriate a share, and usually a disproportionate share, of any kills made. Thus, I suggest that because of lions' social way of life, males have become adapted as fighting animals at the expense of their hunting ability, while the females have adapted as hunters.

Wild dogs and hyenas pursue their prey over a long distance and then pull it down; both these activities probably favor size and strength in the same way as does fighting. Therefore, in these species, to be a good hunter does not make the individual any less effective as a fighter, nor vice versa. Both sexes participate equally in both activities, and natural selection does not "pull" in different directions on the two sexes here. It is not known what selective pressures set the limits of size of any of these species.

Partly as a result of their larger size, female hyenas are usually dominant over males: Kruuk (1972) suggested that large size might be important in helping them to protect their cubs against other cannibalistically minded clan members. I suggest that the very long distances that hyena mothers often have to commute between their food supply and their dependent cubs favor increased size in the traveler. Kruuk (1972) considered that a major cause of hyena cub mortality in the Serengeti was the lack of sufficient food sources within reasonable traveling distance of the hyena dens. Pennycuick (chap. 7) shows that larger size enables an animal to have a larger foraging distance. Male hyenas do not assist in feeding the young (Kruuk 1972), and, therefore, I suggest, are not subject to this selective pressure in favor of larger size.

In wild dog packs, all the adult members travel similar distances because they all hunt together and all return to the young. In addition, with the much higher degrees of relatedness and of cooperation among dog pack members than among hyena clan members, other adults are unlikely to be a threat to the offspring. Thus, in wild dogs, there is little extra selective pressure favoring large size more in one sex than in the other.

*Dominance Relationships and
Competition for Food*

The dominance systems vary among the three social species, depending
on the sizes of the group members and on their degrees of relatedness to
one another. In no species is there a simple linear hierarchy.

As described above, female hyenas are dominant over males; this is
only partly a reflection of their greater size, for a small female is nonetheless dominant over a male who is bigger than she (Kruuk 1972). She
can oust him from resting places, and from pieces of carcasses. Subadults
and cubs, because they are smaller, are lower in rank, and often cannot
feed until after the adults; but a mother will sometimes defend her offspring against others and so enable it to feed unmolested if there are not
too many other competing hyenas around. Kruuk (1972) found no clear
dominance hierarchy within either sex, although some animals seem to
be more dominant than others, and he thought that there probably exists a
complex system of special individual relationships.

At a large carcass, as Kruuk described, hyenas all feed as fast as they
can. Competition occurs by eating faster and therefore obtaining more
than companions, and there is relatively little squabbling. When the
carcass has become dismembered, individual hyenas will carry off separate
pieces. A hyena may try to dispossess another of such pieces. It does so
by chasing, rather than by attacking, the owner and trying to grab the meat;
the owner defends its property by fleeing with it, and only rarely by
threatening the would-be usurper.

Lions appear to be more socially organized, as would be expected in
a species with smaller, more coherent, and longer-lasting social groups.
The following summary comes from the observations of Schaller (1972)
and myself (unpub. obs.). Because of their larger size, males are dominant
over females at carcasses. If a kill is large, all the lions present usually
feed at it together (pl. 30). There is a considerable amount of snarling
and squabbling among the feeders, one of whom may also swat at another
with a paw, or bite at another animal which is feeding too close. Lions are
slow feeders compared with hyenas, and may take several hours to consume a typical large kill.

A small kill, or piece of it, can be "owned" by a single lion. Ownership
seems to be established when a lion has been able to carry away the piece
alone. At this stage, the animal in possession is able to keep away others
equal in size by virtue of its ownership: it threatens another lion which

comes too close, and will attack if the latter persists. The usurper recognizes the extra determination conferred by first possession, and respects the convention. However, the convention works best among equals: larger animals, such as males, often rob and keep away all smaller ones, and females often rob cubs of pieces of food the latter have obtained. Therefore, as in hyenas, the cubs are the first to starve when food is in short supply. However, an adult male, more than a female, will often allow cubs to feed from a carcass at which he is not permitting adult females to feed. A male is usually quite closely related to the cubs in his pride (although he does not know which he has fathered), but is not related to the lionesses. Lion cubs are relatively expendable: they are very small—less than 1 percent of adult weight at birth (pl. 29)—and they can be produced quickly; gestation takes three and a-half months, as against an adult life of fifteen to twenty years. Presumably this reduces the amount of risk and effort which it is worthwhile for their mother to take on their behalf.

Within each sex, in lions, there appears to be absolute equality, with no dominance relationships at all. No male is dominant over the other adult males, nor is any lioness above or below the others. No particular animal consistently leads the way in hunts or other moves, although some —cubs, males, and partially incapacitated females—tend to lag behind.

There is less overt competition in wild dogs than in either of the other two species (Schaller 1972). There is one dominant male and one dominant female; it is not clear how their dominant position is established, but apparently it is a rapid process (Frame and Frame 1976a) and a surprisingly peaceful one. There does not appear to be any further rank order among the other members of the pack. The exception is that puppies are dominant at food (Schaller 1972). They follow the adults on the hunt, drive them from the resulting kill, and feed first. Meanwhile, the adults wait nearby and keep away any hyenas from the carcass. When the puppies have finished, the adults all feed together, fast but without the squabbling so characteristic of feeding groups of lions or hyenas. Adult wild dogs also regurgitate food for one another, which neither of the other two species do. On the whole, wild dog society is remarkably peaceful, which may be partly because the members are so closely related to one another, and partly because there is little for them to compete over. The high mortality rate among adults (Schaller 1972) probably makes young especially "valuable" in a genetic sense, and, thus, it particularly favors behavior such as letting them feed first, which enhances their survival.

Distribution of Reproduction

A hyena clan contains many adult males, among whom there does not appear to be a dominance order, and so it is likely that any of them may mate with an estrous female. Mating has rarely been observed. There seems to be little competition among males for estrous females (Kruuk 1972), perhaps partly because the female, by virtue of her dominance, can easily determine with whom she will mate, rather than leaving it to competition.

Any adult female hyena may reproduce. Almost invariably there are two cubs in a litter, and the female rears them alone. Although several females may have young at the same den, and although these young play together extensively, each female feeds only her own young (Kruuk 1972). The average relatedness between adult hyenas is so low that kin selection could provide little significant selective pressure in favor of helping to rear one another's young, as lions and wild dogs do. A male, even if he is allowed at the den, does not assist in any way. There is little he could usefully do either to guard or to feed them, because the cubs' best defense against enemies is being able to escape far down the small tunnels of the den, and because while at the den they are fed only on milk. It is surprising that a hyena does not regurgitate food to her offspring, as wild dogs do, for it would be an extra and probably more efficient method of bringing food to the cubs; it would also be a method in which males could participate. The low quality of the hyena diet may be responsible.

A lion pride contains fewer adult males than adult females. There is no dominance order among them, and they acquire estrous females with roughly equal frequency and with remarkably little competition (Bertram 1976). Competition over a lioness in heat resembles competition for small amounts of food, in that the same convention over ownership operates—the first male in possession is temporarily dominant over others, but only as long as he is within a few meters of her. There do not appear to be any particular pair bonds within the pride, and a lioness does not have any discernible preference for any particular male (unpub. obs.). Copulation takes place a few times an hour for the two to four days that most estrous periods last; nonetheless, only about one in five of these periods results in the birth of cubs. Thus, mating in lions may be said to be extremely inefficient compared with almost any other species, in the sense that, on average, 1500 matings are needed for each litter born. I have argued elsewhere (Bertram 1976) that this inefficiency may be an important factor in reducing intermale competition by reducing the genetic "value" of each copulation.

Most litters consist of two to four cubs. At about six weeks of age, when the young are mobile, their mother leads them to join the rest of the pride. All the adult lionesses breed, and they tend to synchronize births with other lionesses in their pride (Bertram 1975a), so there are likely to be other cubs of similar age. They are reared communally, and are able to suckle from lactating lionesses other than their own mother. There are probably two factors which contribute to this most unusual situation: first, the fact that the lionesses are quite closely related, so they may have a genetic interest in one anothers' offspring; and second, the fact that the lioness' own young (particularly the males among them) do much better later if they have companions (Bertram 1975a).

In wild dog packs, only the dominant male mates, and only with the dominant female (Frame and Frame 1976a). Mating is rarely observed, and so presumably is not inefficient as in lions. It is not clear whether subordinate females fail to come into estrus, or whether they are prevented from mating, as is the case with wolves (Zimen 1976), but it is rare for a female other than the dominant one to produce puppies. All the animals in a wild dog pack are together almost the whole time, unlike the scattered members of a lion pride, and they are totally dependent on being together in order to catch their prey. For this reason, I suggest that a dominant female wild dog is in a better position than a lioness both to assert her dominance over her subordinates and to inhibit their reproduction.

A wild dog may have up to sixteen puppies in a litter (Schaller 1972). Thus, it may be that a single female can produce as many young as the pack is likely to be able to rear; if so, a pack would do little better if more than one female reproduced. Subordinate males, if they were to leave a pack, would probably be unlikely to find mating opportunities elsewhere. Thus, they do better, in terms of their genes in subsequent generations, to remain and help rear their dominant brother's offspring, each of which contains one quarter of their own genes. By staying, too, there is the possibility that the dominant male might die and leave one of them to reproduce in his stead; Frame and Frame (1976a) documented an instance of this. Subordinate females are just as closely related to the dominant female and to her offspring, which they, too, help to rear. As we have seen, if she leaves the pack, a female's chances of successful reproduction may be higher than a male's, which could account for the unusual transference of females rather than males in this species (Frame and Frame 1976a). But if she remains, she, too, like subordinate males, contributes to the propagation of a proportion of her genes.

All adult wild dogs feed the young by regurgitating meat for them, and a case has been reported of a five-member pack—all males—managing to rear young puppies whose mother had died (Estes and Goddard 1967). They feed other adults in the same way, which makes possible a system of division of labor, with some adults remaining on guard with the puppies while the rest hunt, and receiving a food supply nonetheless when the hunters return.

Conclusion

I have reviewed briefly the large predator species of the Serengeti. It is clear that they have a relatively small effect on the populations of their prey species, for three main reasons: because, on the whole, they take either different species or different sections of the population of the same species; because many of their victims are, in a sense, expendable, being surplus or doomed anyway; and because a high proportion of the prey are migrating in response to changing feeding conditions, and thus are geographically unavailable to most of the less mobile predators. Nonetheless, predation exerts selective pressures on the prey species, forcing them to continue to be alert, swift, and responsive to a range of different kinds of predator attack.

The five predator species exhibit a wide array of different hunting methods and social systems. Three of the five—lions, hyenas, and wild dogs—live in social groups; whereas the other two species, leopards and cheetahs, are solitary.

The same species tend to hunt in groups. As Kruuk (1975) pointed out, the group-hunting predators are able to kill prey larger relative to their own size than the solitary predators. Under certain circumstances, the advantages to be gained from group hunting are likely to be the most important factor contributing to a social way of living.

It is apparent that sociality in these predators is a most complex phenomenon. It does not depend on phylogeny, nor on size, nor on any one particular method of hunting. Nor is it correlated with abundance, either in the Serengeti or elsewhere.

Sociality takes completely different forms in the three social species. Examination of the dynamics of the social systems shows that the differences can be related to a number of interdependent factors. One is the phylogenetic origin of the social group. A second is the way in which the individuals obtain their food, for this may influence the interactions between the sexes and either favor or constrain sexual dimorphism. A third is the transference of individuals from one group to another, influenced indirectly by the probable breeding opportunities in other groups. A fourth is the presence or absence of a dominance system, influencing competition for food and particularly for reproductive opportunities. And a fifth is the degree of relatedness among group members; this is clearly dependent on the third and fourth factors above, but, through the operation of kin selection, it probably also influences them.

Kin selection is implicated in several aspects of the social organization and behavior of the social predators. Guarding, mutual regurgitation, reproductive toleration, and communal suckling and rearing are striking examples of cooperative acts. Such cooperation is most marked in wild dogs and least marked in hyenas, and the average degrees of relatedness show a similar decline from dogs to lions to hyenas. A variety of other important selective pressures are obviously implicated too.

The Serengeti is almost unequaled in having healthy populations of all five of these magnificent predators. It is unique in having long-term data available on both individuals and populations, and it is to be hoped that such long-term records will continue.

References

Bertram, B. C. R. 1973. Lion population regulation. *E. Afr. Wildl. J.* 11:215–25.

Bertram, B. C. R. 1974. Radio-tracking leopards in the Serengeti. African Wildlife Leadership Foundation *Newsletter* 9(2):7–10.

Bertram, B. C. R. 1975a. Social factors influencing reproduction in wild lions. *J. Zool.* (Lond.) 177:463–82.

————. 1975b. Weights and measures of lions. *E. Afr. Wildl. J.* 13: 141–43.

————. 1976. Kin selection in lions and in evolution. In *Growing points in ethology,* ed. P. P. G. Bateson and R. A. Hinde, pp. 281–301, Cambridge: Cambridge Univ. Press.

————. 1978. Living in groups: predators and prey. In *Behavioural Ecology: an evolutionary approach,* ed. J. R. Krebs and N. B. Davies. Oxford: Blackwell.

Bischof, N. 1975. Comparative ethology of incest avoidance. In *Biosocial anthropology,* ed. R. Fox. pp. 37–67. New York: Malaby Press.

Eaton, R. L. 1970. Group interactions, spacing and territoriality in cheetah. *Z. Tierpsychol.* 27:481–91.

Eisenberg, J. F., and Lockhart, M. 1972. An ecological reconnaissance of Wilpattu National Park, Ceylon. *Smithsonian Contrib. Zool.* 101: 1–112.

Elliott, J. P.; Cowan, I. McT.; and Holling, C. S. 1977. Prey capture by the African lion. *Can. J. Zool.* 55:1811–28.

Estes, R. D., and Goddard, J. 1967. Prey selection and hunting behavior of the African wild dog. *J. Wildl. Manage.* 31:52–70.

Frame, G. W., and Frame, L. H. 1977a. Serengeti cheetah. African Wildlife Leadership Foundation *Newsletter* (Nairobi) 12(3):2–6.

———. 1977b. Census of predators and other animals on the Serengeti Plains, May 1977. Serengeti Research Institute Rep. no. 52 (typed). Arusha: Tanzania National Parks.

Frame, L. H., and Frame, G. W. 1976a. Female African wild dogs emigrate. *Nature* (Lond.) 263:227–29.

———. 1976b. Wild dogs of the Serengeti. African Wildlife Leadership Foundation *Newsletter* (Nairobi) 11(3):1–6.

Greenwood, P. J.; Harvey, P. H.; and Perrins, C. M. 1978: Inbreeding and dispersal in the great tit. *Nature* (Lond.) 271:52–54.

Hamilton, W. D. 1964. The genetical evolution of social behaviour, parts 1 and 2. *J. Theoret. Biol.* 7:1–16, 17–52.

Hamilton, P. H. 1976. The movements of leopards in Tsavo National Park, Kenya, as determined by radio-tracking. Master's thesis, University of Nairobi.

Herdman, R. 1973. Cheetah breeding program. In *The world's cats,* ed. R. L. Eaton, 1:255–62. Winston, Ore.: World Wildlife Safari.

Kleiman, D. G., and Eisenberg, J. F. 1973. Comparisons of canid and jelid social systems from an evolutionary perspective. *Anim. Behav.* 21: 637–59.

Kruuk, H. 1972. *The spotted hyena.* Chicago: Univ. of Chicago Press.

———. 1975. Functional aspects of social hunting by carnivores. In *Function and evolution in behaviour,* ed. G. Baerends, C. Beer, and A. Manning, pp. 119–41. Oxford: Oxford University Press.

Makacha, S., and Schaller, G. B. 1969. Observations on lions in the Lake Manyara National Park, Tanzania. *E. Afr. Wild. J.* 7:99–103.

Malcolm, J. R., and van Lawick, H. 1975. Notes on wild dogs (*Lycaon pictus*) hunting zebras. *Mammalia* 39:231–40.

McLaughlin, R. T. 1970. Aspects of the biology of cheetahs, *Acinonyx jubatus* (Schreber), in Nairobi National Park. Master's thesis, Nairobi University.

Muckenhirn, N. A., and Eisenberg, J. F. 1973. Home ranges and predation of the Ceylon leopard (*Panthera perdus fusca*). In *The world's cats,* ed. R. L. Eaton, 1:142–75. Winston, Ore.: World Wildlife Safari.

Rudnai, J. 1974. The pattern of lion predation in Nairobi Park. *E. Afr. Wildl. J.* 12:213–25.

Sadleir, R. M. F. S. 1966. Investigations into the reproduction of larger Felidae in captivity. *J. Reprod. Fert.* 12:411–12.

Schaller, G. B. 1968. Hunting behaviour of the cheetah in the Serengeti National Park, Tanzania. *E. Afr. Wildl. J.* 6:95–100.

———. 1972. *The Serengeti lion.* Chicago: Univ. of Chicago Press.

Sinclair, A. R. E. 1977. *The African buffalo.* Chicago: Univ. of Chicago Press.

Taylor, C. R., and Lyman, C. P. 1972. Heat storage in running antelopes: independence of brain and body temperatures. *Amer. J. Physiol.* 222: 114–17.

Taylor, C. R., and Rowntree, V. J. 1973. Temperature regulation and heat balance in running cheetahs: a strategy for sprinters. *Amer. J. Physiol.* 224:848–51.

Taylor, C. R.; Schmidt-Nielsen, K.; Dmi'el, R.; and Fedak, M. 1971. Effect of hyperthermia on heat balance during running in the African hunting dog. *Amer. J. Phsyiol.* 220:823–27.

Wilson, E. O. 1975. *Sociobiology: the new synthesis.* Cambridge, Mass.: Belknap Press.

Zimen, E. 1976. On the regulation of pack size in wolves. *Z. Tierpsychol.* 40:300–341.

J. P. Hanby
J. D. Bygott

Ten   Population Changes in Lions
and Other Predators

Populations of some large ungulate species have shown a substantial increase during the last two decades (see chap. 4), and one might therefore expect that the numbers of large carnivores have also increased. This study was designed to ascertain the current population trend of lions, for although Schaller (1972) concluded from his three-year study that the Serengeti lion population was stable, subsequent rumours suggested that it might be decreasing.

In order to discover what changes, if any, had occurred between Schaller's study period (1966–69) and our own (1974–77), we reexamined his main study area, about 2500 km² of open grasslands and woodlands. We found that there were more lions resident in the area and that they were now found in areas that were not fully occupied before. These changes resulted from at least two factors acting in combination: an increase in rainfall in the dry season (June through October), and an increase in the numbers of resident prey species.

In this chapter, we shall present data in support of these conclusions, and we shall then briefly consider how the same ecological factors may have affected the other carnivores. It is unfortunate that the main Serengeti predator—the spotted hyena—has not been monitored since the termination of Kruuk's study (Kruuk 1972), but a census of predators on the plains conducted in May 1977 has provided some information on the current number of hyenas. Cheetahs and wild dogs, the rarest species, are currently being studied by George Frame and Lory Herbison-Frame, respectively, who have kindly allowed us to quote their preliminary figures.

The Serengeti Lion Population

The social structure of the Serengeti lion population has been described in detail by Schaller (1972) and Bertram (1975; chap. 9 above). The two basic components of lion society are residents and nomads. The social units of the resident population are prides of two to eleven adult females (presumably related) with their cubs (pl. 30), and groups of one to seven adult males. Males usually attach themselves to a pride and occupy the same range, but large groups (three to seven animals) may maintain a range large enough to span several prides. Resident prides occupy home ranges of 20–300 km² in habitats where prey, water, and cover are available throughout the year. Females defend their ranges from female intruders, males from males.

The nomad population is much smaller (perhaps about 20 percent of the total) and consists of individuals (pl. 28) who have emigrated from resident prides. It is normal for subadults, particularly males, to leave their parent pride at two to four years of age and become nomadic; however, some young females are recruited into the parent pride. Nomads are characterized by having no stable range (or an extremely large range), and they wander through the areas occupied by resident lions, whom they avoid as best they can, converging in places where prey is temporarily abundant. Nomads may become resident if conditions permit; thus, females may settle in any suitable area which is not already occupied by a pride, and males may join a group of females if there are no males already present or if they can displace the resident males.

Schaller showed that the main source of recruitment for both the nomadic and resident lions was through cubs born to resident prides. He concluded that the resident prides produced a surplus of 5.5 percent of the population per year and that most of these lions became nomadic and emigrated from the area or died; thus, the lion population remained stable. We conjectured that if conditions improved, much of this "surplus" could survive and a new population level would be reached.

In order to assess trends in the Sergenti lion population, we worked for three years (September 1974 to September 1977) in Schaller's main study area. This area, illustrated in figure 10.1, included the Serengeti plain and a large strip of woodland at the northern and western borders of the plain. Within this area, we attempted to find and identify individually all lions, and to determine whether they were resident or nomadic.

Each lion over one year old was photographed and sketched, so that all lions found could be positively reidentified by the combination of ear notches, vibrissae spot patterns (Pennycuick and Rudnai 1970), and other natural features. This system of individual recognition was also used by Bertram (1975), and his identity cards for three prides in the study area (the Seronera, Masai, and Loliondo prides) enabled us to

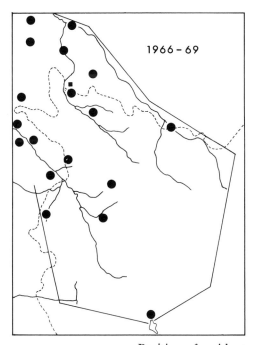

1966 - 69

Figure 10.1   Position of resident prides (*solid circles*) on the plains and surrounding woodlands 1966–69 (after Schaller 1972). *Solid line* indicates study area boundary; *broken line,* edge of plains. Seronera is shown by the square.

recognize members of those prides without difficulty. Schaller had marked all the other prides in the study area by ear-tagging an adult female in each pride, but, unfortunately, all the marked lions had lost their ear-tags or died by the time our study began. We could not, therefore, be certain which of the prides we found corresponded to those which Schaller had named, although, in some cases, reasonable guesses could be made on the basis of geographical location. However, the overall numbers and distribution of the resident prides in the study area could still be compared with Schaller's data for the same area, ten years earlier.

*Changes in the Resident
Population*

Prides. The distribution of resident prides in Schaller's study period and our own are shown in figures 10.1 and 10.2, respectively. There are now more resident prides; we found a minimum of twenty-four distinct prides, where Schaller found eighteen (an increase of 33 percent). We actually found twenty-eight different groups, but we did not see some of these often enough to be able to state definitely whether or not they were separate prides or just subgroups. Note that most of the "new" prides were found on the plains, an area in which Schaller found very few prides living throughout the year.

While the absolute number of prides has increased over the ten-year period, mean pride size has also increased from fifteen to nineteen. Figure

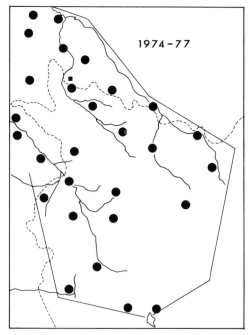

Figure 10.2   Resident prides, 1974–77, in the
                                   same area as that shown in
                                   fig. 10.1.

10.3 shows that the mean number of adults and cubs per pride has not changed significantly, the increase being mainly due to subadults.

Cub survival. In 1966–68, Schaller recorded only ten subadults in fourteen prides, while in 1975–77, we found seventy-one subadults in fifteen prides in the same area (the new plains prides were excluded from this analysis). This difference (statistically significant by sign test, or *"t"* test, at $P < 0.01$) implies either that cub survival has increased significantly during the past few years, or that more cubs are allowed to remain with the pride into subadulthood. Both factors are probably operating.

Increased cub survival is clearly shown by continuous data on the reproductive history of the Seronera and Masai prides, collected by Schaller, Bertram, L. H. Frame (pers. comm.), and ourselves. The sur-

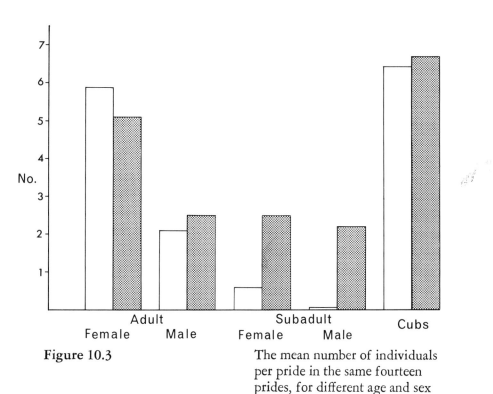

Figure 10.3   The mean number of individuals per pride in the same fourteen prides, for different age and sex classes. *Open column, 1966–68; stippled column, 1975–77.*

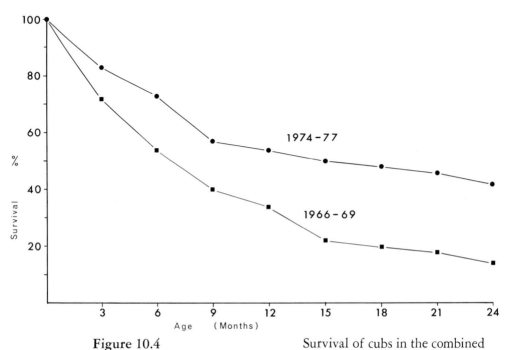

Figure 10.4   Survival of cubs in the combined
Seronera and Masai prides. One
hundred twenty-seven cubs born in
1971–75 (*circles*) have a 44%
chance of surviving to two years of
age. Eighty-seven cubs born in
1966–70 (*squares*) have a 14%
chance of surviving to two years of
age.

vivorship curves in figure 10.4 show a threefold increase in the proportion
of cubs surviving to subadulthood (that is, to two years of age). Once a
cub reaches two years of age, the probability of its death decreases considerably. Schaller calculated an annual mortality of 5.5 percent for all
lions over two years old. We have no direct evidence that the mortality of
subadults is any lower now than it was in the past, but we agree with
Schaller's conclusion that mortality must be higher for nomads than for
residents, and if subadults can remain for longer with their prides, their
chances of survival are presumably improved.

Formation of new prides. When subadults leave their parent pride and become nomadic, littermates of the same sex tend to stay together as a stable group. Male groups may eventually succeed in attaching themselves to established prides and becoming resident breeding males, and female groups, under the right conditions, are capable of establishing new prides.

We have found that new prides may be formed in two ways: (1) a group of subadult females may leave the parent pride and settle in an adjacent area, or (2) one or a few nomadic females, of any age, may settle in a suitable area and build up a pride by recruitment of their offspring.

Five of the eight new prides we have found in the study area appear to have been formed in the first way, and in three of these cases, we have been able to trace the parent pride through photographic records. In all three instances, the females left their prides as subadults and took up residence at the periphery of their mothers' range, attracted new males, and began to breed. The age structure of two other prides implies the same process of formation.

Three other new prides were probably formed in the same way, by a "founder female" who recruited her daughters. These prides, though small, contained females of several different ages, and, in each case, the oldest female had obviously been ear-notched or tagged by Schaller. Only one of these presumed founder females could be positively identified by Schaller (pers. comm.), who gave us a list of his sightings of her. These data showed how she initially wandered as a nomad all over the plains, with a variety of companions, then in 1968 began rearing cubs in the northeastern plains. Seven years later we found her still in that area, with two middle-aged female companions and three cubs.

The formation of both these types of pride would seem to be a response to improved environmental conditions. In the first case, improved cub survival in a resident pride results in a large group of subadults; in the second case, a solitary female nomad at last succeeds in rearing some daughters to maturity.

Changes in Nomad Numbers

Schaller (1972) estimated the numbers of nomadic lions on the plains during each wet season; his estimates varied from 267 in 1967 to 123 in 1969 (mean = 185). We identified on average only sixty-eight nomads per wet season during 1974–77, which appears to be a substantial decrease. The sex ratio also seems to have changed; Schaller recorded an average ratio of eighty-eight males to one hundred females among plains nomads, that is, 53 percent of adults and subadults were females. We found that only 17 percent of nomads were females.

There are several possible reasons for a decrease in nomads on the plains: (1) nomad mortality may have increased (for example, from poaching or starvation); (2) more nomads may now remain in the woodlands during wet seasons; (3) fewer subadults may leave their prides; (4) more subadult females may be forming new prides. An increase in mortality so heavily biased against females seems unlikely. One would expect females to be better able to support themselves than males (see Schaller 1972), while snaring and hunting might be biased against males, since they are more mobile than females. It is conceivable that more nomads are staying in the woodlands, particularly since several large subadult groups (whose members we knew individually) left their prides in the woodlands but were never seen on the plains. But alternatives 3 and 4 could explain the proportional reduction of female nomads, and fit well with the observed data.

During the wet season of 1976–77, our sightings of nomads, plus the new plains prides, totaled 141 individuals; this is a minimum figure but well within the confidence limits of Schaller's estimates of plains nomads in 1968 and 1969. Of the adults and subadults within this sample, 49 percent were females, which more closely approaches Schaller's nomad sex ratio.

Thus, the number of nomads using the plains during the wet season may have decreased because many females and some males, who would formerly have become nomads, now stay on the plains throughout the year in resident prides. These prides live in areas which used to attract nomads during the wet season during Schaller's study. These new resident lions are territorial and tend to keep nomads out of their ranges; thus, to the casual observer, there may appear to be fewer lions in the area.

Climatic Changes and Prey
Availability

The formation of new prides and the improvement in cub survival in established prides are presumably due to an increase in the availability of food. It would be naive to assume that the big increase in the population of wildebeest (the species most frequently eaten by lions) was responsible for the increase in lions, because most of the Serengeti wildebeest are migratory and most of the lions are resident. Any resident pride will have access to migratory prey for only a few months in every year, and must subsist at other times on resident herbivore species. Schaller considered that the major factor limiting the size of the resident lion population was the availability of prey during the leanest time of year; this is the wet season in the northern and western woodlands, when the migratory species (wildebeest, zebra, Thomson's gazelle, and eland) have moved out to the plains, but in our study area on and around the plains, there is least prey during the dry season. If more lions can now survive in this area than formerly, we might expect resident prey to have increased since Schaller's study.

Unfortunately, there are few data on the resident ungulate populations that span the past ten years and include the numbers on the plains. Buffalo have shown a continual increase (see chap. 4) but are only available to lions living in the woodlands or at their edge. Their increase may well have benefited lions in these areas; Schaller found that only 2.4 percent of 545 ungulates eaten by the Seronera and Masai prides during 1966–69 were buffalo, whereas during our study, 20 percent of 101 ungulates eaten by the same prides were buffalo (pl. 30). Our sample is smaller and less systematic, but supports the hypothesis that buffalo now form a greater part of the lions' diet.

Schaller (1972) presents data on the total numbers of resident prey in the Masai pride's home range. Utilizing his data, we found that in 1966–67, the mean biomass of resident prey in that area was about 300 kg/km^2. Our own systematic monthly sample counts from specified points within the same area in 1976–77 gave a figure of 600 kg/km^2. A comparison of the two methods showed that the sample counts underestimated the actual biomass, so resident prey had at least doubled. The increase is mostly due to buffalo, giraffe, topi, and warthog. It is noteworthy that during this ten-year period, the number of adult females in the Masai pride increased from six to sixteen.

Reliable quantitative data on the numbers of ungulates on the plains during the dry seasons have been difficult to obtain. Norton-Griffiths (pers. comm.) calculated a mean biomass minimum of 99 kg/km^2 for resident ungulate species on the plains during the dry season of 1972. Most observers (for example, Schaller 1972; Kruuk 1972; R. M. Bradley and A. R. E. Sinclair, pers. comm.) agree that toward the end of the 1960s there were very few ungulates or large carnivores on the short-grass plains during the dry seasons. Schaller conducted one prey count approximately in the center of the plains in October 1966 and found a prey biomass of 131 kg/km^2. Our sample counts of prey at the Gol Kopjes, where a resident lion pride now lives, showed a mean biomass of 950 kg/km^2 in the dry season of 1975, and 850 kg/km² in 1976. However, in July of both years, prey biomass in this area fell as low as 25 kg/km². Although the overall availability of prey may have increased enough to support a resident pride, this area is still a marginal habitat subject to large fluctuations in prey availability.

The prey species now found on the plains during the dry season include Thomson's and Grant's gazelles, topi, kongoni, and warthog. Warthogs, in particular, seem to be important prey for several plains prides; the lions can stalk them in the daytime, or dig them out of their holes at night. There are no data on the numbers of warthogs on the plains over the past ten years, but G. Frame has notes from 1965 onward that indicate an eastward expansion of warthogs around Olduvai Gorge during the past five years, and, thus, presumably on the short-grass plains as well. R. M. Bradley (pers. comm.), who studied gazelles on the plains during 1970–72, considered that there were virtually no warthogs on the plains at that time. A ground census of the whole plains by transects during May 1977 gave a mean density of 0.76 warthogs/km² (representing about 2000 in our study area), which would suggest an increase.

The greater abundance and wider distribution of prey species suggested a change in the rainfall pattern—in particular, an increase in dry-season rainfall on the plains. Fortunately, this hypothesis could be checked against the Serengeti Ecological Monitoring Programme's rain-gauge data, which covers the entire area and time period under consideration. The mean monthly rainfall for twenty-one rain-gauge stations on the short- and intermediate-grass plains over the last thirteen years is shown in figure 10.5. Wet-season rainfall has been variable with no particular trend, but for the last five years, dry-season rainfall has been higher than for the previous five years. This small increment is sufficient to promote grass growth and, thus, to attract and sustain a greater number of herbivores.

Corresponding increases in dry-season rainfall throughout the park (chap. 4) may account in part for increases in resident herbivores in the woodland areas, and the resulting improvement in cub survival in those areas.

Changes in Other Carnivore Populations

Hyenas

The most important large predator in Serengeti is undoubtedly the spotted hyena (pls. 31, 32), which Kruuk (1972) studied from 1964 to 1968.

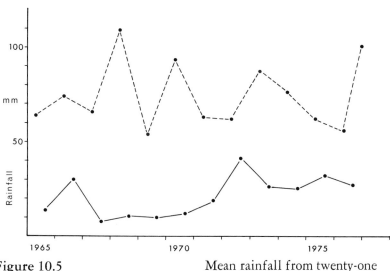

Figure 10.5   Mean rainfall from twenty-one
stations on the plains. Dry-season
rainfall (*solid line*) was highest in
the seventies. Wet season rainfall
(*broken line*) showed no trend.

He found that Serengeti hyena numbers were much lower than would
be expected from the size of the prey populations, and suggested that the
limiting factor for the hyena population was the distance that females
had to travel from their dens to the concentrations of wildebeest, then
main prey. The increase in both the number of wildebeest and the dry-
season rainfall (and, hence, resident prey) may now allow hyenas access
to a greater food supply than before. From this, one would predict an
increase in hyena numbers comparable to the lion increase discussed in
the preceding sections.

   In order to assess the number of hyenas, the Serengeti Research Institute conducted a census of the plains in May 1977. At this time (the end
of the wet season), the wildebeest herds were massed on the plains before
migrating into the woodlands, and the density of hyenas was presumably
maximal. An area of about 3000 km² (bounded by the plains-woodland
border, Olduvai Gorge, and the Gol Mountains) was sampled by ground
transects (Anon 1977). The resulting estimate for hyenas was 3391 (with
95 percent confidence range of 2560–4122). The lower limit still exceeds

Kruuk's estimate of the wet-season hyena population on the plains (2117) by 21 percent, so we conclude that there are now considerably more hyenas on the plains during the wet season. Kruuk found that the density of hyenas in the woodlands was very low, and we have no evidence that this has changed.

The factors which have caused lions and hyenas to increase in numbers or distribution may also have affected other predators, but apart from cheetah and wild dog, there are no long-term records for other species. Records of the resident leopard population (pl. 34) around Seronera over the past eight years, kept by Bertram and ourselves (unpub. data), show no indication of a change in numbers, but the area concerned is small, and the status of leopards elsewhere in the park is unknown.

Neither Schaller nor Kruuk considered lions or hyenas to have a limiting effect on their prey in Serengeti, and we have no data that conflict with this view. However, they may well affect other predators. Hyenas and lions are themselves major competitors, in that both are large, common, hunt similar prey, and scavenge from each others' kills, though in the latter situation, lions are usually dominant. In the past, during the dry season on and around the plains, this competition may have been reduced, since lions were then dependent on resident prey within their pride ranges, while the majority of hyenas commuted to areas where migratory prey species were abundant. However, it is possible that some hyenas now take advantage of the increased resident prey populations, a possibility that deserves further study.

Cheetahs

Potentially, hyenas could be serious competitors of cheetahs (pl. 33), since they can easily steal kills from cheetahs. However, cheetahs are much more diurnal than hyenas, and G. Frame (pers. comm.) has found that most cheetah hunts occur at times when hyenas are seldom active. The hunting technique of cheetahs is also fairly unobtrusive; a long stalk in the cover of grass or herbs is followed by a quick chase and concealment of the prey after capture.

The frequency of encounters between the two species could be further reduced because hyenas tend to stay with the wildebeest concentrations, while cheetahs follow the gazelle herds, and these two ungulate species are usually concentrated in different areas at any one time. The reduction in dry-season grass fires over the past ten years (chap. 13) may have benefited cheetahs by providing more cover for stalking and also for safe concealment of cubs. It is perhaps for these several reasons that the Serengeti cheetah population is flourishing. G. Frame estimates that there are at least five hundred in the ecosystem, a large proportion of which are young.

Wild Dogs

The only large predator which is known to have suffered a decrease over the past ten years is the wild dog (pl. 35) living on the plains. (There are no reliable figures for dogs living in the woodlands.) The number of adults living on the plains year-round has declined from around 110 in 1970 (J. R. Malcolm, pers. comm.; Malcolm and van Lawick 1975) to only thirty at present (L. Herbison-Frame, pers. comm). Packs are fewer and smaller, and pup survival is very low.

It is on the plains that wild dogs must compete most directly with other predators, especially hyenas. While single hyenas are effectively driven off by a dog pack, a group of hyenas can easily steal the dogs' kill (pls. 37, 38), and may also kill puppies that are left unattended. One or more adults must stay at a den to guard the young when hyenas are around, thus diminishing the hunting pack and increasing the number of mouths to be fed. Lions will also steal kills from dogs whenever possible, but they are neither as widespread nor as mobile as hyenas. Competition with an increased hyena population on the plains may well have contributed to the high mortality of wild dog pups during the past three years, but shooting and diseases such as distemper have also taken a toll in the past.

Conclusion

Lions and hyenas have increased in numbers on the Serengeti plains and adjacent woodlands, but the main causal factors seem to be different for the two species. The main factor accounting for the improved cub survival and wider distribution of resident lions is an increase in resident prey. The main factor accounting for the increase in hyenas is probably the increase in migratory prey, though they may also have benefited from the increase in resident prey. The cheetah population may have increased for the same reasons that the lion population has. Wild dogs living on the plains have decreased, probably because of disease combined with intense competition with the enlarged hyena population.

A relatively small change in dry-season rainfall has made a large area of Serengeti available to resident herbivores and carnivores during the years 1972–77. Schaller's and Kruuk's studies (of lions and hyenas, respectively) happened to coincide with a run of dry years, and ours with a run of wet years. Fortunately, the existing long-term records of predator populations and rainfall have enabled us to detect, and suggest explanations for, the changes in predator populations. The importance of long-term studies that monitor populations and climate is well illustrated here, and it is easy to see that a reversion to more severe dry seasons might eliminate the resident lion prides on the plains and reduce the reproductive success of other resident prides.

This study would not have been possible without the financial support of the Science Research Council, England, and the New York Zoological Society. Tony Sinclair and the African Wildlife Leadership Foundation have also contributed substantially to the predator counts. We acknowledge our indebtedness to all the scientists and S.R.I. for their generous help in counting predators and their help in other ways too. We particularly want to thank G. Frame, L. Frame, J. Malcolm, A. Sinclair, and M. Norton-Griffiths for contributing data and comments on the manuscript, as well as for support and inspiration.

References

Anon, 1977. Census of predators and other animals on the Serengeti plains, May 1977, rep. no. 52. Arusha: Tanzania National Parks.

Bertram, B. C. R. 1975. Social factors influencing reproduction in wild lions. *J. Zool.* (Lond.) 177:463–82.

Kruuk, H. 1972. *The spotted hyena.* Chicago: Univ. of Chicago Press.

Malcolm, J. R., and van Lawick, H. 1975. Notes on wild dogs (*Lycaon pictus*) hunting zebras. *Mammalia* 39:231–40.

Pennycuick, C. J., and Rudnai, J. 1970. A method of identifying individual lions (*Panthera leo*) with an analysis of the reliability of identification. *J. Zool.* (Lond.) 160:497–508.

Schaller, G. B. 1972. *The Sergenti lion.* Chicago: Univ. of Chicago Press.

D. C. Houston

Eleven   The Adaptations of
         Scavengers

A few years ago, it was commonly believed that mammalian carnivores could be divided into predators which killed their own prey, such as lions, leopards, and cheetahs; and scavengers, such as hyenas and jackals, which relied on finding dead animals. We now know that this distinction is completely unjustified (Schaller 1972; Kruuk 1972). All mammalian carnivores are primarily predators, and most of them will scavenge food whenever they have the opportunity. The distinction between predatory and scavenging animals is, therefore, not a very useful one, and none of the mammalian carnivores feeds entirely, or even substantially, by scavenging. There are, however, many species of carnivorous birds which are exclusive scavengers, that is, they never kill, but rely entirely on finding dead animals.

The proportion of food that a carnivore obtains by predation or scavenging varies with many factors, but there is no advantage in killing prey if the animal could easily obtain good-quality meat by scavenging. In this chapter, I want to look at why some carnivorous animals are able to feed exclusively by scavenging, while others need to kill. In order to do this, I have examined the nature of the food supply available to carnivorous animals, the way it is apportioned between species, and the adaptations necessary to be an efficient scavenger.

The Food Supply for Carnivorous Animals

Whatever the cause of death, almost all the food for the large carnivorous animals comes from ungulates. Table 11.1 shows an estimate of the food available to carnivorous animals each year from the Serengeti ecosystem, based on population sizes for the main ungulate species, estimates of their

Table 11.1

Estimate of the number of ungulates that die in the Serengeti region each year, and their total weight (kg).

Species	Population size[a]	Body weight[b]	%-age adult mortality	Deaths	Combined weight of dead adults
Wildebeest	720,000	123	12[d]	86,400	10,627,000
Zebra	240,000	200	10	24,000	4,800,000
Buffalo	108,000	450	8[e]	8,640	3,888,000
Thomson's gazelle	981,000	15	20	196,200	2,943,000
Giraffe	17,400	750	7	1,218	914,000
Eland	24,000	340	8	1,920	653,000
Topi	55,500	100	10	5,550	555,000
Impala	119,100	40	10[f]	11,910	476,000
Elephant	4,500	1,725	4	180	311,000
Kongoni	20,700	125	10	2,070	259,000
Warthog	34,200	45	15	5,130	231,000
Hippo	2,400	1,000	6	144	144,000
Waterbuck	3,000	160	12	360	58,000
Rhino	900	816	7	63	51,000
Grant's gazelle	6,000	40	15	900	36,000

Species	Births	1st-year deaths	Weight at death[c]	Combined weight of 1st-year deaths	Total weight of dead
Wildebeest	250,000	150,000	30.75	4,612,000	15,239,000
Zebra	76,000	52,000	50	2,600,000	7,400,000
Buffalo	34,560	26,000	112.5	2,925,000	6,813,000
Thomson's gazelle	500,000	300,000	3.75	1,125,000	4,068,000
Giraffe	5,568	4,350	187.5	816,000	1,730,000
Eland	8,000	6,000	85	510,000	1,163,000
Topi	17,760	12,210	25	300,000	855,000
Impala	48,000	36,090	10	360,000	836,000
Elephant	400	220	431.25	95,000	406,000
Kongoni	6,624	4,600	31.25	144,000	403,000
Warthog	34,200	29,070	11.25	326,000	557,000
Hippo	500	350	250	88,000	232,000
Waterbuck	960	600	40	24,000	82,000
Rhino	225	162	204	33,000	84,000
Grant's gazelle	2,400	1,500	10	15,000	51,000
					40,094,000

(Not including oryx, roan, bushbuck, reedbuck, oribi, duiker, dik-dik, steinbuck, and klipspringer, which collectively contribute less than 0.5% of total food for carnivores.)

a   Norton-Griffiths, pers. comm.
b   Coe, Cumming, and Phillipson 1976
c   Assumed at ¼ adult body weight
d, e   Sinclair, pers. comm.
f   Jarman, pers. comm.

annual mortality rates, and estimates of the numbers and weights of young animals which die. Oryx, roan, bushbuck, reedbuck, oribi, dik-dik, klipspringer, steinbuck, and duiker, which collectively contribute less than 0.5 percent of the total weight of carcasses each year, are not included in these figures. Information on mortality rates is only available for a few species, so there has been some extrapolation and inspired guesswork; this will introduce some errors in the details, but it is only the overall orders of magnitude of the food supply that I am concerned with here. Table 11.1 shows that the Serengeti system is, from a carnivore's point of view, dominated by the migratory ungulates. About 67 percent of the food available to carnivorous animals comes from migratory species, and a single species, the wildebeest, provides about 38 percent of the total.

The Utilization of the Food Supply

There are basically three ways in which a carnivore feeding on large ungulates can obtain food, and we should briefly review the relative importance and features of these methods in the Serengeti. An animal can kill its own prey. This is dangerous, in that there is a risk of injury involved in hunting, attacking, and killing. It also requires considerable time and energy, and, in addition, most hunts are unsuccessful. Predation is also not the major mortality factor among ungulates. Table 11.2 shows estimates for total weight of ungulates killed by large predators. Mammalian carnivores account for only about 36 percent of the total weight of carcasses which arise from annual ungulate mortality, as given in table 11.1. These large predators kill to obtain most of their food, but a proportion of their diet comes from scavenging the carcasses of animals which have not been killed by some other predator. Bertram (table 9.1) shows that this accounts for 33 percent of food intake for hyenas and 15 percent for lions. Predation, therefore, accounts for about 30 percent of the total ungulate mortality each year.

The second method of feeding is to locate a predator kill and feed on any remains after the predator has finished feeding. This is commonly believed to be the food supply for "scavengers," although it is actually of minor importance: firstly, as already mentioned, predation is responsible for a comparatively small proportion of total ungulate mortality; and secondly, predators are efficient feeders, leaving little meat behind for others to scavenge. The feeding behavior of all large predators is adapted to prevent substantial amounts of meat being lost. Leopards carry large kills high up into trees to take them out of the reach of hyenas and to hide

Table 11.2

Estimated weight of ungulates which are taken each year by the large predator population in the Serengeti. Figures include both animals killed and those scavenged, and include parts of carcasses not eaten (usually bones, skin, and contents of digestive tract).

Predator	Estimated number in Serengeti ecosystem	Weight taken per day per animal (kg)	Weight taken per year per species (kg)
Hyena	3,000[a]	3[a]	3,285,000
Lion	2,400[b]	9[b]	7,884,000
Cheetah	250[b]	10[b]	912,500
Wild Dog	300[b]	7.5[b]	821,250
Leopard	1,000[b]	4[b]	1,460,000
		Total	14,362,750 (including wastage)

Figures derived from a Kruuk 1972
b Schaller 1972

them from vultures flying overhead, and a lion pride will guard a large kill for several days until the animals have finished feeding. Packs of hyenas and wild dogs usually consume a whole carcass within a few hours. When these predators abandon a carcass, only small amounts of food remain (Houston 1974a). The cheetah is the only predator which regularly leaves a large proportion of the carcass, but cheetah are so uncommon as to provide an insignificant food supply for scavengers. The only way that a scavenger can obtain substantial amounts of meat from a predator kill is if it can dominate the predator species, drive it away, and take over the carcass.

The third method of feeding is to locate and scavenge from those animals which have died from some cause unconnected with predation. Two-thirds of all ungulate deaths come into this category, making it the major food supply for carnivorous animals. Unfortunately, we have little information on the relative importance and action of these other mortality agents. However, it is clear that most such mortality factors are greatly influenced by malnutrition. During the dry season, the protein content of many grass species declines sharply and may fall below the required level to maintain body condition of herbivores, which then have to rely on stored body reserves. Their condition therefore declines (Jarman, pers. comm.; Sinclair 1974b). Ultimately, some ungulates exhaust their fat reserves (as judged by their bone-marrow condition; Sinclair and Duncan 1972), and die from starvation: this has been recorded during the dry season in wildebeest, zebra, kongoni, and buffalo (Sinclair 1974a, 1977). Apart from direct starvation, animals that are in a weak condition are more susceptible to disease, parasite burdens, and accidents. Diseases, such as sarcoptic mange in gazelle and other animals, are only seen in dry-season conditions, and some ubiquitous diseases, such as *Corybacterium pyogenes* infection in buffalo calves, only cause pathological symptoms in animals that are in poor body condition. Similarly, many internal parasites only reach heavy infection in weak animals (Sinclair 1974a, 1977).

Variations in the Food Supply

Whatever the ultimate cause of death, therefore, it is clear that malnutrition is a major contributing factor—as was suggested by Hirst (1969) in the Transvaal. It follows that mortality is not uniform through the year; during periods of good-quality grazing, there is probably very little mortality. Most mortality of adults occurs at the end of the dry season, and it is probable that over half of all ungulate deaths occur during the period from August to October. There is, therefore, a considerable fluctuation in the quantity of food available for scavengers during the year.

Mortality is also unpredictable, especially when there are not heavy losses from factors associated with malnutrition. Animals, then, die chiefly from unexpected causes. Wildebeest are particularly accident-prone. They stampede at night, falling into gullies or drowning when crossing rivers and lakes. The drowning of several dozen wildebeest provides a massive supply of food for scavengers, but the location and timing are unpredictable, and such food sources are transient and may be widely spaced.

There are also major variations in the distribution of these carcasses. Table 11.1 shows that about 70 percent of the food available from all ungulate deaths comes from migratory species; the activities of these species, therefore, are of prime importance. The main wildebeest population occupies the Serengeti plains during the wet season, moving toward the western woodlands at the beginning of the dry season, and then later traveling north into the northern extension of the Serengeti National Park and into the Kenya Mara Park (see chap. 5 above; Pennycuick 1975). Zebra and Thomson's gazelle undertake similar extensive movements, although in the dry season they remain more dispersed throughout the woodland region than the wildebeest, and some gazelle remain on the western plains. These migrations are rather erratic, and their timing is influenced by variations in the rainfall, but each year the herds travel many hundreds of kilometers. It is worth considering the implications of these movements for the scavengers. When the migratory herds are occupying the plains, there are a total of about 1,970,000 ungulates in the 9000 km² area. This contrasts with about 364,000 ungulates occupying the 21,000 km² woodland area. Assuming that during this time there is a daily mortality rate of 0.02 percent (taking the mean annual mortality rate of 12 percent, of which 6 percent die during the nine months from November to July at a constant rate) then, on average, there is about one carcass every 33 km² in the plains area and one carcass every 412 km² in the woodlands. At the start of the dry season, the herds enter the western woodlands, and the plains are occupied by comparatively small numbers of gazelle. The carcass density on the plains then probably falls to well under one carcass per 300 km². Thus, the arrival and departure of the migratory herds cause major changes in the abundance of carcasses.

These carcasses are only available for a short period. The rate of breakdown of tissues by bacteria and fly larvae is fast and varies with different-sized carcasses. Those of small animals, such as gazelle and impala, usually break down within a week, while those of larger animals, such as buffalo, may remain longer; but little meat remains in an edible condition after ten to fourteen days.

There is also a fluctuation in ungulate mortality over a longer time scale. Sinclair has outlined the population dynamics of the major ungulate species, which show considerable changes in population size. The wildebeest have quintupled over the past fifteen years, and buffalo have doubled (chap. 4 above; Sinclair 1977).

These increases may partly reflect a recovery after the heavy mortality caused by rinderpest, but it is also likely that such fluctuations in ungulate density are a normal feature of savanna regions like the Serengeti. It has long been known that in semiarid regions, there is a linear relationship between primary production and rainfall (Walter 1954), and this applies to the Serengeti (Braun 1973). Since the food supply ultimately limits the size of the ungulate populations, it is likely that ungulate biomass will also show a relationship to rainfall, and this has been demonstrated in a comparative study by Coe, Cumming, and Phillipson (1976). Rainfall does not appear, however, to be constant over long periods of time, and it is becoming apparent that annual rainfall in savanna regions fluctuates over periods of twenty years or so (see Phillipson 1975; also Stoddart 1978, for cycles in Indian Ocean rainfall patterns), which would influence the primary production and, hence, the ungulate biomass. Over a long time scale, there are probably phases when populations of grazing animals are rising and phases when populations are falling. These phases would involve changes in the annual mortality rates of ungulates, and, hence, in the food supply for scavengers.

Food Division

Mammalian Carnivores

The proportion of food that is obtained by predation and by scavenging varies considerably among different species of carnivores. Table 9.1 summarizes information on the large mammalian carnivores and their food. Cheetah and leopard scavenge rarely. All other species scavenge food whenever the opportunity arises, but this accounts for only a small proportion of their food intake. Hyenas scavenge more than other species, showing several adaptations for scavenging—chiefly the habit of solitary individuals spending part of the day wandering extensively, when they are likely to encounter carcasses (Kruuk 1972). Lions also scavenge part of their diet. Unlike hyenas, they do not often find carcasses by actively searching over large areas, but usually steal them from hyenas or vultures, which have led them to the site. The three species of jackals are all chiefly predators on small mammals and insects. They scavenge meat whenever they can, but this is probably not an important part of their diet (3 percent, based on fecal analysis).

Avian Carnivores

Among the carnivorous birds, eleven species feed regularly from carrion. Among these the kite, white-naped raven, bateleur eagle and tawny eagle are comparatively unimportant (Houston 1978): all these species have alternative sources of food, and while carrion may be important as a part of their diet for certain periods of the year, it does not account for a significant proportion. The other seven species feed entirely, or substantially, from carrion, and table 11.3 summarizes the information on these species

Table 11.3 Species of birds commonly seen feeding on carrion in Serengeti. Based on Kruuk (1967), Penny-cuick (1972a, b), and Houston (1975).

Species	Proportion recorded at feeding groups	Body weight (g)	Skull structure	Flight	Food supply	Social organization	Distribution
Ruppell's griffon vulture	18%	7400	Head almost bare. Long neck also sparsely feathered. Bill long and with a sharp cutting edge. Tongue barbed for gripping soft tissue.	High wing loadings. Rely on soaring and travel long distances each day. Fly at compara-tively high altitudes.	Muscle and viscera of large-mammal carcasses.	Nonterritorial. Search over wide areas and collect in large numbers at suitable carcasses	Highland and hilly country—only found in Serengeti near to migratory herds.
White-backed vulture	70%	5300					Lowland savanna. Found throughout Serengeti, but concentrated over the migratory herds.
Lappet-faced vulture	8%	6200	Large head, sparsely feathered. Bill deep and power-ful with strong hook at the tip.	Low wing loadings. Do not travel long distances. Fly comparatively low.	Skin, tendons, and meat from large-mammal carcasses. May occasionally kill its own prey.	Appear to hold group terri-tories. Unusual to see more than six birds, even at carcasses where several hundred griffon vultures.	Occurs in arid and savanna conditions. In Serengeti in both plains and woodlands.
White-headed vulture	0.3%	4000				Appear to be territorial, each pair defending a feeding range.	Savanna conditions. In Serengeti found throughout the woodlands.

Species	Proportion recorded at feeding groups	Body weight (g)	Skull structure	Flight	Food supply	Social organization	Distribution
Egyptian vulture	0.03%	1800	Face region bare. Bill long, thin and comparatively weak.	Lowest wing loadings. Will make long journeys, but not regularly. Usually fly at low altitudes.	A wide variety of items: termites, beetles, snakes, lizards, scraps of meat from carcasses, dung of carnivores, human refuse.	Nonterritorial but tend to be far more resident than the griffon vultures.	Arid and semiarid areas. Only common in eastern Serengeti plains.
Hooded vulture	3%	1900					Savanna and forest conditions. Common throughout Serengeti.
Lammergeier	Rarely seen	5000	Head and neck covered by more feathers than other vulture species. Bill can tear meat, and tongue is adapted for removing marrow from long bones.	Relies almost entirely on soaring in rising air over hills and cliff faces.	Carrion from large mammal carcasses. Will drop long bones onto rock slabs to break open the marrow cavity.	Territorial. Pair of birds patrol a range.	Precipitous mountain ranges. Only seen in Serengeti region near Gol Mountains and rift valley.
Marabou stork	Very variable Av. 3%	5500	Bill very long, deep, and broad. Unable to tear meat from carcass. Relies on other birds to do this.	As for griffon vultures, but few birds travel long distances regularly.	In moist areas feed on fish, frogs, and other small vertebrates. Also at rubbish tips and pirating food from vultures.	Nonterritorial.	Throughout plains and woodland areas of Serengeti.

in the Serengeti. Kruuk (1967) showed that these birds can be divided conveniently into categories according to their feeding strategies.

Hooded and Egyptian vultures: small birds with slim bills, which peck at scraps around a carcass and also eat a variety of other foods. Nonterritorial.

Two species of griffon vultures (white-backed and Ruppell's griffon): large birds with long necks, which feed only at large carcasses, taking the meat and viscera. Nonterritorial. (pl. 39)

Lammergeier, lappet-faced and white-headed: birds with large bills which feed by taking skin and tough meat from large carcasses, or feeding on smaller carcasses. Territorial.

Marabou stork: a large bird which feeds on fish and frogs and other small vertebrate prey in some areas of East Africa. In the Serengeti, they feed chiefly from carcasses. They cannot tear meat with their long bills, and they are dependent on pirating food away from griffon vultures.

Considering these birds in terms of their relative food consumption, the lammergeier is so rare in the Serengeti as to be insignificant, and the two smaller vultures—hooded and Egyptian—are unimportant because they have several other sources of food; they obtain only part of their food from carcasses, are not very abundant, and have a small body size and, hence, small daily food requirements. The other four species of large vultures (white-backed, Ruppel's griffon, white-headed, and lappet-faced) and the marabou stork together account for about 98 percent of food consumption by the bird scavengers (Houston 1978), and these species all feed almost exclusively by scavenging.

White-headed vultures are quite unlike all the other large scavenging birds. They are the only species which rarely feed beside other species of vultures at large feeding parties around carcasses. They are timid birds which usually feed by themselves, locating the carcasses of smaller animals, such as hares, and they are also probably the only vulture species which obtains a significant amount of food from predator kills. They are comparatively uncommon at carcasses, representing about 0.3 percent of the bird scavengers.

The lappet-faced vulture has a totally different feeding behavior. It is highly aggressive and congregates at ungulate carcasses, where it can normally dominate and drive away griffon vultures. Despite its aggressive behavior, it is far less numerous than the two species of griffon vulture, and it is unusual to see more than six birds, even at a large carcass, where griffon vulture numbers may reach several hundred. This species represents about 8 percent of the birds present at carcasses.

The two species of griffon vultures are by far the most important of the scavenging birds—about 88 percent of bird species recorded feeding at carcasses belong to this genus (pl. 39). They are large birds with correspondingly high food requirements (although the lappet-faced vulture appears to be the largest of the vultures because of its intimidating threat display, it is considerably lighter than the Ruppell's griffon). Both species have similar methods of food searching: they cover large distances and rely heavily on the activities of neighboring birds to locate food sources.

Marabou storks sometimes feed in association with the griffon vultures, and search for food in the same way. But in the Serengeti, they are commonest near village rubbish dumps and abbatoirs, and birds are only occasionally found far away from settlements. Compared with griffon vultures, they are unimportant.

Feeding Strategies of Scavengers

These figures of the relative abundance of the species, and the proportion of food they obtain from ungulate carcasses, are what one would expect, given the nature of the scavengers' food supply and the different feeding strategies adopted by each species. The amount of food an animal can obtain by scavenging from ungulate carcasses will depend on its searching methods. Species which hold territories are unlikely to obtain a substantial part of their food by scavenging because migratory herds are only in their feeding range for short periods during the year. At other seasons, carcasses are infrequent, and it is doubtful if a solitary searcher could both locate such widely spread food sources, and also prevent other species from displacing it before it had made full use of the carcass. The only vulture species which uses this strategy, the white-headed, obtains comparatively little food from ungulate carcasses, scavenging mainly from smaller animals. It is also suspected of killing some of its food. It maintains a fixed hunting range which it knows intimately, and which it presumably patrols regularly for food. Leopards have a similar strategy; they feed mostly by killing smaller animals and rarely scavenge.

Species which hold group territories will also have migratory herds within their range for part of the year only. But we might expect that searching for isolated and unpredictable carcasses would be more efficient in a social animal; and a group is also more likely to dominate and make full use of any carcasses it finds. Lappet-faced vultures seem to hold group territories, and they undoubtedly obtain a large proportion of their food by scavenging ungulate carcasses: they also feed on smaller animals and may occasionally kill their prey. Lion and hyena also hold group territories and scavenge a small proportion of their food.

It is the nonterritorial species which can travel widely and remain with migratory ungulates throughout the year. Thus, they can not only stay in areas where carcasses are likely to occur, but also can search communally, which is an efficient strategy to locate widely spaced carcasses. The two species of griffon vultures are both nonterritorial, feed exclusively from ungulate carcasses, and are by far the commonest of all the scavengers. They congregate over migratory herds and follow their movements. There are two mammalian carnivores, cheetah and wild dogs, which are also nonterritorial and which might therefore be expected to scavenge regularly, but both are rather specialized (see chap. 9). Kruuk (1967) has suggested that wild dogs are adapted for a specialized method of hunting, making them far less versatile in food searching than hyenas. Both species rarely obtain food by scavenging.

Invertebrate Scavengers

Mammalian and avian carnivores are not the only scavengers. Any carrion which is not eaten by them will be taken by invertebrates, chiefly fly larvae (*Sarcophaga* sp.) and some beetle species, for example, *Thanatophilus* sp., and by bacteria. It is almost impossible to estimate what proportion of food goes to the invertebrates because although carcasses where large scavengers are feeding are obvious, those left to invertebrate scavengers are easily overlooked. It is unusual to find a carcass that has been left long enough for bacteria and fly maggots to break down the tissues; in most cases, a vulture or mammalian scavenger finds it long before the meat has decomposed. Invertebrates will sometimes consume some of the soft tissues of carcasses where avian scavengers have been feeding: in the dry season, when carcasses are abundant, vultures remove most of the soft tissues, except skin and tendons, which are broken down by insect larvae and bacteria. I suspect that invertebrates account for only a few complete carcasses—such as those in dense undergrowth (where vultures cannot find them) and for some at the end of the dry season, when carrion is superabundant.

The Numbers of Avian Scavengers

Since most of the available carrion is eaten by avian scavengers, I have calculated the number of vultures that would be needed to consume this amount, to see if such a figure is realistic. There are about 26 million kg of dead animals each year (from table 11.1) not eaten by mammalian carnivores (from table 11.2). Assuming that 45 percent of the carcass weight is inedible for vultures (approximately 15 percent for the contents of the digestive tract; 15 percent for skin, hooves, tail, and so on; and 15 percent for bones; Ledger 1968), then about 14 million kg of soft tissues remain. If we assume that 85 percent of this is taken by vultures (the rest going to invertebrates and bacteria), we have a food supply of 12 million kg a year, or 33,000 kg for an average day.

Under caged conditions, the amount of food needed to maintain body weight for a white-backed and Ruppell's griffon vulture is about 375 and 470 g of meat, respectively (Houston 1973). These captive birds took little exercise; active, uncaged birds should have greater food requirements. There is little information on the effect of activity on energy requirements in birds, but in man there is a two- to sevenfold increase from resting energy requirements to various unexacting forms of activity (Brobeck 1974). Apart from brief periods of fighting over food, vultures spend most of the day perched in trees or soaring. Flying requires little additional energy: Pennycuick (1972a) calculated that the energy demands of flight would be equivalent to an increase of about 50 percent of standard metabolic rate. Assuming wild birds require three times the food intake of captive birds, one would expect the white-backed and Ruppell's griffons to obtain a minimum of 1125 and 1410 g, respectively, per day. The ratio of abundance of these two species is 4 to 1, so the two species combined would consume about 5910 g per day.

Assuming the vulture population as a whole consumes an average of 33,000 kg a day, griffon vultures would account for 29,000 kg per day, since they comprise about 88 percent of the vulture population. This consumption rate would require an average of 25,000 griffon vultures per day. This figure is consistent with other, tentative estimates of total bird numbers based on nesting density. It is difficult to estimate vulture numbers accurately because of their clumped distribution, and wide-ranging movements. However, it seems there are sufficient birds to account for the proposed food intake.

Adaptations for Scavenging

Searching Efficiency

The total meat consumption of avian scavengers is substantial, probably greater than the combined food intake of the mammalian carnivores. This is true despite their relatively small body size and timid behavior, which enables most mammalian carnivores to dominate them and to chase them away from any food source. Why, then, are avian scavengers so much more efficient than mammals at exploiting the scavenger food supply?

The ability to search for food from the air is an obvious advantage. A ground observer has its view restricted by surrounding vegetation, local depressions, and small hills, while, from the air, the view is unobstructed over large areas. Not only does this enable birds to locate food more easily, but they can also clearly see all other birds in the same area, and so can search communally: once one of them starts to descend to food, all other birds within sight converge, even if they themselves cannot see the source of food. Birds out of sight of the first bird, but within sight of these followers, also notice the change in flight behavior and start to follow, and so a chain reaction radiates out from the food site. This is an efficient method of food searching, whereby several hundred birds can congregate within an hour (Houston 1974b).

Birds can also travel fast to reach food. A vulture can probably descend in a steep dive at a horizontal speed of about 70 km/h, while the running speed of a hyena is about 40 km/h, so if both locate food the same distance away, the vulture will arrive there a few minutes before the ground scavengers (Pennycuick 1971). But these are not major advantages. The obvious advantage that the birds gain in being first to spot a carcass is only useful if they can prevent other scavengers from finding it. Once one vulture has located food and started to descend, the neighboring birds immediately follow; and most birds find food by watching the activities of other birds. But both lions and hyenas also watch for signs of vultures descending, and they will run toward any such site, even if the source of food is out of their line of sight. In a hypothetical situation where mammalian and avian scavengers are equally abundant in an area, the vultures would arrive slightly earlier at each carcass, but they would be chased away a few minutes later when the larger ground carnivores arrived. This is not, however, the way the system actually works. Sixty-four carcasses were watched from the arrival of the first birds until all food had been eaten, and in 84 percent of those cases, no mammalian carnivore appeared—although the birds usually fed for several hours (Houston 1974a). This suggests that there is little competition from the mammals and explains why avian scavengers have become the most numerous carnivore: they are able to exploit a food supply that the mammals are unable to utilize efficiently.

An avian scavenger has an even more important advantage, in that it can travel long distances with relatively little energy expenditure. As a method of locomotion, even-powered flying requires less energy than running (Schmidt-Nielson 1972). But vultures rely on gliding flight, which enables them to take advantage of surrounding air movements to keep airborne. They normally search by thermal soaring: relying on thermal upcurrents of air for gaining altitude and then gliding, slowly losing height, until they contact another thermal. Their energy requirements are small and require only that the birds hold their wings open for long periods. There are specialized tonic muscles used for this (Kuroda 1961), which require little energy compared with the fast-acting muscles used for powered flight, and Pennycuick (1972b) has suggested that gliding flight requires only about one-thirtieth of the energy necessary for flapping flight. The energy required for flying may therefore be only slightly greater than that required when the bird is resting on the ground.

Pennycuick (1972a) has made detailed studies of the gliding flight of vultures by following them in a glider and has recorded an average cross-country speed of 47 km/h for a Ruppell's griffon, documenting the fact that birds can cover large distances rapidly. A feature of the scavenger food supply is that it is transient and irregular. To locate food, an animal must search over large areas. An avian scavenger is at a considerable advantage in reaching isolated and widely separated food sources. But the major advantage that low-energy flight brings is that it enables vultures to follow the migratory ungulates throughout the year, even while the birds are breeding.

Seasonal Movements

During the year, the large herds of migratory ungulates travel many hundreds of kilometers. These herds supply the majority of the food for carnivores, and any animal which is going to feed extensively by scavenging will be most likely to succeed if it remains with these herds. The daily movements of the herds are not particularly arduous—individual wildebeest have been recorded moving up to 18 km per day, although the average distance traveled was under 4 km per day in the wet season

(Inglis 1976) and 10 km per day over the year (D. Kreulen, in chap. 7 above), and mammalian scavengers would have little difficulty covering such distances. But over a period of months, the herds move several hundred kilometers. This becomes a major problem during the breeding season of the chief mammalian scavengers, lion and hyena, for their young cannot cover such large distances (chaps. 9, 10). These species are territorial, and when the migratory herds leave their area, they have to rely on finding food from the resident ungulate populations. Most hyena dens are located on the plains, and from there the adults have to commute, often long distances, to reach the herds and obtain food for the young. There is an obvious limit to the distance that a female can travel from her den to reach the herds. When the wildebeest enter the northern extension of the Serengeti and the Kenya Mara Park, few hyena follow them north of the Grumeti River. It is difficult to see why, for this area is occupied by the wildebeest at the end of the dry season, when their mortality is heaviest. Ungulate carcasses in this area are abundant then, but there are few lions or hyenas, so the vultures have virtually no competition.

Unlike mammalian scavengers, the avian scavengers can cope with long distances comparatively easily. Vultures have unusually long fledgling periods for birds of prey; the eggs and young of the Ruppell's griffon are in the nest for up to five months. When the breeding season starts in January, the herds are within a few kilometers of the nest sites, but later on in the year, the herds move several hundred kilometers away. Birds travel regularly to these migratory herds, taking perhaps two or three hours to reach them, feed, and return to feed the chicks (Houston 1976). Until they are almost fully grown, the young are always guarded by one of the parents. Such long-distance commuting would be impossible for a mammal.

Long-Distance Movements

Griffon vultures travel long distances not only while searching for food. In southern Africa, where ringing programs on the Cape vultures have been continued for over twenty years, birds have been recorded moving over 1100 km from the nesting site at which they fledged (Houston 1975c). There is little direct information on the distances traveled by birds in East Africa, where bird ringing is unlikely to yield many recoveries. Among the birds marked in the Serengeti, the longest movement recorded was that of a white-backed vulture which, in six days, traveled from Lake Ndutu to Maji Moto, a distance of 180 km. But far greater movements, comparable to those undertaken by griffon vultures in southern Africa, also occur in East Africa. There are two pieces of evidence which suggest this. A noticeable movement occurs with immature Ruppell's griffon vultures, which are present in the Serengeti only during the nonbreeding season. As soon as the breeding season starts, these immature birds are virtually never seen. It is unlikely that this is due to a sudden heavy mortality, and the most probable explanation is that the young birds move away from the Serengeti ecosystem to avoid competition with adult birds. They presumably move to other feeding areas which are out of reach of the breeding colonies, where they can feed without competing with the more experienced birds.

Another indication of long-distance migration comes from the sightings of marked birds. One hundred and sixty-two griffon vultures were caught and marked with conspicuous plastic leg rings so that each individual could be recognized through field glasses. Only thirteen individuals were resighted, some on more than one occasion, making a total of twenty-two resightings. If there had been an equal probability of each marked bird's being seen, then twenty-one different individuals should have been seen. The number recorded is significantly less than this, which suggests that not all the birds that were originally marked remained available for resighting. This is probably because many of the birds which were first caught in the Serengeti later moved on to other areas of East Africa (the only other explanation would be heavy mortality or loss of the rings, both of which are considered unlikely). Therefore, there does not appear to be a discrete "population" of vultures in the Serengeti; most likely, the birds present there are only part of a continuous blanket over East Africa, and the density of birds in any particular area varies seasonally with their movements.

It is interesting to speculate on the purpose of the long-distance movements. It is clear that the period of food abundance for scavengers comes at the end of the dry season, when ungulates are in worst body condition. The time at which the dry season occurs varies in different parts of Africa, due to the timing of the movements of the rain belts which move north and south through Africa during the year (Thomson 1965). Hence, periods of food abundance occur in different months at different latitudes. In the Serengeti, the end of the dry season occurs between September and October. But in the southern Sudan, where there is still a migratory ungulate concentration comparable to the Serengeti's in the Boma plateau region, the dry season ends between January and February. By traveling north and south along the savanna belt during the year, the birds could move from one ungulate population to another and so have a favorable food supply for much of the year.

It has long been suspected that many small savanna bird species undertake comparable movements. Many birds occur within one area only for a brief period in the year, a period when feeding conditions are optimal. Ward (1971) has suggested that the populations of *Quelea quelea* in Africa undertake large seasonal migrations, following the rain belt, where conditions are suitable for feeding and breeding. Similar movements have recently been shown in a wide range of savanna bird species in Nigeria (P. Ward and P. Jones, pers. comm.).

I would expect vultures to preceed the rain belt, arriving in areas at the height of the dry season. Presumably, only a proportion of the vultures undertake these movements, for breeding birds are restricted to one area for about nine months of the year. But large-distance movements between different populations of migratory ungulates is probably a strategy used by many immature and nonbreeding individuals. This seasonal immigration and emigration of birds enables griffon vultures to exploit effectively the large seasonal variations in the abundance of carcasses in any one area.

These movements may be far less extensive today than they were in the past, for the conditions under which these birds evolved have now changed considerably. Today there are few areas in Africa where large migratory systems, such as those in the Serengeti, remain intact. In many other savanna areas, it is clear that seasonal movements of large ungulate herds occurred in the past, but these herds have been either exterminated or drastically reduced during the last hundred years.

Evolution of Exclusive Scavengers

For birds to travel such large distances, it is important that they be efficient in flight and so travel with low energy expenditure. This may explain why vultures have lost the ability to kill—I presume they evolved from predatory ancestors. Most species of vulture obtain their food exclusively by scavenging. By contrast, there is, as far as I know, no mammal which feeds only by scavenging. A carnivorous mammal is more likely to survive if it can kill its own food when none is available from scavenging. For a carnivore, there is always more "live" food in an area than dead. It would gain no advantage by losing the ability to kill, but would suffer a considerable disadvantage. Griffon vultures, however, do have the disadvantage that when they are hungry, they cannot go out and kill something—they just remain hungry.

There must, therefore, have been a large selective advantage in abandoning predation, and this was probably an adaptation for efficient gliding. Vultures could only achieve this by losing the agility, accuracy of landing, and ability to take off rapidly which are necessary for predators. For example, their wings have a large surface area, which prevents agile maneuvers. A large body size is also of considerable advantage to scavenging birds with an irregular food supply, for they can withstand longer periods between feeds and have a larger food intake at each meal. They are also able to dominate smaller scavengers when feeding. Indeed, some vultures are among the heaviest flying birds in the world. Such a large body size, which results in clumsy and awkward landing and take-off, could probably only be achieved by abandoning predation.

Long-term Variations
in Food Supply

So far I have considered adaptations evolved by vultures to exploit a food supply which varies seasonally in distribution and abundance. The food supply also varies over a longer time scale corresponding to phases of ungulate population increase and decline. Bird scavengers have an advantage over mammals in this situation as well because of their greater ability to survive a long period of poor food supply. Birds, despite their small body size, have a greater longevity than mammals. As a result of many long-term ringing studies, it is becoming clear that among tropical passerines, the annual mortality rates are usually of the order of 10 percent per year. Assuming mortality rates to be constant with age, life spans of up to ten years should be considered normal for many individuals (Fry 1978). There is, unfortunately, no similarly detailed information for larger land birds, but there are many species of large, tropical sea birds which have been studied by long-term ringing programs. The results of these studies suggest that adult survival rates of 95 to 97 percent may not be uncommon. If one assumes that mortality remains constant with age, then these high survival figures imply that, for most species studied, a substantial number of birds which reach breeding age will survive for well over fifty years (Ashmole 1971). Large vultures, together with albatrosses, have the lowest reproductive rate of any bird species in the world

(Lack 1968). It follows, then, that their mortality rates must be among the lowest of any bird species. I assume that among those vultures which reach breeding age, some will live for at least sixty or seventy years.

This contrasts markedly with the survival of large mammalian carnivores. Kruuk estimated adult mortality of hyenas of one year and over at 17 percent in Ngorongoro Crater, but they may have slightly lower annual mortality in the Serengeti. Bertram (1975) suggests that few lionesses live beyond fifteen years, and males considerably less. Considering that their food supply fluctuates considerably over a long time scale due to declines and increases in ungulate populations, scavengers face times of low food availability. In such a situation, a long life span would be advantageous, since animals can breed only if food is relatively abundant. Breeding is demanding in griffon vultures, and probably only those in peak condition can rear young (Houston 1976). One can envisage a situation when food for animals which feed exclusively by scavenging would be so scarce that breeding would be impossible for a number of years. A small population of bird scavengers could survive such conditions for perhaps forty or fifty years. When food conditions again became favorable, the population could reproduce, while any mammalian competitor would have joined the fossil record.

Effect of Scavenging Birds on Predators

Lions, hyenas, and other mammals could obtain food from many of the carcasses which are now taken by vultures if the food was present long enough for the mammals to locate it with their relatively inefficient searching methods. Predators would then need to kill less often. However, there is an advantage for predators in watching the activity of vultures and using them to locate carcasses. Lion obtain at least 11 percent of the carcasses that they scavenge directly by watching the activity of vultures (Schaller 1972). But it is unlikely that vultures influence the food supply of predators to any large extent. The density of both lions and hyenas in the Serengeti is probably determined by the food supply during the worst part of the year—which corresponds to the season when the migratory ungulate herds are out of the predators' range (chap. 10 above). At these seasons, griffon vultures are uncommon because they are concentrated over the migratory herds. Hence, avian scavengers are unlikely to remove substantial amounts of food from the predators—although any reduction in food supply could be of some importance.

Conclusion

Virtually all predators will scavenge food if they have the chance. Exclusive scavengers, which never kill, are at an obvious disadvantage because their sources of food are far more limited. They evolve in situations in which there is both a large potential food supply and also a strong selective advantage in abandoning the ability to kill. In savanna conditions in Africa, such scavengers have developed because of the dominant influence of the migratory ungulates on the food supply for carnivorous animals. In the Serengeti, most ungulates are migratory, while their major predators have restricted feeding ranges. This prevents the predators from reaching a population size where they could cause heavy mortality among migratory ungulates. Among the migratory herds, most animals die from causes other than predation, and so provide a large food supply for scavengers. This supply is transient, infrequent, and irregular in distribution, but each carcass contains sufficient food to feed a large number of scavengers. The mammalian carnivores are unable to cope effectively with these variations in the temporal and spacial distribution of the food, but, for a variety of reasons, the avian scavengers have major advantages, and the griffon vultures, in particular, have become adapted to exploit this large food supply and have become the major carnivorous animals in the Serengeti system.

Migratory systems were once found in most savanna areas in Africa, but the widespread destruction of animal populations has left few such systems intact. In most other areas of Africa today, the ungulates are resident, and such communities usually support a higher predator-prey ratio than is found in the Serengeti (Schaller 1972). Scavengers are not influenced by predator density as such, but by the effect of these predators on the prey species; and in areas of resident ungulates, it is clear that vultures play a far less important role. For example, Kruuk (1972) found that in a population of resident ungulates in Ngorongoro, the predators (chiefly hyenas) were responsible for most ungulate deaths and that few animals died from other causes—and those which did were probably quickly scavenged by hyenas or lions ,both of which live at a comparatively high density. In such a situation, there is little food available for scavenging birds, and vultures are rarely seen feeding there.

One can therefore predict to some extent the situations in which scavengers will be abundant. In addition to migratory systems, scavengers might be expected to play an important role in communities where the biomass is dominated by elephants—such as in Tsavo, Kenya, and in some

Ugandan National Parks. Here the major herbivores are too large for predators to exert much impact and this may provide suitable conditions for scavengers.

Similarly, the presence or absence of exclusive scavengers may enable us to predict some features of a past community. Animal communities in the Indian subcontinent have been so severely restricted by man's activities that we know very little of their original interactions. But the presence there of a range of species of large avian scavengers indicates that in the past predators must have been responsible for a relatively small proportion of ungulate mortality. The American prairies, however, once contained large migratory populations of bison and pronghorn antelope, but there were no large avian scavengers. This suggests that predators were responsible for a higher proportion of mortality among these ungulates—we know that their major predator, the wolf, can indeed control prey numbers in some modern ungulate communities (Mech 1970).

References

Ashmole, N. P. 1971. Sea bird ecology and the marine environment. In *Avian biology,* ed. D. S. Farner and J. H. King, pp. 224–86. London: Academic Press.

Bertram, B. C. R. 1975. Social factors influencing reproduction in wild lions. *J. Zool.* (Lond.) 177:463–82.

Braun, H. M. H. 1973. Primary production in the Serengeti; purpose, methods and some results of research. Ann. Univ. d'Abidjan, ser. E, no. 6, pp. 171–88.

Brobeck, J. H. 1974. Energy balance and food intake. In *Medical physiology,* 13th ed., ed. V. B. Mountcastle, pp. 1253–72. New York: Mosby.

Coe, M. J.; Cumming, D. H.; and Phillipson, J. 1976. Biomass and production of large African herbivores in relation to rainfall and primary production. *Oecologia* 22:341–54.

Fry, C. H. 1979. Mortality rates in tropical birds. *Ostrich,* in press.

Hirst, S. M. 1969. Predation as a regulating factor of wild ungulate populations in a Transvaal low veld nature reserve. *Zool. Afr.* 4:199–231.

Houston, D. C. 1973. The ecology of Serengeti vultures. D. Phil. dissertation, Oxford University.

————. 1974a. The role of griffon vultures as scavengers. *J. Zool.* (Lond.) 172:35–46.

————. 1974b. Food searching behaviour in griffon vultures. *E. Afr. Wildl. J.* 12:63–77.

————. 1974c. Mortality of the cape vulture. *Ostrich* 45:57–62.

————. 1975. Ecological isolation of African scavenging birds. *Ardea* 63:55–64.

————. 1976. Breeding of the white-backed and Ruppell's griffon vultures, *Gyps africanus* and *G. rueppellii. Ibis* 118:14–40.

————. 1979. Inter-relations of African scavenging animals. *Ostrich,* in press.

Inglis, J. M. 1976. Wet-season movements of individual wildebeests of the Serengeti migratory herd. *E. Afr. Wildl. J.* 14:17–34.

Kruuk, H. 1967. Competition for food between vultures in East Africa. *Ardea* 55:171–93.

————. 1972. *The spotted hyena.* Chicago: Univ. of Chicago Press.

Kuroda, N. 1961. A note on the pectoral muscles of birds. *Auk* 78: 261–63.

Lack, D. 1968. *Ecological adaptations for breeding in birds.* London: Methuen.

Ledger, H. P. 1968. Body composition as a basis for a comparative study of some East African mammals. *Symp. Zool. Soc. Lond.* 21:289–310.

Mech, D. M. 1970. *The wolf.* New York: Natural History Press.

Pennycuick, C. J. 1971. The soaring flight of vultures. *Sci. Am.* 229: 102–9.

————. 1972a. Soaring behaviour and performance of some East African birds observed from a motor glider. *Ibis* 114:178–218.

————. 1972b. *Animal flight.* London: Arnold.

Pennycuick, L. 1975. Movements of the migratory wildebeest population in the Serengeti area between 1960 and 1973. *E. Afr. Wildl. J.* 13: 65–87.

Phillipson, J. 1975. Rainfall, primary production and "carrying capacity" of Tsavo National Park, Kenya. *E. Afr. Wildl. J.* 13:65–87.

Pienaar, U de V. 1969. Predator-prey relations amongst the larger mammals of the Kruger National Park. *Koedoe* 12:108–76.

Plowright, W., and McCulloch, B. 1967. Investigations on the incidence of rinderpest virus infection in game animals in N. Tanganyika and S. Kenya. *J. Hyg.* (Camb.) 65:343–58.

Schaller, G. B. 1968. Hunting behaviour of the cheetah in the Serengeti National Park, Tanzania. *E. Afr. Wildl. J.* 6:95–100.

————. 1972. *The Serengeti lion.* Chicago: Univ. of Chicago Press.

Schmidt-Nielson, K. 1972. Energy cost of swimming, running and flying. *Science* 177:222–28.

Sinclair, A. R. E. 1973. Population increases of buffalo and wildebeest in the Serengeti. *E. Afr. Wildl. J.* 11:93–107.

———. 1974a. The natural regulation of buffalo populations in East Africa. Part 3. Population trends and mortality. *E. Afr. Wildl. J.* 12: 185–200.

———. 1974b. The natural regulation of buffalo populations in East Africa. Part 4. The food supply as a regulating factor, and competition. *E. Afr. Wildl. J.* 12:291–311.

———. 1977. *The African buffalo.* Chicago: Univ. of Chicago Press.

Sinclair, A. R. E., and Duncan, P. 1972. Indices of condition in tropical ruminants. *E. Afr. Wildl. J.* 10:143–49.

Stoddart, D. R. 1978. Long-term climatic change in the western Indian Ocean (pers. comm. See also *Phil. Trans. Roy. Soc.* A. Vol. 291).

Thomson, P. D. 1965. *The climate of Africa.* London: Oxford University Press.

Walter, H. 1954. Le facteur eau dans les régions arides et la signification pour l'organisation de la végétation dans les contrées sub-tropicales. In *Les Divisions Ecologiques du Monde,* pp. 27–39. Paris: Centre Nat. Res. Sci.

Ward, P. 1971. The migration patterns of *Quelea quelea* in Africa. *Ibis* 113:275–97.

Wyman, J. 1967. The jackals of the Serengeti. *Animals* 10:79–83.

Ray Hilborn
A. R. E. Sinclair

Twelve   A Simulation of the
         Wildebeest Population,
         Other Ungulates,
         and Their Predators

It is clear from the preceding chapters that the dynamics of the Serengeti ecosystem involve interaction among many species, and that our understanding of these interactions is far from complete. The major perturbation of increasing wildebeest numbers suggests several interesting questions, which we hope can be answered with the available data. Firstly, at what size will the wildebeest population cease to grow if the dry-season rainfall continues at the recent level of 250 mm? In particular, will the population reach equilibrium, and how does the equilibrium depend upon the rainfall? With respect to the time dynamics of wildebeest numbers, will they reach equilibrium rapidly, or so slowly that the numbers will not be able to track rainfall? When a dry year comes, as it will inevitably, how much will the population decline?

The resident prey in the north must bear the full brunt of lion predation in the wet season, but increasing numbers of wildebeest (chap. 9) provide a buffer against dry-season predation. Therefore, our second question is What has been the net effect of the wildebeest increase upon predation pressure on the resident woodland ungulates?

Predators now take a small proportion of wildebeest in the Serengeti, but they take a large proportion of ungulates in other areas, particularly Ngorongoro Crater (Kruuk 1972). It is possible that a substantial wildebeest crash might reduce the wildebeest numbers to a level where predators could have a major effect for a short time. It is remotely possible that, as a result of several years of poor rain, wildebeest numbers would drop so low that they would be driven extinct by predators. For example, the few wildebeest of Lake Manyara, once they were cut off from the migrating herds of Tarangire by the rise in the lake level in 1962–63, were then eliminated by predators (I. Douglas-Hamilton, pers. comm.), and we would like to explore the possibility of such an outcome in the Serengeti. Local extinctions of major ungulates, like caribou, precipitated by poor food conditions, and aggravated by temporarily high predation are thought to be a frequent occurrence in Alaska (Haber 1977).

In this chapter we will attempt to synthesize the available data on African grassland ecosystems and will suggest a possible interaction between rainfall, food, ungulates, and predators in the Serengeti in order to answer the questions we have posed above. If the goal of the Serengeti work has been to understand the dynamics of the ecosystem, then some synthetic model should be constructed as an explicit hypothesis of what we believe to be the important processes, and how these processes interact. Without this step the research would be incomplete. By explicitly stating hypotheses, we highlight the areas where data are lacking. As a result we will be wrong in places, but we invite criticism because this is the most rapid means of improving our understanding of Serengeti dynamics.

Data for African Grassland Ecosystems

Spatial and Temporal Resolution

Our attempt at synthesis will rely heavily upon the construction of models. These models represent our hypotheses about the structure of the Serengeti ecosystem and its behavior, and could be formulated at several levels of temporal and spatial resolution. For our first question, What will happen to the wildebeest under different rainfall regimes? the appropriate temporal resolution is annual, with the seasonal changes implicitly considered. The most appropriate spatial scale is the entire Serengeti ecological unit. Such a hypothesis (model) is similar to Sinclair's (1977) model of the buffalo population. For our questions concerning predators and other ungulates, different space-time scales seem necessary because wildebeest migrate, while predators do not. We have chosen a monthly scale, so that the different timings of the three migratory ungulates—wildebeest, zebra, and Thomson's gazelle—can be handled easily.

The Serengeti is usually divided into four areas: the plains, the western corridor, the central woodlands, and the northern woodlands. Our feeling, based upon long familiarity with the region, is that the central and northern areas are sufficiently similar in structure and animal usage to warrant combining them. Thus, our model considered each species in each of three areas on a monthly scale.

The Migration

The spatial and temporal scales described above were largely dictated by our desire to examine the phenomena affected by the migration (chap. 5), which is largely dictated by food availability on the plains. We made no attempt to simulate the dynamics of the migration, and we have assumed that the movement of the ungulate species depends upon the time of year. The proportions of each migratory species in each area for each month are presented in figure 5.4. We use these data as presented, with one exception. Maddock shows a small proportion of wildebeest remaining in the north. These are resident animals on the Loita plains, well outside the area considered in this model, and we assume that they are not available to the predators in our model. Hence, for purposes of our model, there are no wildebeest, zebra, or gazelle in the northern and central areas during the wet season.

Grass Production

An apparently unique feature of the Serengeti ecosystem, as it has been observed in the last twenty years, is that grass production appears to depend solely on rainfall. McNaughton (chap. 3) has shown that some grazing can stimulate growth, as can burning, but rainfall is the overwhelming influence and predicts grass growth well. For our purposes, we consider food to be any green grass with a crude protein content averaging 8 percent. Such high-protein grass is a potential limiting factor for wildebeest: low-protein plant material, which can fill the gut but does not provide enough nutrition for body maintenance, is always available.

Wildebeest are affected by food limitation in the dry season, when they are in the north and central woodlands. Sinclair (1977) has described the effect of rainfall on grass production in these areas. These data are modified for a monthly dry-season grass production curve shown in figure 12.1. We specify the dry-season rainfall, and the model calculates the appropriate amount of grass. Since dry-season rainfall occurs periodically, we divide the dry season into four separate monthly rainfalls and calculate the grass production on a monthly basis.

Wildebeest Reproduction and Calf Survival

Female wildebeest older than two years have a constant fertility, such that nearly every female gives birth in February every year. There is evidence

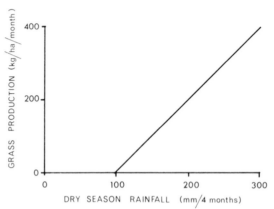

Figure 12.1   Dry-season grass production
plotted against dry-season rainfall.
Taken from Sinclair (1977).

for some reduced yearling fertility at low food availabilities, but, since yearlings constitute a small proportion of the breeding population, small changes in their fertility will have little, if any, effect upon the entire population, and we have assumed this effect to be negligible.

There is good evidence that calf survival to one year depends upon dry-season food availability (calculated from chap. 4). Figure 12.2 shows this survival plotted against the food available per animal per month of the dry season, with the curve fitted by eye. The survival of calves through the dry season was calculated from the food available per individual, as per figure 12.2.

Adult Wildebeest Survival

There is no evidence to suggest that adult survival rates changed during the wildebeest increase. Therefore, we concluded that, within the range of past observations of environmental conditions, adult survival is likely to remain constant at 95 percent per year. However, with the present high wildebeest population, one year of poor dry-season rain (100–150 mm) could push the system well beyond the range of past observations. Although years of 100–150 mm were commonly observed in the 1960s, today such low rainfall, coupled with the present wildebeest population increase, would lead to much lower food availability per individual than

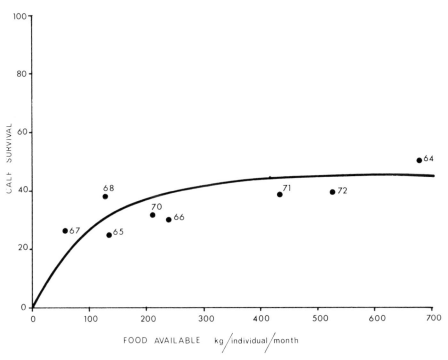

Figure 12.2   Calf survival plotted against food
availability during the dry season.
Points calculated from data in
chap. 4, curve fitted by eye.

has previously been observed. Hence, large-scale adult mortality might
result from such a dry year, if it occurs in the near future.

To present adult mortality as a function of food availability in the dry
season, other information must be included: (1) wildebeest eat 4.2 kg/
day of high-protein grass when it is plentiful (Sinclair 1975); (2) a
ruminant can lose up to 30 percent of its body weight and still survive,
but it is likely to be susceptible to disease after a 20 percent loss (H. Nor-
dan, pers. comm.); (3) green grass has an 8 percent crude protein
content, and low-quality dead forage averages 2 percent (Sinclair 1977);
(4) from studies of cattle (Dradu and Harrington 1972), we know how
weight loss depends upon protein consumption. Combining this information, figure 12.3 shows the number of days required to lose 26 kg (20

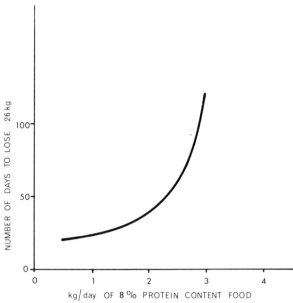

Figure 12.3   Number of days required for an adult wildebeest to lose 20 kg, plotted against kg/day of high-protein food. Calculated from Dradu and Harrington (1972).

percent of body weight), relative to the number of kg/day of green food (8 percent crude protein) available. Since the dry season is 120 days long, we would expect to see little adult mortality when there are 3 or more kg/day of food per animal. In 1967, there was a little more than 2 kg/day of food, yet there was no pronounced increase in adult mortality, indicating that 2 kg/day is sufficient. But at 1 kg/day, we would expect most individuals to have lost 20 percent of their body weight and to see high mortality.

The first month of the dry season is the most crucial, because if large numbers of individuals die at that stage, there will be fewer animals to eat the next months' production, resulting in more food per animal. Based on figure 12.3, we have guessed that if food were as low as 1 kg/day, we would see 50 percent adult mortality. This function is drawn in figure 12.4. We assume that 50 percent per month is the maximum mortality for several reasons. First, there will always be spatial variability in food conditions; low-lying soils are wetter and will have better grass in poor years. Second, some animals are hardier than others; older and younger animals are likely to die before three- to eight-year-olds. In our opinion, some animals will always find enough food to survive. This is, of course, largely

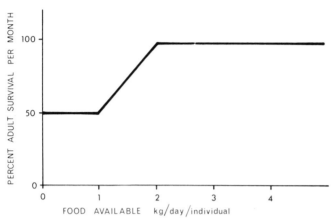

Figure 12.4   Assumed relationship between
monthly adult survival and food
availability.

assumption, but in different runs of the model, we tried different assumptions and found the results insensitive to this one.

In summary, each month all the wildebeest in the model are partitioned in areas according to Maddock's data. The adult and calf survival is calculated as a function of food availability and the density-independent survival rates described above. During February the calves are born. The effects of predators are described below.

Predators

The principal predators in the Serengeti are lions and hyenas. Their diet consists primarily of wildebeest, zebra, and gazelle (Schaller 1972; Kruuk 1972), except during the wet season in the central and northern woodlands, where the migratory species are not present. Under these conditions, the predators must sustain themselves on the resident prey, which are much less abundant.

We consider two questions about predation. First, what is its likely impact during a dry year? Is it possible that predation (which has been a minor source of mortality in the past) will become important when wildebeest numbers are reduced? Second, what has been the effect of predators on resident prey in the woodlands as the wildebeest have increased?

An increase in predation could have occurred by the following sequence. During the dry season, predators feed almost exclusively on wildebeest. The increase in wildebeest numbers does not reduce predation pressure on nonmigratory prey species in the dry season, but it could lead to increased predator numbers that would cause more predation during the wet season. An alternative hypothesis is that the increased numbers of wildebeest have not led to increased numbers of predators because of the low prey availability in the dry season. A third hypothesis holds that the increased rainfall leads to wildebeest's spending more time in the plains, thus producing less of a buffering effect for resident prey in the woodlands. A fourth alternative is that increased wildebeest numbers have led to wildebeest's occupying a larger portion of the north and central areas. This might lead to reduced predation and resident prey, since wildebeest would not be present in a higher proportion of predator territories in the woodlands. We explore this jungle of conflicting ideas with the available data.

The Functional Response of Predators

Any understanding of predator-prey interactions must include some hypothesis about how the number of prey taken per predator depends upon the number of prey available, a relationship known as the functional response. The derivation is primarily due to Holling (1959, 1965, 1966) and Ivlev (1961). The simplest functional response hypothesis that proved adequate for our purposes assumes the predator searches at random, encounters prey in relation to their abundance, and captures a proportion of those encountered that depends upon the prey species. It then spends a fixed amount of time eating and digesting the prey before it resumes hunting (Holling 1959, 1965, 1966).

The simple functional response, known as the disc equation, can be written as follows:

$$E = \frac{spN}{1 + hspN} \tag{1}$$

where E is the number eaten per unit time by each predator; s is the rate of search by the predator in an area per unit time; p is the probability that a prey encountered will be successfully attacked and captured; N is the abundance of prey in numbers per unit area; h is the amount of time taken to capture, eat, and digest an individual prey item.

Of these variables, E and N are frequently observed in the field; h can

be estimated by calculating the maximum number of prey a predator eats
in a year, and assuming that when he is eating that many prey, he spends
nearly all his time capturing, eating, and digesting (resting) and has little
time for anything else. The remaining parameters can be estimated from
field data if the N's have been observed over a range of conditions. This
estimation procedure requires a modification of the above formulas and
the use of some nonlinear curve fitting computer routines.

Equation 1 can be rewritten as follows:

$$E = \frac{aN}{1 + abN} \tag{2}$$

where a is the product of s and p from equation 1.

We use this particular equation for the predator functional response
primarily because it is well adapted to any type of behavior that saturates
at high prey densities and diminishes smoothly at lower densities. The
available data do not permit an evaluation of the quality of the fit, but we
believe that the disc equation is an appropriate approximation for the
qualitative form of the predator functional response.

The most useful data on predator consumption rates and prey availability come from censuses of Nairobi National Park in Kenya (Foster and
Kearney 1967; Foster and McLaughlin 1968). They present data showing
prey densities and proportion of different prey items in the diet of resident
lions over a period of six years during which a drought greatly altered prey
abundance. In particular, lions' principal prey, wildebeest, declined from
2000 to 200. The Nairobi data included four species: wildebeest, zebra,
impala, and kongoni. A modified version of equation 2 was required to
accommodate multispecies data. This modified equation, known as the
multiprey disc equation, was developed principally by Charnov (1974).

$$E_i = \frac{a_i N_i}{1 + \sum_{j=1}^{4} h_j a_j N_j} \tag{3}$$

All symbols are the same as in equation 2, except for subscripts of species
i or j.

The h parameter was estimated as follows: Schaller estimated that a
lion kills about 2,469 kg/year. A zebra weighs an average of 200 kg, so
that an adult lion must eat 12.5 zebra to get his full ration. Assuming that
all time is devoted to capturing, eating, and digesting (resting), h is
equal to 1/12.5, or 0.08. This is an overestimate, since some time was

devoted to searching. Similar calculations were made for the handling times of wildebeest, kongoni, and impala. The a parameter is calculated using nonlinear least squares to get the best fit to the Nairobi Park data. However, since the Nairobi park data contain only the proportion of prey species in the kill, the actual magnitude of the a estimates have little meaning; we have assumed maximum consumption at all prey densities and therefore cannot estimate the levels at which low prey density is reflected in lower prey consumption.

The Nairobi park data do tell us two useful things. The relative values of the a values can be estimated using the procedure described above, or by direct computation for an individual year (N. Gilbert, pers. comm.). The data show wildebeest are preferred to zebra, which are preferred to impala and kongoni. Since a is the product of effective rate of search (area/time) and probability of successful capture, the differences in the values must reflect differences between the species in probabilities of detection and successful capture.

Elliott, Cowan, and Holling (1977) present data on this process for lions attacking wildebeest, zebra, and Thomson's gazelle. They found that the probability of successful capture, once a chase was begun, was high for wildebeest, lower for zebra, and quite low for Thomson's gazelle. If we consider impala and kongoni to be similar to gazelles in capture success, then our estimated parameters seem to be properly ordered.

The second useful information we can derive from the Nairobi park data concerns the absolute magnitude of the a values. In the lowest year of prey observed, all indications were that lions were able to obtain near maximal food intake. The density of wildebeest was 2/km², zebra 4/km². Thus, the lions saw a combined density of these preferred prey of 6/km². Therefore, the a parameter should be large enough so that near maximal consumption is obtained when combined wildebeest-zebra densities are as small as 6/km². The actual value is only important when prey density is low. The lowest prey density considered later in this paper is 100,000 wildebeest for the Serengeti, which is about 15/km² on the plains—far above the density at which we suspect food might become scarce for lions. Thus, for purposes of this exercise, we assume that predators always get near full diet.

The major qualitative observations, that wildebeest numbers in the diet of lions dropped significantly but not nearly as much as the tenfold drop in wildebeest numbers, are reflected well by the fit. Similarly, the number of kongoni in the diet increased dramatically, since kongoni numbers were not affected by the drought. Hence, we consider a simple multiprey disc equation sufficient to explain the Nairobi Park data. Since we were fitting sixteen data points with four parameters, the model should fit reasonably well, and it would take a much larger data set to get good evidence for more complex processes such as "switching."

Predator Population Dynamics

The population dynamics of lions and hyenas have been studied in the field by Schaller (1972), Bertram (1975), and Kruuk (1972). It is clear from this work that social behavior is an important factor in predator dynamics; but, for the lions, it seems that an equally important determinant of population change is the availability of food, as it affects cub survival. Both Hanby and Bygott (chap. 10) and Bertram (1973) have demonstrated major changes in cub survival due to increased food availability. Figure 12.5 is a plot of the average monthly survival rate of cubs in two lion prides against the numbers of months for which there are at least 1000 prey present per km² (from Bertram 1973, table 1). Hanby and

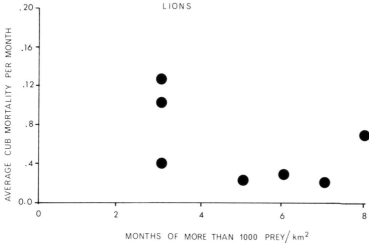

Figure 12.5   Lion cub survival as a function of food availability. From Bertram (1973).

Bygott also observed that cub survival increased dramatically as food became more available over several years, and that this, in turn, led to dramatic increases in the abundance of lions. Unfortunately, no such information about time dynamics of hyenas is available.

Thus, we believe that there is a qualitative relationship between food availability and cub survival. However, the quantitative nature of this relationship remains uncertain. To simulate the increase of lions on the plains as the prey numbers increased from 1969 to 1976, we used a linear relationship of cub survival to food availability. We assumed that when food availability was low (1000 kg/km²), cub mortality was high (15 percent per month); and when food available was high (10,000 kg/km²), cub mortality was low (5 percent per month). There is no evidence that predators increased in the woodlands, but data are insufficient, and major changes could have occurred without being detected.

Population dynamics and feeding data were not as complete for hyenas as for lions. We have assumed that those for hyenas are the same as for lions, but have counted each hyena as two-fifths of a lion.

Alternative Prey

Wildebeest constitute the largest proportion of predator diet, 56 percent for Serengeti lions on the plains (Schaller 1972) and 75 percent for Ngorongoro hyenas (Kruuk 1972). However, in the Serengeti, alternative prey species become important for predators when the migratory species are not present. Table 12.1 presents estimates from various sources of the number of nonmigratory prey present in each of the three areas of our model. It appears that most of the prey species on the plains have increased, while only topi and impala have increased in the west, and none of the species have increased in the center and north (appendix A).

Model Results and Discussion

Rain-Grass-Wildebeest Interaction

The eventual equilibrium population size of wildebeest depends primarily on the dry-season rainfall. Since it takes several decades to reach an equilibrium, it is unlikely that the populations will ever do so before the climate changes. Instead, the population will track the current rainfall up and down. Nevertheless, it is interesting to examine the potential equilib-

Table 12.1

Estimated animal numbers in
Serengeti (Dry Season)

	1960s	1970s
Plains		
Topi	100	1,000
Kongoni	100	1,000
Warthog	0	2,000
Grant's Gazelle	2,500	10,000
Lion (100 kg)	800	1,350
Hyena (40 kg)	2,000	3,400
West		
Topi	20,000	50,000
Kongoni	500	500
Impala	30,000	30,000
Lion (100 kg)	800	800
Hyena (40 kg)	500	500
North and Central		
Topi	10,000	10,000
Kongoni	7,500	7,500
Impala	50,000	50,000
Lion (100 kg)	800	800
Hyena (40 kg)	500	500

rium as a function of rainfall (fig. 12.6). One should note that the
equilibria are large, and that, even with the low dry-season rainfall observed in the 1960s, the wildebeest population will be about the same as in
1977 (1.4 million). Thus, a return to 1960s rainfall levels would possibly
not lead to a catastrophic decline at 1977 levels.

Figure 12.7 shows a twenty-year projection of wildebeest numbers, the
solid line indicating increasing numbers with constant 250 mm dry-season
rain. The broken line represents numbers with a drought of 125 mm dry-
season rain beginning in the sixteenth year, the population reaching 2.5
million, and then declining rapidly to 500,000 due to high juvenile mortality. The food per adult never became less than 2 kg/day, so there was
never any additional adult mortality. This, in turn, prevents a possible
undershooting of the total population below its equilibrium for the amount
of rainfall. It seems likely that if there was a sixty-day period without

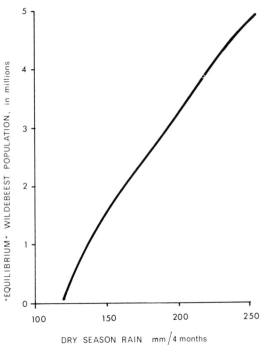

Figure 12.6   The relationship between a theoretical wildebeest equilibrium population size and dry-season rainfall.

rainfall, severe adult mortality would result, but such periods are infrequent (Schaller 1972, table 1; Pennycuick and Norton-Griffiths 1976).

Alternative Prey

In the predator-prey interaction, the wildebeest are the most preferred prey (highest a value, equation 2) and are also by far the most abundant. Therefore, if wildebeest are present in an area, they constitute most of the kills, as was found by Schaller and Kruuk. Since those studies, wildebeest numbers have increased greatly, so they should now constitute an even higher percentage of the kills. Under the assumption of our model of fixed migration timing, we find that the total predation on alternative prey in

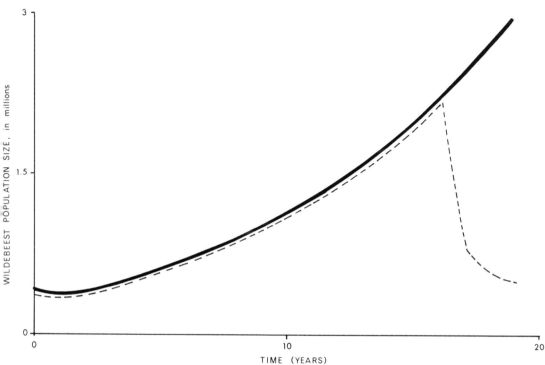

Figure 12.7

Projection of wildebeest numbers
for twenty years, given 250 mm
dry-season rainfall (*solid line*) and
a drought of 125 mm dry-season
rain from years 16–20 (*broken
line*).

the woodlands should have increased gradually as the predators increased.
This is due only to increased predator numbers, leading to increased predation in the wet season. Figure 12.8 shows the number of topi eaten by
predators during the wildebeest increase and decline depicted in figure
12.7. Figure 12.8 also shows a predicted increase in the number of adult
predators in the north. This is in disagreement with the estimates in
table 12.1.

   We are unhappy with this result. Our conclusions depend upon two
assumptions: (1) that the migration timing is independent of rainfall
and wildebeest numbers, and (2) that the migratory prey are spread
uniformly within the central and north in the dry season. Maddock (chap.
5) has shown both these assumptions to be false, but a realistic model of

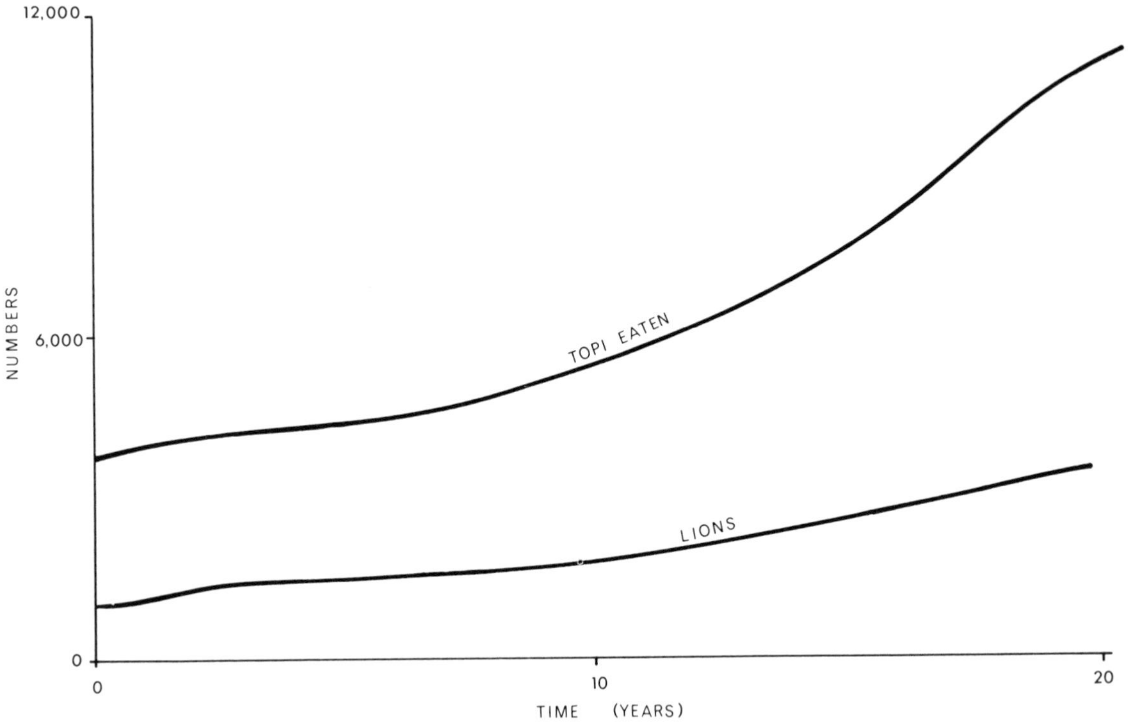

Figure 12.8   Number of topi eaten and number
of predators present in the north-
central area for a twenty-year
projection of 250 mm dry-season
rainfall (years 1–15) and 125 mm
rainfall for years 16–20.

the migration in relation to weather would require substantial data that
may not exist. At present, we place little confidence in our results regarding
the impact of the wildebeest increase upon the alternative prey.

Wildebeest-Predator Interaction

Field studies (chaps. 9, 10) show that predators have little effect upon
wildebeest in the Serengeti. However, with resident prey increase has come
a predator increase, at least in the plains. Also, there is the potential for
a rapid drop in wildebeest numbers in the event of a severe dry season. To
examine the interaction between predation, rain, and wildebeest numbers
and to clarify the dominant processes, we constructed a simplified model

of this interaction. This model assumes that predators and alternative prey
are constant, and wildebeest can be represented by a single age class for
simplicity's sake. Alternative prey consists of a single type with predator-
prey parameters similar to those of zebra. The equations of this model can
be written as follows:

$$\text{Food per individual wildebeest} = \frac{\dfrac{\text{Food}}{\text{ha}} \times \text{available area}}{\text{number of wildebeest}}$$

Food/ha $= +200 + (2 \times \text{rainfall})$ (regression from Sinclair 1975)
Available area $=$ approx. 1,000,000 ha

$$\text{Calf survival rate} = \frac{\text{Food} \times 0.5}{75 + \text{Food}} \qquad (\text{Curve fit to fig. 12.2})$$

Wildebeest next year $=$ wildebeest surviving $+$ calves produced and
surviving

Wildebeest surviving $=$ Wildebeest $\times 0.95$
Calves produced $=$ Wildebeest $\times 0.5$
Calves surviving $=$ calves produced $\times$ calf survival rate
Wildebeest eaten $=$ number of predators $\times$ wildebeest eaten per predator

$$\text{Wildebeest eaten/predator} = \frac{a_1 \dfrac{\text{Wildebeest}}{\text{ha}}}{1 + (a_1 \times h_1 \times \text{Widlebeest/ha}) + (a_2 \times h_2 \times \text{alternative prey/ha})} \quad (\text{from equation 3})$$

$$\text{Wildebeest eaten/predator} = \frac{317 \times \dfrac{\text{Wildebeest}}{\text{ha}}}{1 + (.05 \times 317 \times \text{Wildebeest/ha}) + (.08 \times 100 \times \text{alternative prey/ha})} \quad (\text{from table 12.1})$$

From this simplified model, we can describe the major interactions.
Figure 12.9 plots the wildebeest population size in a given year against the
population size the previous year for a dry-season rainfall of 250 mm.
The 45-degree line represents exact replacement of population. When
the wildbeest curve (known as the recruitment curve) is above the 45-
degree line, the population increases; when below, it decreases. The upper
crossover at approximately 7 million represents the upper equilibrium,
which is higher than that in figure 12.6, due to the simplifications of the

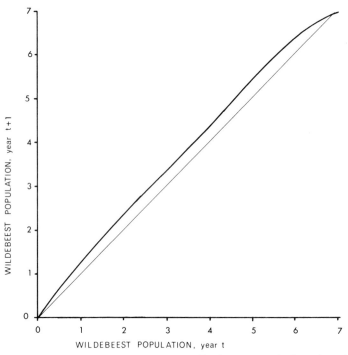

Figure 12.9   Wildebeest population size plotted
against population size of the
previous year for years with 250
mm dry-season rainfall. Results
from simplified model.

model. These results should be interpreted in a general sense, for there is
no reason to believe that exact quantitative predictions will be accurate.

In figure 12.10, the ratio of the population size in a given year to that in
the previous year $\left(\dfrac{N_{t+1}}{N_t} \right)$ is plotted against the population size N_t to
emphasize the biological relationships. The intersection of the curves and
the 1.0 line in figure 12.10 is the same as that with the 45-degree line in
figure 12.9. When the population is very low, it will decline due to predation pressure. Predators are efficient at capturing wildebeest, even when
the latter are very scarce, as was observed at Nairobi Park (Foster and
Kearney 1967). Below a certain density, the predators can capture them

Figure 12.10   Rate of change of wildebeest
                                         population plotted against wilde-
                                         beest population size for two
                                         dry-season rainfall regimes.

at a higher rate than they can reproduce. This assumes that the predators
and alternative prey numbers remain constant, which is somewhat unlikely. However, predators did not decrease in Nairobi Park, so we believe
that this figure represents reasonably accurately potential behavior over
several years.

Figure 12.10 shows that high dry-season rainfall leads to more productive wildebeest populations, so that the lower crossover gets lower
and the higher crossover gets higher.

That the lower crossover is due to predation pressure becoming greater
than reproductive rate of wildebeest can be seen in figure 12.11. We have
plotted the recruitment curve resulting from half the present number of
predators, and it can be seen that the crossover becomes much lower.

The Effect of Rinderpest

Sinclair (chap. 4) has hypothesized that wildebeest were held at much
lower densities in the past because of the additional juvenile mortality due
to rinderpest. We simulated this possibility by artificially doubling the
present juvenile mortality rate to represent the effect of rinderpest. This
generates the recruitment curve labeled *rinderpest* in figure 12.11. The

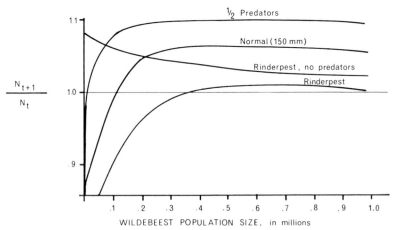

$$\frac{N_{t+1}}{N_t}$$

WILDEBEEST POPULATION SIZE, in millions

Figure 12.11   Low-density recruitment curve for wildebeest population under four assumptions: (1) normal predation conditions and 150 mm dry-season rainfall; (2) predator numbers reduced to half; (3) rinderpest simulated by doubling calf mortality; and (4) rinderpest included, but predators eliminated.

doubled juvenile mortality greatly lowers the overall recruitment curve and dramatically increases the point of crossover. In fact, the crossover is at approximately 400,000 wildebeest, which is twice the number of wildebeest that occurred in the Serengeti in the rinderpest era. It is tempting to say that rinderpest was responsible for holding down the wildebeest population, but we should be cautious. First, most of the animals dying of rinderpest would have provided food as easy kills or scavenge for the predators in the north, so there would have been some compensatory effect: calves dying would have reduced the predation pressure on healthy animals.

Second, we mentioned earlier that this simplified model is dependent upon a number of assumptions that may be in error. We could thus easily believe that the crossover at 400,000 wildebeest could be at 100,000 or less. However, we do believe that such a crossover exists, at least on a short time scale. The predator numerical response would greatly affect the long-term dynamics. A third objection to believing that this explains the "rinderpest story" is that our model, taken literally, says the wildebeest go extinct below the crossover, yet the wildebeest did manage to survive in spite of rinderpest.

However, we feel that several conclusions can be made. In the presence of rinderpest, wildebeest numbers were much lower than their absolute limit determined by food; when rinderpest was removed, they increased rapidly. Therefore, it does seem probable that wildebeest were limited by a combination of predation, food, and rinderpest mortality.

Conclusion: What Is Missing?

Perhaps the most useful product of a modeling analysis is the perspective it provides on what type of data would be most useful. We have found that an understanding of the predator numerical responses to changing prey abundance is most important to the predator-prey dynamics, but these data are largely unknown. Apparently less important is the food partitioning among various ungulates, which Bell (1970) has postulated as being the key influence in the interaction between the ungulates and their food.

For each major component of our model, we found one element that seemed to need further work. In the wildebeest dynamics, it was the temporal distribution of wildebeest in relation to predator territories and alternative prey. For predators, the weakest areas seem to be the relationships between prey availability, territory size, and infant survival. We have assumed that infant survival is a function of prey availability, yet we know that predators have persisted in situations with substantially fewer prey than currently available. It seems probable that infant survival is related to prey available per territory, and that territories would expand in low prey situations. Perhaps a comparative study of many lion prides in greatly different prey regimes would be most useful.

We feel confident that the functional response of the predators (at least for lions) is sufficiently understood to make some predictions about predator-prey interactions. It is clear from our model that the timing, rather than amount of prey available, is the most important factor in the Serengeti in determining predator consumption. This information may be available, but more analysis is needed.

The dynamics of alternative prey were ignored in our model, due to a lack of information. We have nevertheless been able to make some tentative statements about the effect of the wildebeest increase on these other populations. The most important information that must be obtained in the future is that on population dynamics of alternative prey species, particularly impala, topi, and kongoni.

In summary, we can say that,

1. In the past, wildebeest were probably held at low numbers by a combination of rinderpest mortality and predation.
2. With the elimination of rinderpest as a mortality agent, the wildebeest were able to escape the predation pressure at low densities and greatly increase in numbers; they will apparently increase up to limits determined by food shortages acting on calf survival.
3. If a serious drought occurs in the Serengeti, the wildebeest population may simply stop increasing, or may decrease, depending upon the severity of the drought. If rainfall reverts to its 1960–70 level, the wildebeest population will most likely stabilize around 1.0–1.5 million animals.
4. The data were insufficient to make any firm statements about the effect of the wildebeest increase on the alternative prey.

We wish to thank Neil Gilbert for the very useful comments he made on an early draft of this chapter.

References

Bell, R. H. V. 1970. The use of the herb layer by grazing ungulates in the Serengeti. In *Animal populations in relation to their food resources,* ed. A. Watson, pp. 111–23. Oxford: Blackwell.

Bertram, B. C. R. 1973. Lion population regulation. *E. Afr. Wildl. J.* 11:215–25.

Bertram, B. C. R. 1975. Social factors influencing reproduction in wild lions. *J. Zool.* (Lond.) 177:463–82.

Charnov, E. L. 1973. Optimal foraging: some theoretical explorations. Ph.D. dissertation, University of Washington.

Dradu, E. A. A., and Harrington, G. N. 1972. Seasonal crude protein obtained from a tropical range pasture using oesophageal fistulated steers. *Trop. Agric.* (Trinidad) 49(1):15–21.

Elliott, J. P.; McTaggart Cowan, I.; and Holling, C. S. 1977. Prey capture by the African lion. *Can. J. Zool.* 55:1811–28.

Foster, J. B., and Kearney, D. 1967. Nairobi National Park game census, 1966. *E. Afr. Wildl. J.* 5:112–20.

Foster, J. B., and McLaughlin, R. 1968. Nairobi National Park game census, 1967. *E. Afr. Wildl. J.* 6:152–54.

Haber, G. C. 1977. Socio-ecological dynamics of wolves and prey in a subarctic ecosystem. Ph.D. dissertation, University of British Columbia.

Holling, C. S. 1959. The components of predation, as revealed by a study of small mammal predation of the European pine sawfly. *Can. Ent.* 91: 293–320.

———. 1965. The functional response of predators to prey density and its role in mimicry and population regulation. *Mem. Ent. Soc. Can.* 45: 1–60.

———. 1966. The functional response of invertebrate predators to prey density. *Mem. Ent. Soc. Can.* 48:1–86.

Ivlev, V. S. 1961. *Experimental ecology of the feeding of fishes.* New Haven, Conn.: Yale University Press.

Kruuk, H. 1972. *The spotted hyena.* Chicago: Univ. of Chicago Press.

Pennycuick, L., and Norton-Griffiths, M. 1976. Fluctuations in the rainfall of the Serengeti ecosystem, Tanzania. *J. Biogeog.* 3:125–40.

Schaller, G. B. 1972. *The Serengeti lion.* Chicago: Univ. of Chicago Press.

Sinclair, A. R. E. 1975. The resource limitation of trophic levels in tropical grassland ecosystems. *J. Anim. Ecol.* 44:497–520.

———. 1977. *The African buffalo.* Chicago: Univ. of Chicago Press.

Starfield, A. M.; Smuts, G. L.; and Shiell, J. D. 1976. A simple wildebeest population model and its application. *S. Afr. J. Wildl. Res.* 6:95–98.

M. Norton-Griffiths

Thirteen The Influence of Grazing,
Browsing, and Fire on the
Vegetation Dynamics of
the Serengeti

Within the National Parks, Game Reserves, and similar conservation areas of Eastern, Central, and Southern Africa, the course of plant succession is being halted and reversed, and there is a general trend for woodland vegetation types to revert to more open, grassland forms (for example, in Tanzania: Croze 1974a, b; Douglas-Hamilton 1972; Vesey-Fitzgerald 1973; in Kenya: Corfield 1973; Leuthold and Sale 1973; Western and van Praet 1973; in Uganda: Laws 1970; Laws et al. 1975; in Zambia: Caughley 1976). In the recent past, this trend has been viewed with alarm, for the preservationist outlook did not appreciate that these areas are (semi-) natural ecosystems in which change is an integral part. Even today, change of any sort is still viewed as undesirable, and the response of most wildlife biologists and conservationists alike is to apportion "blame," especially to elephants, sometimes to fire, and occasionally to both.

"Elephant problems," as these reversals of plant succession have come to be called, have a certain sameness about them. They appeared simultaneously in the early sixties in many different types of habitat. In most cases, a "sudden and devastating" increase in woodland "destruction" was reported, usually first from around the main tourist areas, where much capital had been invested in lodges and tourist circuits. Two consistent characteristics of these reports were a lack of firm, quantitative information on what exactly was happening, and dire and somewhat silly predictions of total deforestation and even desertification.

The "elephant problem" in the Serengeti National Park, with which this chapter is concerned, developed along the now classical lines. Stands of large, overmature trees of the greatest esthetic value, situated in and around the best tourist areas, suddenly fell prey to marauding elephant. The role of fire in preventing recruitment to the tree populations was soon recognized, and fire control was implemented; and despite much evidence about their minor role in reducing tree cover, a number of elephant were shot. The fact that some 50 percent of the woodlands in the north of the Serengeti National Park had meanwhile vanished seemed to have escaped everyone's notice.

Croze (1974a, b), in his analysis of the Seronera elephant problem, pointed out that this crises approach lead to misleading conclusions. The careful approach of Western and van Praet (1973) to the woodlands of the Amboseli National Park underlines this. Here, what was apparently the most clear-cut example of elephants' destroying a woodland turned out to be a matter of rising water tables. Of equal interest was the time elapsing between the establishment of these woodlands and their final senescence, which was on the order of sixty to seventy years.

The main difficulty in interpreting the ecological significance of any of these "elephant problems" has been the lack of long-term data. It has not been possible to compare contemporary events against long-term trends, or local patterns against overall patterns; and without such macroscale information in both time and space, uninformed opinion has clashed, inevitably, with uninformed opinion.

In the Serengeti, long-term data on many aspects of vegetation change was routinely collected through the Serengeti Ecological Monitoring Programme, and an analysis of the interactions between the agents of vegetation change (namely browsing, fire, grazing, and climate) can be attempted. I will show that the vegetation has been undergoing a rapid shift toward a more open grassland phase; but the relevance of this must be seen against the perspective of an ecosystem in a state of perturbation and readjustment following the eradication of rinderpest (chap. 1), the subsequent eruption of ungulate populations (chap. 4), and a consequent change in fire regime; and a simultaneous expansion of range by elephant. Neither the ecological significance of the vegetation changes, nor their implications for the management of the Serengeti, can be judged without this macroscale picture of trends and interactions within the ecosystem.

Methods

The ecosystem is described in chapter 2. For present purposes, I consider three areas: the north, the center and west, and the plains. I refer to the grasslands and woodlands of the Serengeti park alone, for the most important data were derived from airphoto analysis, and the quality and extent of coverage outside the park boundaries was not adequate. Neither were there reliable data on the frequency and extent of fires. Important areas have, therefore, been omitted, including the Maswa Game Reserve to the southwest of the Serengeti, the Grumeti Controlled Area to the northwest, the Loliondo Controlled Area to the northeast, and the Masai Mara Park in Kenya, north of the Serengeti (see fig. 1.1).

Photointerpretation

In January 1972, a large section of northern Tanzania, including most of the Serengeti ecosystem, was photographed at an average scale of 1:60,000 as part of a government survey and mapping project. The cover density (canopy cover) of the woodlands, measured from these photographs, was compared against that obtained from photographs dating back some fourteen to twenty years. Direct tree counts, which have been used in similar studies (Buechner and Dawkins 1961; Western and van Praet 1973; Harrington and Ross 1974; Croze 1974b), could not be used here because of the very small scale of the 1972 photography.

All the photocenters of the 1972 photography that fell within the woodlands of the park were treated as sample plots at which the cover density was measured by means of a dot-grid. The grid was a 10 x 10 regular array of black dots, 0.075 mm in diameter and 1 mm apart, on a transparent acetate sheet. Cover density was measured by counting, under a x 8 magnifying stereoscope, the number of dots touching or covering woody vegetation. Five counts were made at each sample plot (photocenter), with the grid centered over it each time, but oriented randomly.

The 1972 sample plots were exactly located on earlier sets of small-scale photography which were of varied dates and scales. The majority of the plots (53%) could be located on the 1958 photography (fourteen years earlier) which had a mean scale of 1:35,000. The remainder were located on photography dating back four years (4%), seven years (6%), ten years (3%), fifteen years (25%), and nineteen years (21%). Five counts were again made at each sample plot.

Differences in scale between the 1972 and the earlier photography were adjusted exactly by making a wide range of dot-grids from the same template. The size and spacing of the dots relative to the ground were, thus, the same in all photographs. In effect, each sample plot was 45 ha with one hundred "dots" of 5 m diameter spaced 69 m apart.

Photography of scale 1:60,000 is by no means ideal for this type of work; scales of 1:30,000 or 1:15,000 are preferable. The cover density is therefore undoubtedly overestimated, even though the dot-grid had been chosen to give the best results possible with this scale. The bias will, however, be the same for both sets of photography, so estimates of the magnitude and direction of change should be unbiased.

The two sets of five counts on each sample plot, five 1972 counts and five pre-1972 counts, were transformed to arcsines for all statistical evaluations. The absolute rate of change at a plot was found by dividing the difference between the two means by the number of years separating them, yielding the percentage of tree cover change per year. Relative change, however, was measured over the ten-year period 1962 to 1972; the change was expressed as a percentage of the 1962 value, that is, it is not a measure of cover but a measure of change. This standardized all the plots to a common time scale. The implicit assumption that the rate of change was constant seems to be upheld by the little data available (table 13.2).

In 1972, large-scale (1:15,000) photographic sampling was carried out on a restricted basis in the northern and central woodlands using the Serengeti Research Institute aircraft and cameras (Super Cub, F 24 camera). The large scale enabled an analysis within different woodland vegetation types, but the highly restricted nature of the sampling made it impossible to use this photography to correct the bias in the interpretation of the smaller-scale, 1972 photography. A few transects were also photographed in the north in 1967, giving three measurements of cover density between 1958 and 1972.

Fire

All fires within the park have been mapped from the air on a more or less systematic basis since 1963, with the exception of 1965, for which complete data are unavailable. This mapping has been done regularly once a month during the burning season since 1969, and all fires within the park are now mapped accurately at a scale of 1:250,000.

The frequency of burning in different areas of the park was found by counting the number of times that the center point of each 10 x 10 km grid square had burned between 1963 and 1972. Figure 13.3 shows a trend surface fitted to these frequency counts. Fires occur with low frequency on the plains compared with the northern stratum.

The area burned each year in each stratum was measured with a planimeter. These areas were expressed as km^2 in all statistical manipulations, although they are expressed as a percentage in the figures.

Elephant Utilization

Between August 1969 and July 1972, a series of twenty-eight survey flights were made across 32,000 km² of the Serengeti ecosystem (Norton-Griffiths 1973; Pennycuick 1975; Norton-Griffiths, Herlocker and Pennycuick 1975). The densities of some thirteen species of large mammal were recorded in each of 320 grid squares of 10 x 10 km, the flight lines being completely systematic. Densities were usually estimated on a log scale, but for some species, including elephants, the total number seen in each square was also recorded. The annual, wet-season and dry-season indices of elephant density were calculated from these records (table 13.3). I assume that temporal and spatial variations in these indices reflect changing patterns of utilization by elephant.

Rainfall and Climate

Norton-Griffiths, Herlocker, and Pennycuick (1975) give the details of rainfall and climate within the Serengeti. The source data were monthly rainfall totals from sixty-four storage gauges within the ecosystem.

Results

Photointerpretation

The woodlands of the park have undergone an overall decrease in cover density (table 13.1). The mean absolute and relative rates of change are −0.47 percent and −1.8 percent per annum, respectively. Some 13 percent of the cover density has been lost between 1962 and 1972.

However, the northern dry subhumid woodlands are changing some three times as fast as the semiarid *Acacia* savannas of the central stratum. The north lost 26 percent of its cover density over the decade, compared with 7 percent in the central woodlands.

The spatial patterning of the change also differs between the two woodland types. The northern areas (fig. 13.1) show a blanket change, with almost all sample plots showing a decrease. The central woodlands are strikingly different; there is a mosaic of change, with some plots showing increase and some decrease. The absolute rate of change increases toward the north (fig.13.1) and the relative change shows a similar pattern (fig. 13.2), with the most northern areas losing up to 50 percent of their cover density.

Table 13.1   Results from the dot counts of the Serengeti woodlands

	% cover density		
	North	Central	Overall
1962	33	39	37
1972	24	36	32
Absolute rate of change (% cover/year)			
per annum	−0.86	−0.28	−0.47
s.e.	0.027	0.039	0.035
d*	32	7.2	13.3
N	39	88	127
P	<0.001	<0.001	<0.001
Relative rate of change (as percent of 1962 value)			
Per decade	−26	−7	−13
Per annum	−3.13	−0.8	−1.44

* paired comparisons test

The larger-scale photography was used to study the rates of change within the main vegetation types of the northern woodlands. Sample plots were chosen within riverine forest, bush thicket, and open *Acacia* or *Terminalia* woodlands, the major components of the vegetation there. The rates of change were fastest in thickets and slowest in riverine forest (table 13.2), and they seem to have remained fairly constant between 1958 and 1972, at least in the woodlands. This is an important point, for there was a marked immigration of elephant into this area in the mid-sixties (see below).

Croze (pers. comm.) took similar scale photography within a small part of the central *Acacia* woodlands. Plots in *Acacia/Balanites/Albizia* wooded grasslands showed a decrease in cover density from 13 percent to 10 percent over an eighteen-year period. In contrast, plots within *A. clavigera* and *A. xanthophloea* woodland nearby showed an *increase* in cover density of 7 percent and 17 percent, respectively, over the same time period.

These admittedly restricted samples reinforce the initial impression of a mosaic of change in the *Acacia* woodlands and a blanket decrease in the north.

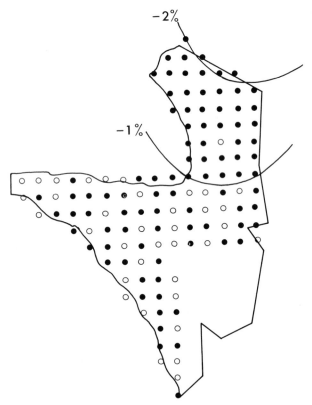

Figure 13.1

Grid squares in which there has been an increase (*open circles*) or decrease (*solid circles*) in cover density between 1962 and 1972. Contours show mean annual absolute change in cover density (drawn from surface of best fit [cubic], accounting for 35% of the variation). Note the blanket decrease in the northern areas, compared with the mosaic of change in the central woodlands.

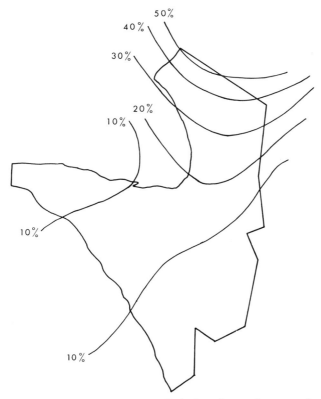

Figure 13.2

Relative change in cover density
between 1962 and 1972. Contours
drawn from surface of best fit
(*cubic*) accounting for 35% of the
variation. The northern areas
have lost up to 50% of their cover
density.

Multiple Regression Analysis

Multiple regression analysis was used to investigate the association between the direction and extent of change measured at the photographic
sample plots and a number of ecological variables. The six variables

Table 13.2   Results of dot counts within
                                                    different vegetation types in the
                                                    Serengeti's northern woodlands

| | | % cover density | | |
		Woodland	Bush thicket	Riverine forest
1958	n	370	121	43
	x	23%	78%	94%
	s.e.	1.6%	7%	3.7%
1967	n	63		
	x	16%		
	s.e.	1.3%		
1972	n	370	121	43
	x	11%	58%	88%
	s.e.	0.9%	7.1%	2.2%
Absolute change per annum				
1958–1967		−0.78% (s.e. 0.11)		
1967–1972		−1.0% (s.e. 0.13)		
1958–1972		−0.86% (s.e. 0.09)	−1.43%	−0.47%

chosen for this regression analysis are described in table 13.3, and all of them were estimated on the basis of 10 x 10 km grid squares. Accordingly, the absolute rates of change had to be converted into a similar format. This was done by averaging the values from sample plots falling within the same grid square. The dependent variable was, therefore, the average absolute rate of change at the sample plots falling in each of the 127 grid squares that make up the woodlands of the park.

When the whole data set was analyzed, the only variables significantly associated with changes in cover density were rainfall, climate, and the dry-season elephant index (table 13.4). This merely restates what has already been observed, namely, that change is more extensive in the higher rainfall, dry subhumid woodlands of the north.

The analysis was therefore repeated on two subsets of data, those squares falling in the central woodlands and those in the northern woodlands. This effectively stratified out the influence of rainfall and climate, so these variables were dropped from the analysis. The results (table 13.4) were considerably more revealing: in the central woodlands, where a mosaic of change was observed, two elephant indices followed by fire were the variables most closely associated with decreases in cover density, the

Table 13.3   Variables used in the multiple regression analysis. Each variable refers to values within individual grid squares of 10 x 10 km²

Y	mean absolute rate of change in cover density
$X1$	mean annual rainfall, from Norton-Griffiths et al. (1975, fig. 2a)
$X2$	Thornthwaite's climatic index, from Norton-Griffiths et al. (1975, fig. 4)
$X3$	number of times the sample plot had burned between 1963–1974 (fig. 13.3 below)
$X4$	annual elephant index; mean number of elephants seen per flight (fig. 13.4 below)
$X5$	wet-season elephant index; mean number of elephants seen per wet-season flight (November-May, fig. 13.5 below)
$X6$	dry-season elephant index; mean number of elephants seen per dry-season flight (June-October, fig. 13.6 below)

elephant indices being by far the most important. In contrast, in the dry subhumid woodlands of the north, where there has been a blanket decrease in cover density, fire was the most important variable, accounting for 44 percent of the variation in changes in cover density. The wet-season elephant index was the only other variable significantly associated with changes in cover density.

Influences of Elephants and Fire

Elephants

Croze (1974a, b) agrees with many other authors that elephants are primarily grass eaters (Laws 1970; Field 1971; Wing and Buss 1970; Wyatt and Eltringham 1974). He demonstrates further that in the Seren-

Table 13.4

Results of multiple regression analysis. R^2 gives the proportion of the total variance accounted for by the best multiple regression equation; the proportion explained by each significantly associated variable is expressed as a percentage.

	Overall	North	Central
1st variable	X2 (climate)	X3 (fire)	X6 (dry season—elephants)
	21%	44%	26%
2d variable	X1 (rainfall)	X5 (wet season—elephants	X5 (wet season—elephants)
	17%	23%	19%
3d variable	X6 (dry season—elephants)		X3 (fire)
	10%		13%
R^2	0.48	0.67	0.58
N	127	96	31

geti the elephants were utilizing the woody vegetation largely in proportion to its composition, there being relatively little selection for either species or size classes, with the exception that very small trees of less than one meter in height were rarely eaten. Nonetheless the browsing pressure was by no means spread evenly across the vegetation. High-impact utilization occurred locally, and up to 6 percent of the mature trees could be removed from a feeding site, compared with an overall mortality rate from all causes of 2.7 percent per year.

Fire

Herlocker (pers. comm.) considers that the most important influence of fire is the inhibition, and even complete suppression of tree recruitment, for young trees of less than 3 m in height are very susceptible to fire damage. For example, in one large sample from a mixed *Acacia* woodland that had been burned, Herlocker found that 92 percent of trees less than one meter in height had been burned back to ground level. This mortality rate had dropped to 68 percent for trees of 1–2 m; to 28 percent for trees

of 2–3 m; and to 1 percent for trees of 3–4 m. Fire has a negligible effect
on trees higher than this, and Herlocker considers that trees of over 3 m
in height are effectively safe from the effects of fire and can therefore be
considered recruited into the mature tree population.

Results similar to these have been obtained from *Combretum/
Terminalia* woodland in the northern Serengeti (pers. obs.). In a sample
of 273 trees, fire had burned back 94 percent of those less than 1 m in
height, 68 percent of those between 1 and 2 m in height, and 45 percent
of those between 2 and 3 m in height.

Fire can also completely destroy some of the woodland vegetation types,
especially in the north, where the bush thickets are particularly susceptible
(pls. 44, 45, table 13.5). Fires burn back the edge regeneration, penetrate
the thicket, and burn back the vegetation within the thicket. These effects
are more marked once the thicket has been opened up (by previous fires,
or by the activities of elephant, buffalo, and rhinoceros), and hot fires
late in the dry season are more damaging than the early, cooler fires. These
bush thickets used to make up a large proportion of the vegetation of the
northern woodlands. The rapid rates of change in these woodlands may
therefore be associated with the extreme vulnerability of these thickets to
fire. And fire in the north has been more frequent than in other areas of
the park (fig. 13.3).

Table 13.5	Effect of early (cool) and late (hot) burns on the evergreen/ semievergreen bush thickets in the Serengeti's northern woodlands.	
	Early Burn	Late Burn
Number of thickets sampled	6	3
Number of sample points	85	51
Significant damage to edge regeneration	64%	98%
Penetration by fire		
Main body of thicket	0.5m	4m
Previously affected areas*	12m	19m
Significant internal damage		
Main body of thicket	6%	50%
Previously affected areas*	52%	85%

* previously affected and opened up by
fire

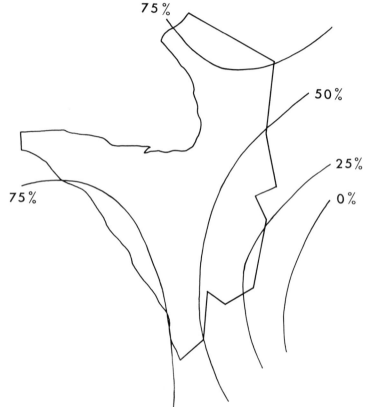

Figure 13.3

Frequency of burning between 1962 and 1972. Contours from quadratic surface of best fit (66% of the variation explained).

Combined Influences

The data presented here are by no means comprehensive or complete. Nevertheless, they represent the impressions and results of many workers in the Serengeti. Elephants are undoubtedly utilizing and removing large, mature trees from the canopy (pl. 43), which, after all, is what elephants have evolved to do. But fire is having a more pernicious effect, in that it is effectively holding back the recruitment of trees into the mature age classes. The young trees are not necessarily killed, but they are prevented from replacing those lost to elephants and other causes. The end result of fire is the production of a woodland mosaic, each patch having an even-aged stand of trees resulting from a chance escape from fire; the published age structures (Croze 1974a, b; Lamprey et al. 1967; Glover 1968; Vesey-Fitzgerald 1972, 1973) demonstrate this well. Such stands are ecologically unstable (Horn 1974).

Elephants are undoubtedly interacting synergistically with fire in the northern woodlands (Buechner and Dawkins 1961), for, by opening up thickets, they encourage the growth of grass within them, which further fuels the fires (table 13.5). However, in the central woodlands, it would appear that it is the local, high-impact utilization by elephants that sets the observed mosaic of change; fire merely tips the balance so that the overall trend is toward a decrease in cover density.

Dynamics of the Elephant Populations

There are two populations of Serengeti elephants, a northern one and a southern one (figs. 13.4–13.6), which move into the park during the rains and out again during the dry season. These seasonal movements described by Croze (pers. comm.) are of a similar nature and magnitude to those described elsewhere (Leuthold and Sale 1973; Caughley and Goddard 1975; Laws et al. 1975).

Croze considers that these two populations increased in size between 1962 and 1968 as a result of immigration. This phase of immigration seems now to have ceased, and the populations appear stable at a mean density of 0.2 km^{-2} in the north and 0.13 km^{-2} in the south. Immigration was ultimately caused by expansion of agricultural settlement in the areas bordering the Serengeti. Kurji (1976) has documented these rates of expansion from aerial photographs and from population censuses. There was a wave of settlement spreading northeastward toward the Serengeti from Sukumaland (see also Ford 1971) and eastward from Lake Victoria, and the entire western boundary of the park is now heavily settled (Norton-Griffiths et al. 1975). Kurji demonstrates that the population density in these areas has approximately doubled over the last decade. This expansion could have caused the immigration of elephant, especially in the southern parts of the Serengeti, where large areas of former elephant range are now densely settled. The expansion of settlement along the

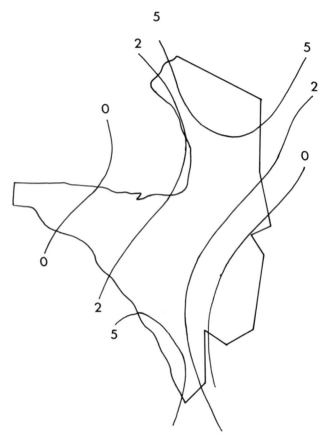

Figure 13.4

Annual elephant index across
the Serengeti. Contours from
surface of best fit (quartic, 38%
explained).

Serengeti's northwestern boundaries, as well as in the northern parts of
Narok District in Kenya (the northern boundary of the Serengeti ecosystem), could also explain the immigration of elephant in the north.

It is too early to say whether these populations will increase through
natural recruitment, or whether another phase of immigration will occur.
There is certainly the potential for further immigration, especially in the
south, where the important dry-season areas of the Maswa Game Reserve

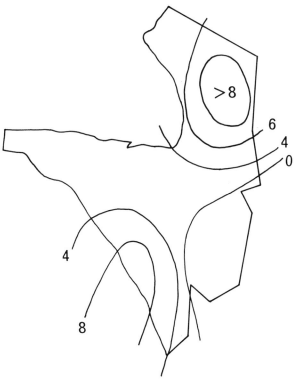

Figure 13.5

Wet-season index of elephant utilization across the Serengeti. Contours from surface of best fit (quintic, 48% explained).

are seriously threatened by illegal settlement. One encouraging trend is that the widespread settlement along the western boundaries of the park is contracting into population centers under a government resettlement program. Ranching schemes are being established in the vacated land, thus forming a potential buffer zone, where the man-wildlife interface will be less acute, around the park. The resettlement program may prevent further inroads into present wildlife habitat. However, these gains may well be

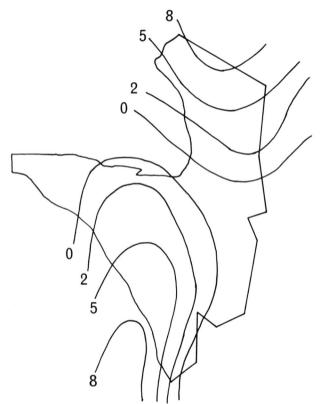

Figure 13.6

Dry-season index of elephant utilization across the Serengeti. Contours from surface of best fit (quintic, 31% explained). Note demarcation of northern and southern populations.

offset by the planned agricultural development of the Loliondo highlands to the northeast of the Serengeti (see fig. 1.1). The elephants there, unless they are eradicated, can only be displaced further into the Serengeti ecosystem, thus setting off another ripple of immigration. Potentially, therefore, further expansion of settlement could increase elephant densities by 50 percent in a very short time.

Dynamics of Fire

Causes of Fire

The Serengeti experiences grass fires rather than canopy fires, and their frequency is high (fig. 13.3). There are no quantitative data on the causes of fire in the Serengeti, although some of them are well known. Lightning starts fires (M. Turner, pers. comm.), although the intense rainstorms that normally accompany lightning make this an infrequent occurrence. Most of the fires are started by people (Buechner and Dawkins 1961). The pastoralists and agriculturalists around the park burn their rangelands annually, and these fires sweep wild into the park; cattle raiders light fires within the park to obscure their tracks; poachers and honey hunters light fires; park wardens light fires to make it easier to apprehend poachers and honey hunters; and wildlife biologists light fires as part of long-term experiments. All in all, the Serengeti is an area of high fire risk.

Fire and Rainfall

Rainfall has a major influence on the timing, frequency, and extent of grass fires in the Serengeti. Firstly, fires only start once the rains are over and the grass begins to dry out (fig. 13.7); July, the driest month of the

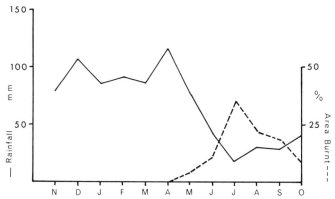

Figure 13.7   The relationship between mean
monthly rainfall (*solid line*) and
the area of the Serengeti burned
(*broken line*). Fires start only
when the rains are over and the
grass begins to dry out.

year, is the month with most burning. Secondly, the strong correlation between the frequency of fires in each grid square (fig. 13.3) and the mean annual rainfall shows that fires are more frequent in the higher rainfall areas ($r = 0.73$, $n = 150$).

There is also a within-year relationship between fire and rainfall, in that the extent of burning in a year is positively associated with the wet-season rainfall (fig. 13.8). The higher the wet-season rainfall, the more extensive are the fires in the following dry season. This relationship holds for the whole park, and for each stratum, with the exception of the north, where at least the direction of the relationship is the same. The dry-season rainfall has a weak negative influence on the extent of fire. Data subsequent to 1972 and not included here show that the negative relationship is stronger than shown.

These results show the relationship between wet-season rainfall, the amount of fuel provided by the growth of grass, and fire. The production of grass, and therefore of fuel, is greater in the higher rainfall areas, and fires are consequently more frequent and more extensive there. The lack of a significant relationship between the wet-season rainfall and the extent of burning in the north may be evidence of a threshold effect. This stratum has the highest rainfall, and the greatest production of grass (table 13.6), which may always provide enough fuel to mask any moderating effect of rainfall.

Trends in Extent of Burning

There has been a marked decrease in the extent of burning throughout the park since 1963 (fig. 13.9). This trend is highly significant within each stratum and for the whole park; it is in the order of −5 percent per year. In view of the close relationship between rainfall and fire, one might assume that there has been some major change in rainfall amounts or distribution. This is not the case. Trends in rainfall are apparent, but none are significant (Pennycuick and Norton-Griffiths 1976). Wet-season rainfall shows a slight negative trend, and dry-season rainfall shows a slight positive trend up to 1972. A multiple regression analysis between the decrease in fire and the trends in wet- and dry-season rainfall indicates that, on a parkwide basis, only 23 percent of the change in the extent of fire can be attributed to them. In the three strata, the proportions range between 0 percent and 20 percent.

It is unlikely that there has been a change in the agents setting fires. Pastoralists, agriculturalists, poachers, and cattle raiders are all as active as

Wet Season

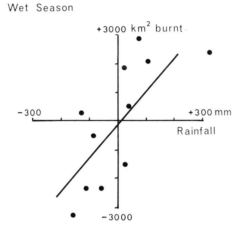

Dry Season

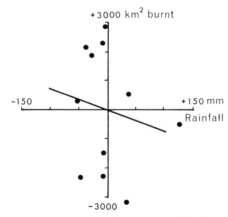

Figure 13.8

The relationship between wet- and
dry-season rainfall and the area of
the Serengeti burned that same
year. Data have been detrended,
and the points represent deviations
from the fitted trend lines.

Table 13.6   Changes to the biomass and pri-
                                    mary consumption of large
                                    mammal grazers, 1962–72, within
                                    the park as a whole and within
                                    each stratum

	North	Central	Plains	Park
Biomass of large mammal grazers $(\text{kg} \cdot \text{ha}^{-1})$				
Buffalo, 1962	6.01	3.02	—	3.01
Buffalo, 1972	32.27	16.25	—	16.18
Wildebeest, 1962	4.56	6.83	13.67	8.20
Wildebeest, 1972	20.66	25.83	43.30	29.52
All other grazers*	33.46	30.76	33.50	32.41
Total biomass, 1962	44.03	40.61	47.17	43.62
Total biomass, 1972	86.39	72.84	76.80	78.11
% increase, 1962–72	96%	79%	63%	79%
Primary production $\text{kg}(\text{ha} \cdot \text{d})^{-1}$	19.07	16.38	12.88	13.82
Consumption $\text{kg}(\text{ha} \cdot \text{d})^{-1}$				
Large mammals, 1962	1.43	1.31	2.98	1.89
Large mammals, 1972	2.92	2.45	5.02	3.36
% increase, 1962–72	104%	87%	68%	77%
Small mammals and insects*	1.44	1.44	0.54	1.17
Total consumption, 1962	2.87	2.75	3.52	3.06
Total consumption, 1972	4.36	3.89	5.56	4.53
% increase, 1962–72	52%	42%	58%	48%
% burned annually (km²)				
1962	91%	90%	56%	86%
1972	48%	42%	15%	38%
% decrease 1962–72	−47%	−53%	−73%	−56%

* values assumed constant between
1962 and 1972

they ever were; and while it is true that the park wardens no longer set fires as frequently and systematically as they used to, fire control, as such, is practiced only in two small areas of the park. A large number of fires still occur. What has changed is that they now go out, instead of blazing unchecked over hundreds of square kilometers. Some other factor is re-

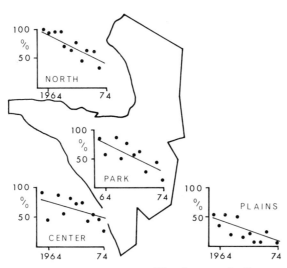

Figure 13.9   The decrease in the area burned
each year between 1962 and 1974.

sponsible for reducing the fuel supply to a level below that at which wild
fires can be maintained.

*Eruptions of Wildebeest and
Buffalo*

The Serengeti's populations of wildebeest and buffalo have erupted over
the last decade (chap. 4). Buffalo have increased from some 20,000 to
their present 70,000; wildebeest from some 200,000 to a 1972 level of
850,000, and presently to 1.3 million (Norton-Griffiths 1973). Buffalo are
resident (Sinclair 1977), the herds living year-round in well-defined home
ranges. They are absent from the plains, and their densities in the central
and northern strata are about 4 km^{-2} and 7 km^{-2}, respectively. The increase in density has been of the same order in each stratum. Wildebeest,
on the other hand, are migratory (chap. 5), and they now use more of
their total range than in previous years, especially the northern parts.
While this is undoubtedly a function of increasing population size, it may
also be correlated with the slight downward trend in wet-season rainfall

and the slight upward trend in dry-season rainfall; Maddock demonstrates that in dry years, the northward movement begins earlier and lasts longer than in wet years. The mean density of wildebeest within the plains, central and north in 1972 was 35 km^{-2}, 21 km^{-2} and 17 km^{-2}, respectively.

Sinclair (chap. 4) argues that the exotic virus rinderpest maintained these two populations at well below the carrying capacity of the ecosystem. The populations erupted once the rinderpest was eradicated from northern Tanzania in the early 1960s. The buffalo population stabilized in the face of density-dependent mortality acting through the dry-season food supply (Sinclair 1977), and similar effects appear to be operating in wildebeest, although they are still increasing (chap. 4).

Other factors may have been involved with this rapid increase in density. More grassland has been created as a result of the high frequency of burning. The change in the north is of particular significance for wildebeest, for it is here that they find their vital supplies of food during the dry season (Pennycuick 1975; chap. 4 above). The heavy burning, again especially in the north, may also have favored the spread of nutritious grasses, such as *Themeda triandra,* at the expense of coarser species, *Hyparrhenia rufa,* for example. Burning may therefore have led directly and indirectly to an increase in primary production (Hadley 1970; Hulbert 1969), while the increase in grazing pressure may in itself have further stimulated primary production (chap. 3). Sinclair (1975) demonstrates the fallacy of equating primary production with food; nonetheless, the changes implied here suggest that the food supply, and therefore the carrying capacity, of these two populations may have increased slightly over the last decade.

Grazing Impact and Fire

The inverse relationship between grazing impact and fire is well documented (Kozlowski and Ahlgren 1974; Daubenmire 1968; Vesey-Fitzgerald 1972; Lock 1967; Lock and Milburn 1971; Olivier and Laurie 1974). Vesey-Fitzgerald (1970), for example, demonstrated that fire always burned through the wooded grasslands that bordered the alkaline flats of Lake Rukwa, Tanzania, except in very dry years, when these grasslands were heavily grazed. Offtake in these years was so high that there was insufficient fuel to maintain fires.

Table 13.6 shows the increase in grazing impact in the Serengeti between 1962 and 1972. The methods and data of Sinclair (1975) were used for these calculations, along with the relevant census results (Norton-Griffiths 1973; Sinclair 1977; Pennycuick 1975). It has been assumed here that the only large mammal populations to change during this period were the wildebeest and buffalo, because the low numbers of other species have little grazing impact. It has also been assumed that the level of primary production is the same, even though some increase is almost certain to have occurred. Data are insufficient to calculate the increase, but it is unlikely to be large.

The biomass of large, grazing mammals has increased 79 percent over the ecosystem as a whole, with the greatest increase in the north (96 percent), which is now more extensively used by the migratory wildebeest. Consumption has risen by 48 percent overall, with the greatest increase occurring on the plains (58 percent), where the (lactating) wildebeest have high food requirements (Sinclair 1975). Within each stratum, the increase in consumption matches exactly the observed decrease in burning.

These calculations strongly support the hypothesis that it is this increase in grazing impact, following the eruption of the buffalo and wildebeest populations, that has led to the observed decrease in burning. McNaughton (1976) has recently demonstrated the effect of the wildebeest's grazing. Locally, up to 85 percent of the standing crop can be removed in a few hours. This creates a mosaic of ungrazed and heavily grazed patches through which fires have great difficulty in spreading; and it is this effect of grazing, rather than the overall impact, which is probably of most significance in influencing the extent of burning.

A Simulation Model and Short-Term Management Implications

The management authorities of the Serengeti National Park became somewhat alarmed at the rapid rate of change in the woodlands, and they expressed fears about the possibility of losing much of the floristic diversity of the park, and some of the habitat types on which a number of large mammals depended. It was thus necessary to predict, however crudely, the probable outcome of the present trends within the woodlands.

Simple, linear extrapolations of observed trends often give highly misleading indications of future events, even in the short-term, and systems modeling is far preferable. A simulation model of the Serengeti's woodlands was developed to incorporate a number of the more important variables. The model is based on the analyses and data already put forward, and it simulates the response of the size structure of tree populations to various mortality and growth-inhibiting agents. The model seems to give results that are intelligible and also meaningful biologically (figs. 13.10, 13.11).

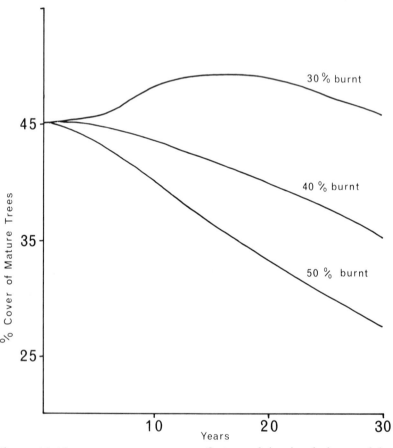

Figure 13.10   Output of the simulation model when the extent of burning is varied, but elephant and giraffe impact is maintained at contemporary levels. Graphs show densities of mature trees over 3 m.

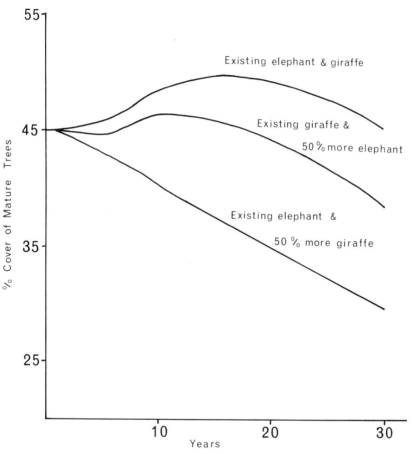

Figure 13.11

Output of the simulation model
when burning is maintained at
30% per annum, but first elephant
and then giraffe impact is increased
by 50%. Graphs show the
densities of mature trees over 3 m.

On the basis of Herlocker's work and the work of Croze (1974b), a
height of 3 m is taken to distinguish between young (Y) and mature (M)
trees. The young trees are further divided into three equal height classes
($Y1$, $Y2$, and $Y3$) to accommodate the observed height-specific sensitivity

to fire. $Y1$, $Y2$, $Y3$, and M thus refer to the numbers of individuals in each height class. The rates $R1 \ldots R6$ refer to the proportions moving from one height class into another in any given period of time. They are influenced by the auxilliary variables: burning (B), survival (SV), giraffes (G), elephants (E), and natural mortality agents (NM).

B represents the proportion burned by fire, which is here made proportional to the total area of the woodlands burned. Thus, if 50 percent of the woodlands burn, then 50 percent of the trees in any height class are affected by fire. Although the majority of trees scorched by fire are burned back to ground level, some escape and continue growth unchecked. This proportion surviving is given by $SV1 \ldots SV3$, taken from Herlocker's data (see above). Burning does not affect trees larger than 3 m in height in this simulation.

G represents the effect of browsing by giraffe and other browsers, elephants excluded. Browsing retards the rate at which young trees grow, so it is here simulated as a time delay that increases the exposure of young trees to a size-specific fire sensitivity. Herlocker (pers. comm.) shows that it takes, on average, some ten years for a young tree to reach 3 m in height in the Serengeti. A three-year delay $(G = 3)$ is therefore built into each stage of the model. Since G influences the proportions that survive, its effect is exponential rather than multiplicative (see table 13.7).

$E2 \ldots E4$ represent the number from each size class destroyed by elephants. This impact is made proportional to the age structure of the model tree population, for Croze (1974b) shows that the Serengeti's elephants feed proportionally on all heights of trees, with the exception of small trees of less than 1 m in height, which are ignored. The total elephant impact is, of course, kept constant. Thus, whereas B and SV represent the proportions of an age class burned or surviving, the Es represent the actual number destroyed by elephant.

Finally, NM represents the proportion dying from natural mortality agents. This rate has been estimated for mature trees by Croze (1974b). There are no estimates for the smaller trees, so it is assumed that natural mortality among trees less than 3 m is negligible. The young trees that are affected by burning are not killed, but are burned back to ground level. They therefore rejoin the smallest size class, $Y1$. In contrast, losses from elephants and from natural mortality are considered permanent.

It will be noticed that there is no provision in the model for seedling establishment. This is because there are as yet no valid data from the Serengeti concerning the rate at which this occurs. The model is therefore "pessimistic," in that, in the absence of any regeneration, the model tree

Table 13.7	Definitions and rate equations for woodland simulation model

$R1$ = size 1 trees that become size 2
$R2$ = size 2 trees that become size 3
$R3$ = size 3 trees that become mature
$R4$ = loss rate of mature trees
$R5$ = size 2 burned
$R6$ = size 3 burned
$R7$ = no. of size 2 destroyed by elephants
$R8$ = no. of size 3 destroyed by elephants
$R9$ = no. of mature trees destroyed by elephants
$E2$ = proportion of elephant killed trees from size 2
$E3$ = proportion of elephant killed trees from size 3
$E4$ = proportion of elephant killed trees from mature class.
$R1 = Y1 * ((1-B) + B*SV1) ** G$
$R2 = (Y2 - R7) * ((1-B) + B * SV2) ** G$
$R3 = (Y3 - R8) * ((1-B) + B * SV3) ** G$
$R4 = (M-R9) * NM * G$
$R5 = Y2 - R2$
$R6 = Y3 - R3$
$R7 = E * E2 * G$
$R8 = E * E3 * G$
$R9 = E * E4 * G$
$E2 = Y2/\Sigma(Y2+Y3+M)$
$E3 = Y3/\Sigma(Y2+Y3+M)$
$E4 = M/\Sigma(Y2+Y3+M)$
$Y1 = (Y1 - R1) + R5 + R6$
$Y2 = R1$
$Y3 = R2$
$M = M - R4 + R3$

population must inevitably decline. However, over short time periods, twenty to thirty years, the simulations are probably little influenced by this.

On the basis of Glover (1968), Lamprey et al. (1967) and Croze (1974b) the initial age structure of the simulated tree population was set at 40 percent, 10 percent, 5 percent, and 45 percent for the size classes $Y1$, $Y2$, $Y3$ and M, respectively. The survival rates from burning among $Y1$, $Y2$, and $Y3$ were set at 8 percent, 32 percent, and 72 percent, respectively, on the basis of the data presented above. The total elephant impact was calculated from figures given by Croze (1974b), who quotes an overall mortality rate among mature trees of 1 percent per year. Since elephants feed proportionally on all age classes except the smallest, this 1 percent represents 75 percent of their impact on this initial simulated population. Their total impact was then distributed in the ratio 2:1:9 among $Y2$, $Y3$, and M. The rate of natural mortality (NM) among mature trees was set at 1.5 percent per year (Croze 1974b).

Two simulations were run on this model population. Firstly, the amount of burning was varied between 30 percent and 50 percent per year (fig. 13.10). Secondly, the amount of burning was kept at 30 percent, while the impact of giraffe and then elephant was increased by 50 percent (fig. 13.11). Each simulation was run for thirty years, which is about the maximum time for which the assumptions of the model might be assumed to hold true. The equations for calculating the rates, and so on, are given in table 13.7.

The simulation in which burning was varied (fig. 13.10) shows that, with moderate burning, recruitment can offset the loss of mature trees through natural mortality and elephant impact. Under conditions of heavy burning, however, the numbers of mature trees decline sharply. It is apparent from figure 13.9 that the level of burning within the park has already dropped to 30–35 percent per year.

In the second simulation, the level of burning was set at the moderate 30 percent per year (fig. 13.11). The increased recruitment could offset a 50 percent increase in elephant impact—at least over the short term—but it was quite inadequate to offset a 50 percent increase in browsing by giraffe.

Although this model is somewhat simplistic, it does indicate that the Serengeti's woodlands are not in a state of inevitable decline in the face of the current rates of mortality. The model suggests that the woodlands are capable of stabilizing under the conditions of 1972 and, furthermore, that they can withstand an increase in elephant utilization, given the degree of protection from fire present in 1972. The woodlands are, however, very susceptible to factors that influence the growth of the young trees, especially fire and browsing by giraffe.

More weight is lent to these conclusions through the fact that the model is constructed so that woodland stabilization is unlikely. Instability in the model stems from a lack of data and observations. For example, no provision has been made for seedling establishment and growth. The effect of burning is also likely to be overestimated, for the survival (SV) of young trees scorched by the flames may well vary with the total area burned. Fewer trees may survive scorching when there is grass adequate to burn through 50 percent of the woodlands than when only 30 percent of the woodlands can burn. In addition, young trees tend to grow in coppices, which further reduces their sensitivity to fire.

The Serengeti ecosystem is obviously in a highly unstable state following the eradication of rinderpest, and it is still readjusting to this massive perturbation. It would appear that the effect of the subsequent increase in grazing impact on fire is now adequate to redress the present drift in vegetation toward a more open, grassland phase. It would therefore seem inadvisable for Tanzania National Parks to encourage a commercial cropping scheme of the migratory wildebeest.

Long-Term Woodland Dynamics

The hypothesis presented here is that rinderpest so reduced the populations of grazing ungulates in the Serengeti that the level of grassland utilization fell to well below the carrying capacity of the rangelands. Unutilized grass thus provided fuel for the wild fires that annually swept across the ecosystem. Woodland recruitment was effectively inhibited, and some woodland types—in particular, the bush thickets—were gradually destroyed by the fires. The woodlands declined because recruitment was no longer adequate to replace mature trees lost to elephants and other causes.

With the eradication of rinderpest from northern Tanzania in the early 1960s, the Serengeti's ungulate populations, especially wildebeest and buffalo, erupted. Large-mammal biomass increased by 79 percent during the decade 1962–72, and grazing impact increased by 48 percent. This change in grazing impact reduced the extent of grass fires. The buffalo population stabilized in the face of density-dependent mortality acting through their dry-season food supply, and perhaps through competition with the wildebeest, whose numbers were still increasing by the time the buffalo had stabilized.

During the early to mid 1960s, a wave of settlement spreading eastward away from the shore of Lake Victoria reached the western boundaries of the Serengeti. This expansion triggered a gradual immigration of elephant, which has now ceased. The rate of woodland change, at least in the north of the park, did not appear to increase as a result of this immigration, and simulation models based on the existing, observed rates of mortality from fire and elephants suggest that the woodlands will now stabilize, providing that the low level of burning seen in 1972 and later is maintained.

Longer-Term Implications
for the Serengeti

Fire plays a number of roles in savanna ecosystems such as the Serengeti, the most important of which are directly beneficial to populations of grazing ungulates. Fire can influence the spread of more palatable grasses, such as *Themeda triandra,* at the expense of coarser species. Fire also enhances primary production of grasslands by removing the overmature, standing crop and by removing litter (Hadley 1970; Hadley and Kieckhefer 1963; Hulbert 1969; Kucera and Ehrenreich 1962; Lemon 1968; Penfound and Kelting 1950), all of which helps maintain the grasslands in a more palatable and productive, submature phase of growth. Fire also directly inhibits the spread of woody plants by destroying seedlings and stunting the growth of immature plants, and under conditions of moderate to heavy burning, woody vegetation types tend to revert to more open grassland forms, thus enhancing the area of grassland available for grazers (Cooper 1961; Glover 1968; Jeffrey 1961; Pickford 1932; Pratt 1966; Strang 1973; Trapnell 1959; Wells 1970). These aspects of the influence of fire are well documented (Daubenmire 1968; Kozlowski and Ahlgren 1974) and they have been very noticeable in the Serengeti over the past decades.

With the eradication of rinderpest and the subsequent eruption of the populations of grazing ungulates, burning is being replaced by grazing as the main pathway for recycling mature herbage. It is far from clear how this will effect the structure and chemistry of the soils, or the pathways of nutrient cycling. In some respects, however, grazing has similar influences to fires. In particular, the mature, standing crop is removed, and the sward is maintained in a submature phase of growth, thus stimulating primary production and maintaining palatability (Hulbert 1969; Vesey-Fitzgerald 1974).

However, the major difference between the effects of burning and those of grazing is that woody plants are no longer inhibited by grazers, except by trampling. Woody plants tend to reestablish themselves under these conditions. The classical conditions for "bush encroachment" are heavy grazing and little or no burning (Jeffrey 1961). The simple simulation model of the Serengeti's woodlands (table 13.7) suggests that they will reestablish themselves over the short term, given the present low level of burning. If the germination and growth of seedlings, for which this model had no provision, are taken into account, a long-term increase in the cover and density of woody plants is inevitable in the Serengeti.

The reinvasion of fire-derived grasslands by woody plants has predictable effects. Grass cover and production decrease because of shading and competition for nutrients and water. Of more interest is that the regenerating young trees physically inhibit ungulates from grazing. This is clearly shown by Olivier and Laurie (1974) in their work on hippopotamus in the northern Serengeti. They reported that the degree of utilization of grasslands by hippo decreased as a function of distance from the riverbank. They further reported that the regeneration and growth of young trees was enhanced in areas heavily grazed by the hippos, and, furthermore, that there were areas of dense young trees where the hippo could no longer graze. A multiple regression analysis of the data presented in their paper showed that a highly significant 18 percent of the variation in the degree of grassland utilization by hippo could be accounted for by the density of young, regenerating trees of less than two meters in height.

The reestablishment of the Serengeti's woodlands will, therefore, erode the resource base of the populations of grazing ungulates. This is particularly true for the migratory wildebeest, who find their vital dry-season food supplies in the grasslands derived from the northern woodlands. Sinclair (1975, 1977, chap. 4 above) has demonstrated that the wildebeest and buffalo populations in the Serengeti are resource-limited. These two populations, at least, can be expected to decline in numbers in the long term.

Buffalo and wildebeest account for over 60 percent of the biomass of the grazing ungulates (table 13.6). Any reduction in their numbers will, therefore, greatly affect the grazing impact on the grasslands, and will lead to a situation where the grasslands will once more be underutilized. As has already been demonstrated, under these conditions, fire becomes a major influence in the ecosystem, and woodlands tend to revert to more open, grassland types of vegetation.

It is feasible, then, that the vegetation of the Serengeti could oscillate between more-woodland and more-grassland phases. This, in effect, may constitute a stable limits cycle (Holling 1973), mediated by fire and grazing. This, however, overlooks the effect of browsers, especially the ways in which their populations might respond to changes in the physionomic characteristics of the vegetation. Recent work in East Africa has highlighted three important aspects in the relationship between grazing ungulates and their food supply. Firstly, grazing enhances primary production, even though it reduces the standing crop of herbage at any particular time. Secondly, populations are resource-limited through their food supply, and density-dependent feedback is adequate to bring about population regulation. Thirdly, and most importantly, this regulation takes place without any permanent reduction in the level of primary production of the grasslands. The grazers die before they begin to erode their own resource base through "overgrazing." Unfortunately, these relationship have yet to be studied with any rigor for the browsing species.

If populations of browsers respond to fluctuations in their food supply in a way qualitatively similar to that demonstrated for some grazing species, and, in particular, if they die before reducing the level of browse production through overutilization, then browsing populations will track woodland-grassland cycles, albeit with a short time lag. Browsers would thus provide a negative feedback on such cycles, perhaps of a magnitude adequate to give the appearance of stability. Furthermore, browsers will speed up the homeostatic responses of ecosystems to massive perturbations of the type described here for the Serengeti. These assumptions are implicit in Caughley's (1976) model of elephant-woodland cycles.

If, on the other hand, browsers do not have such a relationship, and, in particular, if they are capable of reducing the capacity for primary production of browse through overutilization before dying from lack of resources, then populations of browsers will have a positive feedback on such cycles, and the vegetation will eventually become temporarily trapped in a grassland phase. Browsing populations would obviously collapse under these conditions, permitting a reestablishment of woody vegetation. Instead of cycles, there would be a series of precipitous emergence and elimination of woodlands.

Elephants further complicate the issue by their ability to switch from grazing to browsing on a seasonal basis, and, therefore, on a more permanent basis as well. In this way, elephants could possibly override any feedback from a diminishing production of browse, and thus permanently trap the vegetation in a grassland phase. This, of course, can only happen if the total primary production remains the same, irrespective of the particular browse-grass mix at any one moment. This assumption, which will be discussed below, does not always hold true.

The grassland-woodland-fire-grazing interactions in the Serengeti are, if not clear, at least discernible, but the woodland-browsing and woodland-elephant interactions remain obscure. Rigorous, quantitative data are required on the standing crop and primary production of browse, on the offtake by elephant and other browsers, on the effect of this offtake on the standing crop and primary production, and on food-related, seasonal patterns of condition and mortality.

The little data available are fragmentary. That elephants die when they run out of resources is well documented by Corfield (1973), and it is interesting that the pattern of mortality described by him has many characteristics of density independence. Elephant populations can respond more slowly to a gradually diminishing resource base (Laws et al. 1975). Until rigorous, quantitative data is obtained on the relationship between elephant and woody vegetation, only simplistic and somewhat superficial predictions can be made on future patterns of vegetation dynamics. This applies equally to the Serengeti as elsewhere.

Wider Implications

It is tempting to extrapolate from the Serengeti to other areas where the same components of fire, encroachment by settlement, vegetation change, and fluctuations in the numbers of grazers and browsers can be identified. But the responses of elephant populations to apparently similar events are so different that it appears unlikely that the relationships between these components are ever the same. Thus, the Kabalega National Park, Uganda (formerly the Murchison Falls National Park); the Serengeti National Park, Tanzania; and the Tsavo National Park, Kenya, have all undergone widespread, elephant-associated woodland changes, with the complete elimination of woodlands occurring locally. In Uganda, Laws et al. (1975) describe an elephant population declining slowly through reduced reproduction. In the Serengeti, Croze (pers. comm.) describes a vigorous and potentially expanding population, while in Tsavo, Corfield (1973) describes a population experiencing massive, resource-linked mortalities with density-independent characteristics.

These differences highlight yet another aspect of elephant-woodland dynamics where there are insufficient data. This concerns the productivity of the different components of the vegetation (grass, shrubs, trees), how this varies as their relative abundances change, and, finally, how these interrelate in different climatic areas. On this information will depend predictions about the responses of grazing and browsing populations to gross, physionomic changes in the vegetation.

In the Serengeti, for example, one can argue that the primary production of grasslands has increased with the thinning and local eradication of woodlands, first through the formation of more extensive grassland, and secondly through the impacts of fire and, more recently, intense grazing. Although the populations of grazers have been able to harness this increase in production, a more important consequence may be that it provides an alternative resource for elephant, thus damping any potential negative feedback on them from the woodland depletion.

In Kabalega, elephant browsing pressure, aided by fire and possibly by competition between long grass and young regenerating trees (Eggeling 1947) has trapped the vegetation at an early, open-grassland stage of succession. Grass forms a large proportion of the elephants' diet there, and Laws et al. (1975) attribute the fall-off in reproductive activity to the effect of a nutritionally poor diet. Be that as it may, in the absence of a grazing impact of a magnitude similar to that seen in the Serengeti, the vegetation will remain trapped until the slowly declining elephant population falls to a level below that at which succession can once more continue.

The arid bushlands of Tsavo offer a striking contrast. Once again, heavy browsing pressure has trapped the vegetation in an early stage of succession, but there is no compensating growth of grassland, although the cover and production of shrubs may well have increased significantly (Agnew 1968). The elephant population experienced resource depletion, and massive drought-related mortalities became a regular occurrence in the early 1970s dry years. These mortalities have density-independent characteristics (Corfield 1973), in contrast to the more typically density-dependent mortalities seen in Uganda. Eventually Tsavo's elephant population(s) will be reduced to a level low enough for succession to continue, but until then, the vegetation will remain trapped in its present, semidesert form.

Elephants, Fire, Droughts, and Secondary Succession

Late successional and climax communties are ecologically senile, compared with early and mid successional stages (Horn 1974; Odum 1969). They are characterized by species which are adapted to facing severe competition from well-established neighbors; species that are long-lived, highly efficient competitors, hoarding resources and producing few, well-provided-for offspring. In contrast, earlier successional communities are characterized by fast-growing and quickly maturing species, producing large numbers of young, which become widely dispersed. Species diversity and richness, initially low, rises quickly in mid succession, and then levels off (Nicholson and Monk 1975; Shugart and Hett 1973), or even declines (Auclair and Goff 1971; Shafi and Yarranton 1973) in late succession. Of more relevance to these semiarid ecosystems, production outstrips respiration in early successional communtities, leading to an accumulation of biomass. In later stages, there is a gradual decline in gross production (Kira and Shidei 1967; Whittaker 1966) as energy resources are reallocated from production to maintenance. The successional process is characterized by a shift from dynamic, highly diverse production to steady, sustained maintenance.

Since communities are permanently changing toward climax forms, diversity and production can only be maintained by disruptive forces halting, reversing, and even trapping the successional process. Loucks (1970), in his study of Wisconsin forest communities, demonstrates that periodic, drastic perturbations in these communities are essential for maintaining waves of peak diversity and production. Furthermore, he argues that contemporary management policies that prevent these perturbations—in this example, extensive fire-control policies—are highly detrimental to these communities in the long run.

There is a striking analogy here between the processes of succession and those of seasonal growth. Utilization by grazers and browsers traps and maintains vegetation communities at an early, submature, highly palatable, and highly productive stage of growth. Disruptive forces act similarly to recycle and trap successional processes and thus maintain communities of peak diversity and production. And it is of interest here that Cates and Orians (1975) have shown that early successional plant species are more palatable to generalized herbivores than are late successional or climax species.

There are many disruptive forces acting on the successional processes of these East African savanna ecosystems. Diseases such as rinderpest, heavy poaching, and encroachment by settlement all disrupt successional processes in large-mammal faunas. Management policies that suddenly reverse succession by removing significant components of the large-mammal biomass are also disruptive. Eltringham (1974) reports that the effects of completely removing hippo, the dominant grazer, from the Mweya Peninsular in the Rwenzori National Park (Uganda) were similar to those seen in plant communities when the dominant is removed. In ten years the density and biomass of large mammals increased by 54 percent and 20 percent, respectively, while diversity increased from 1.26 to 1.78 (calculated from the biomass figures quoted by Eltringham). Successional advances were also observed in the grasslands (Thornton 1972), even though the density and biomass of the grazers had increased. This reversal of succession led to an increase in both primary and secondary productivity.

On the Serengeti plains, grazing exclosures that have been closed for almost a decade showed that grazing pressure maintains a highly diverse, highly productive grassland community. The ungrazed climax within these exclosures consisted of rank, unpalatable grasses with low cover and production (chap. 3). Grazing impact here maintains an earlier, more diverse and productive successional grassland community.

Elephants and fire are probably the best-documented disruptive forces acting on vegetation succession, and I have suggested how they can reverse and trap succession, as well as create a mosaic of successional stages. Drought too, has disruptive effects, not only through direct selection on species, but also by effectively increasing animal impact by lowered production. In the Serengeti, droughts occur at random intervals, for spectral analyses of long-term climatic records (Pennycuick and Norton-Griffths 1976) show power spectra akin to white noise. There is undoubtedly a damping effect of Lake Victoria on Serengeti weather patterns, for east of the Rift Valley at Machakos (Pennycuick and Norton-Griffiths 1976) and at Voi in the Tsavo National Park (S. Cobb and L. Maddock, pers. comm.) highly significant, low frequency cycles can be found in the rainfall data.

Other potent forces of disruption to vegetation succession are generated when new national parks are declared. Recently in Kenya, the Amboseli National Park and the East Rudolf National Park have been gazetted, and in both areas the human impact (cattle grazing) has been suddenly removed. Cattle made up some 75 percent of the grazing biomass in Amboseli (Western 1973) and 64 percent of the grazing biomass in East Rudolf (pers. obs.). Rapid successional changes are bound to occur in these two parks, and this raises the interesting point that merely "protecting" an area is quite enough to set off successional changes adequate to alter completely its ecological nature.

Rinderpest, poaching, park management, elephants, fire, and droughts are all having disruptive impacts on plant and animal succession in these East African savanna ecosystems. This is not to say that they are having destructive impacts. They may well be maintaining a fine-grained mosaic of successional stages, thus maximizing the diversity, productivity, and resilience of these systems. Holling (1973) discusses the concepts of resilience and stability as they apply to ecological systems. He defines stability as the ability of a system to return to an equilibrium state after a temporary disturbance; and he defines resilience as a measure of the persistence of a system and of its ability to absorb change and disturbance while still maintaining the same relationships between populations and equilibrium-state variables. Highly stable systems, such as climax rain forest, may be able to absorb local disturbances, such as natural tree mortality, efficiently and quickly. But major disturbances, such as heavy logging, fire, or colonization by settlement rapidly destroys the entire system so that it cannot recover. This has happened to large areas of Amazon rain forest, which had been stable for eons until subject to heavy impact by man. In contrast, savanna ecosystems are highly unstable, as the changes taking place in Kabalega, Serengeti, Tsavo, and Amboseli show all too clearly. But their capacity to absorb change and disturbance is equally great, and they may be highly resilient, a point brought out in Noy-Meir's (1975) analyses of arid and desert ecosystems.

It is unfortunate that the conservation "ethic" is so often couched in terms of maintaining the existing diversity of plant and animal life, for ecologically this is a contradition in terms and is unsuitable for highly dynamic and highly unstable semiarid ecosystems. Such a goal might be attainable in mesic, climax communities such as rain forest, but it is unrealistic (with our existing knowledge) for areas that experience either random climatic perturbations (Serengeti) or cyclical climatic perturbations (Tsavo). Stability has no place in systems such as these.

Conclusion

Analysis of small-scale aerial photography indicates that the Serengeti's vegetation is moving toward a more open, grassland phase. There has been a significant decrease in the woodland canopy cover between 1962 and 1972. The central, semiarid woodlands show a mosaic of change, with some areas increasing in canopy cover and some areas decreasing. In contrast, the northern dry subhumid woodlands show a blanket decrease everywhere, and rates of change are some three times those in the central parts. Multiple regression analysis suggests that elephant utilization is most closely associated with the mosaic of change in the central woodlands, while fire is most closely associated with the blanket decrease in the northern area.

Elephants have a high but local impact on woodlands. Fire has a more pernicious, overall effect. In particular, it inhibits the recruitment of young trees into the canpoy. This results in tree populations with uneven age distributions; such populations are highly unstable.

The Serengeti's two elephant populations have recently undergone a phase of immigration from outside the national park, brought about by a wave of settlement spreading eastward, away from the shores of Lake

Victoria. This immigration has now ceased, and the populations appear stable. There is still a great potential for increase, both from further immigration and from natural recruitment.

Fire is widespread in the Serengeti, the timing and extent being closely related to rainfall. Fires start when the rains have finished, and fires are less frequent and extensive in dry years, compared with wet years. There was a marked decrease in the extent of burning in 1962–72, which is not associated with any contemporary changes in rainfall amount or distribution. During this same period, the populations of wildebeest and buffalo have erupted, following the eradication of rinderpest. The biomass of large mammals have increased by some 80 percent, and grazing offtake has increased by over 50 percent. Grazing pressure is now removing so much of the primary production that there is no longer fuel adequate to maintain extensive grass fires.

A simple model investigates the effects of elephants, fire, giraffe browsing, and natural mortality on the Serengeti's woodlands. The present reduction in fire is adequate to offset the existing rate of offtake by elephant. Given that fire remains at the present low level, the woodlands could withstand a 50 percent increase in elephant impact. A 50 percent increase in giraffe impact, however, would lead to their speedy decline. The model suggests that these woodlands are resilient to impacts on mature trees (elephants, natural mortalities) but are sensitive to impacts on regeneration (fire, giraffe).

The role of disruptive forces (elephants, fire, droughts, poaching, park management) may function to provide random perturbations into the natural successional processes, thus maintaining a mosaic of communities of peak diversity, productivity, and resilience. I suggest that the prevention of these disruptions could be highly detrimental to these ecosystems in the long run.

References

Agnew, A. D. Q. 1968. Observations on the changing vegetation of Tsavo National Park (East). *E. Afri. Wildl. J.* 6:75–80.

Auclair, A. N., and Goff, F. G. 1971. Diversity relations of upland forests in the western Great Lakes area. *Am. Nat.* 105:499–528.

Buechner, H. K., and Dawkins, H. C. 1961. Vegetation change induced by elephant and fire in the Murchison Falls National Park, Uganda. *Ecology* 42:752–66.

Cates, R. G., and Orians, G. H. 1975. Successional status and the palatability of plants to generalized herbivores. *Ecology* 56:410–18.

Caughley, G. 1976. The elephant problem—an alternative hypothesis. *E. Afr. Wildl. J.* 14:265–84.

Caughley, G., and Goddard, J. 1975. Abundance and distribution of elephants in the Luangwa Valley, Zambia. *E. Afr. Wildl. J.* 13:39–48.

Cooper, C. F. 1961. The ecology of fire. *Sci. Amer.* 204:150–60.

Corfield, T. F. 1973. Elephant mortality in Tsavo National Park, Kenya, *E. Afr. Wildl. J.* 11:339–68.

Croze, H. 1974a. The Seronera bull problem. Part 1. The elephants. *E. Afr. Wildl. J.* 12:1–28.

———. 1974b. The Seronera bull problem. Part 2. The trees. *E. Afr. Wildl. J.* 12:29–48.

Daubenmire, R. F. 1968. Ecology of fire in grasslands. *Adv. Ecol. Res.* 5:209–66.

Douglas-Hamilton, I. 1972. On the ecology and behaviour of the African Elephant. D. Phil. thesis, Oxford University.

Eggeling, W. J. 1947. Observations on the ecology of the Budongo rain forest, Uganda. *J. Ecol.* 34:20–87.

Eltringham, S. K. 1974. Changes in the large mammal community of Mweya Peninsula, Rwensori National Park, Uganda, following removal of hippopotamus. *J. Appl. Ecol.* 11:855–66.

Field, C. R. 1971. Elephant ecology in the Queen Elizabeth National Park, Uganda. *E. Afr. Wildl. J.* 9:99–123.

Ford, J. 1971. *The role of the trypanosomiases in African ecology.* Oxford: Clarendon Press.

Glover, P. E. 1968. The role of fire and other influences on the savannah habitat with suggestions for further research. *E. Afr. Wildl. J.* 6: 131–37.

Hadley, E. B. 1970. Net productivity and burning responses of native eastern North Dakota prairie communities. *Amer. Midl. Nat.* 84: 121–35.

Hadley, E. B., and Kieckhefer, B. J. 1963. Productivity of two prairie grasses in relation to fire frequency. *Ecology* 44:389–95.

Harrington, G. N., and Ross, I. C. 1974. The savanna ecology of Kidepo Valley National Park. Part 1. The effects of burning and browsing on the vegetation. *E. Afr. Wildl. J.* 12:93–105.

Holling, C. S. 1973. Resilience and stability of ecological systems. *Ann. Rev. Ecol. Syst.* 4:1–23.

Horn, H. S. 1974. The ecology of secondary succession. *Ann. Rev. Ecol. Syst.* 5:25–37.

Hulbert, L. C. 1969. Fire and litter effects in undisturbed bluestem prairie in Kansas. *Ecology* 50:874–77.

Jeffrey, W. W. 1961. A prairie to forest succession in Wood Buffalo Park, Alberta. *Ecology* 42:442–44.

Kira, T., and Shidei, T. 1967. Primary production and turnover of organic matter in different forest ecosystems of the western Pacific. *Jap. J. Ecol.* 17:70–87.

Kozlowski, T. T., and Ahlgren, C. E., eds. 1974. *Fire and ecosystems.* New York: Academic Press.

Kucera, C. L., and Ehrenreich, J. H. 1962. Some effects of annual burning on central Missouri prairie. *Ecology* 43:334–36.

Kurji, F. 1976. *Human ecology.* Serengeti Research Institute, Annual Report, 1974–75, pp. 12–31. Arusha: Tanzania National Parks.

Lamprey, H. F.; Glover, P. E.; Turner, M. I. M.; and Bell, R. H. V. 1967. Invasion of the Serengeti National Park by elephants. *E. Afr. Wildl. J.* 5:151–66.

Laws, R. M. 1970. Elephants as agents of habitat and landscape change in East Africa. *Oikos* 21:1–15.

Laws, R. M.; Parker, I. S. C.; and Johnstone, R. C. B. 1975. *Elephants and their habitats.* Oxford: Oxford Univ. Press.

Lemon, P. C. 1968. Effects of fire on an African plateau grassland. *Ecology* 49:316–22.

Leuthold, W., and Sale, J. B. 1973. Movements and patterns of habitat utilization in Tsavo National Park, Kenya. *E. Afr. Wildl. J.* 11:369–84.

Lock, J. M. 1967. Vegetation in relation to grazing and soils in Queen Elizabeth National Park, Uganda. Ph.D. dissertation, Cambridge University.

Lock, J. M., and Milburn, T. R. 1971. The seed biology of *Themeda triandra* Forsk. in relation to fire. *Symp. Brit. Ecol. Soc.* 11:337–49.

Loucks, O. L. 1970. Evolution of diversity, efficiency, and community stability. *Am. Zool.* 10:17–25.

McNaughton, S. J. 1976. Serengeti migratory wildebeest: facilitation of energy flow by grazing. *Science* 191:92–94.

Nicholson, S. A., and Monk, C. D. 1975. Changes in several community characteristics associated with forest formation in secondary succession. *Amer. Midl. Nat.* 93:302–10.

Norton-Griffiths, M. 1973. Counting the Serengeti migratory wildebeest using two-stage sampling. *E. Afr. Wildl. J.* 11:135–49.

Norton-Griffiths, M.; Herlocker, D.; and Pennycuick, L. 1975. The patterns of rainfall in the Serengeti ecosystem, Tanzania. *E. Afr. Wildl. J.* 13:347–74.

Noy-Meir, I. 1975. Stability of grazing systems: an application of predator-prey graphs. *J. Ecol.* 63:459–81.

Odum, E. P. 1969. The strategy of ecosystem development. *Science* 164: 262–70.

Olivier, R. C. D., and Laurie, W. A. 1974. Habitat utilization by hippopotamus in the Mara river. *E. Afr. Wildl. J.* 12:249–72.

Penfound, W. T., and Kelting, R. W. 1950. Some effects of winter burning on a moderately grazed pasture. *Ecology* 31:554–60.

Pennycuick, L. 1975. Movements of the migratory wildebeest population in the Serengeti area between 1960 and 1973. *E. Afr. Wildl. J.* 13:65–87.

Pennycuick, L., and Norton-Griffiths, M. 1976. Fluctuations in the rainfall of the Serengeti ecosystem, Tanzania. *J. Biogeog.* 3:125–40.

Phillipson, J. 1975. Rainfall, primary production and "carrying capacity" of Tsavo National Park (East), Kenya. *E. Afr. Wildl. J.* 13:171–202.

Pickford, G. D. 1932. The influence of continued heavy grazing and of promiscuous burning on spring-fall ranges in Utah. *Ecology* 12: 159–71.

Pratt, D. J. 1966. Bush control studies in the drier areas of Kenya. *J. Appl. Ecol.* 3:97–115.

Shafi, M. I., and Yarranton, G. A. 1973. Diversity, floristic richness, and species evenness during a secondary (post-fire) succession. *Ecology* 54:897–902.

Shugart, H. H., and Hett, J. M. 1973. Succession: similarities of species turnover rates. *Science* 180:1379–81.

Sinclair, A. R. E. 1975. The resource limitation of trophic levels in tropical grassland ecosystems. *J. Anim. Ecol.* 44:497–520.

———. 1977. *The African buffalo.* Chicago: Univ. of Chicago Press.

Strang, R. M. 1973. Bush encroachment and veld management in south-central Africa: the need for a reappraisal. *Biol. Conserv.* 5:96–104.

Thornton, D. D. 1971. The effect of complete removal of Hippopotamus on grassland in the Queen Elizabeth National Park, Uganda. *E. Afr. Wildl. J.* 9:47–55.

Trapnell, C. G. 1959. Ecological research of woodland burning experiments in Northern Rhodesia. *J. Ecol.* 47:129–68.

Vesey-Fitzgerald, D. F. 1970. The origin and distribution of valley grasslands in East Africa. *J. Ecol.* 58:51–75.

————. 1972. Fire and animal impact on vegetation in Tanzania National Parks. *Proc. Tall Timbers Fire Ecol. Conf.* 11:297–317.

————. 1973. Browse production and utilization in Tarangire National Park. *E. Afr. Wildl. J.* 11:291–306.

————. 1974. Utilization of the grazing resources by buffaloes in the Arusha National Park, Tanzania. *E. Afr. Wildl. J.* 12:107–34.

Wells, P. V. 1970. Post-glacial vegetational history of the Great Plains, new evidence reopens the question of the origin of treeless grasslands. *Science* 167:1574–82.

Western, D. 1973. Structure, dynamics and changes of the Amboseli Basin ecosystem. Ph.D. dissertation, Nairobi University.

Western, D., and van Praet, C. 1973. Cyclical changes in the habitat and climate of an East African ecosystem. *Nature* (Lond.) 241:104–6.

Whittaker, R. H. 1966. Forest dimensions and production in the Great Smoky Mountains. *Ecology* 47:103–21.

Wing, L. D., and Buss, I. O. 1970. Elephants and forests. Wildl. Monogr. no. 19. The Wildlife Society.

Wyatt, J. R., and Eltringham, S. K. 1974. The daily activity of the elephant in the Rwenzori National Park, Uganda. *E. Afr. Wildl. J.* 12: 273–90.

J. J. R. Grimsdell

Appendix A   Changes in Populations of Resident Ungulates

A census was carried out in August 1976 to see if populations of resident ungulates had changed since a similar census in July 1971 by Sinclair (1972). During the last twenty years, considerable ecological change has taken place within the Serengeti ecosystem: human populations have changed in number and distribution, elephants have invaded the park from surrounding areas, the numbers of wildebeest and buffalo have increased, the incidence of wildfires has generally declined, woodlands have been progressively thinned in the north, and the distribution of rainfall has recently been particularly favorable to large herbivores. Against this background of ecological change, it is of interest to know how some of the resident species of the park have been affected.

The species counted were impala, topi, kongoni, giraffe, buffalo bachelor males, and eland. These species, except eland, are resident animals, showing relatively small-scale seasonal movements when compared with the migrants.

Methods

The census was conducted by an aerial random sampling method, essentially similar to the 1971 census. One important difference between the two counts is that in stratum 1, transects were oriented east-west in 1971, but north-south in 1976. The latter orientation was considered to be preferable because the transects cut across the major drainage systems of the stratum and, hence, sampled the various habitat types more consistently than with east-west transect orientation.

Aircraft availability limited the census to strata 1 and 2 (see map in Sinclair 1972); stratum 3, the smallest, was omitted. Stratum 1 covered the western woodlands north of the Nyaraboro hills, south of the northern boundary of the corridor, and west of the Seronera-Kimerishi road. Stratum 2 covered the central woodlands, east of this road to the eastern boundary of the park, north of the plains and south of the Grumeti River (see fig. 1.1). Stratum 3 extended north of the Grumeti to the Mara River.

Details of the census method are given in table A.1. Photographs were taken of the large groups of animals, chiefly impala and topi.

Jolly's method no. 2 (Jolly 1969) was used to analyze the census data. Before analyses, the animal counts were corrected, if necessary, according to photographic counts and correction factors derived from them. In order to reduce the variance of the population estimates, each species was stratified separately for each stratum. This usually involved omitting certain portions of the strata which were devoid of a particular species. On three occasions, the stratum was divided into two secondary strata.

Results and Interpretation

The census results are summarized in table A.2, while a comparison of the 1971 and 1976 results is shown in table A.3. An interpretation of the changes for each species is given below:

Impala

No significant change in impala numbers has occurred in stratum 2. In contrast, there appears to have been a large increase in stratum 1, the

Table A.1   Details of the census method (August 1976)

	Stratum 1	Stratum 2
Aircraft type	Cessna 182	Cessna 180
Height control	Shadowmeter[a]	Radar altimeter
Height above ground	400 feet	400 feet
Transect orientation	North-south	East-west
No. transects	27	24
Mean transect width	353 m	400 m
Area searched	367 km²	479 km²
No. possible transects	290	178
Stratum area	3,862 km²	3,487 km²
Sampling fraction	9.5%	13.7%

[a] See Pennycuick (1973).

Table A.2   Census results, August 1976

Species	Total	S.E.	No. transects	95% C.L.	C.L. as % of Total
STRATUM 1 (3862 km²)					
Impala	30,440	4,550	27	9,100	30
Topi	51,633	7,281	27	14,562	28
Kongoni	934	209	11	460	49
Giraffe	2,240	474	25	995	44
Buffalo males	1,853	350	18	735	40
Waterbuck	1,554	417	22	876	56
Eland	1,526	468	26	983	64
STRATUM 2 (3,487 km²)					
Impala	33,738	4,503	24	9,456	28
Topi	9,651	1,530	23	3,213	33
Kongoni	6,591	1,042	22	2,188	33
Giraffe	4,970	436	24	916	18
Buffalo males	1,646	271	24	569	35
Waterbuck	352	176	22	370	105
Eland	1,468	451	22	947	65

western Serengeti. To some extent, this apparent change may be due to the methods used. In 1971, east-west transects were used for stratum 1; these ran parallel to the main drainage lines, and it is possible that major areas of impala habitat were, by chance, unsampled. North-south transects run in 1976 provided a more effective sampling method with less chance of missing high-density areas.

However, other lines of evidence suggest that there has been a real increase in impala numbers; and if this is so, then the 1971 figure may not be a big underestimate. One line of evidence comes from the age structure of the impala population of the western Serengeti. It was observed by Duncan (pers. comm.) in 1972–73 that the population contained large numbers of young compared to the population of the central woodlands, this being suggestive of different rates of population growth.

A further, and stronger, line of evidence comes from other counts of impala in the western Serengeti. An estimate of 5181 was made in 1966 by Bell (1969), who carried out an aerial total count of an area somewhat smaller than stratum 1. His census zone excluded the Musabi plains and

Table A.3   Comparison of 1971 and 1976
                                             totals with 95% confidence limits

Species	1971	1976	Significant change ($P < 0.05$)
STRATUM 1			
Impala	6,759±1,788	30,440±9,100	Yes
Topi	17,072±1,976	51,633±14,562	Yes
Kongoni	461±335	934±460	No
Giraffe	2,855±430	2,240±995	No
Buffalo males	995±324	1,853±735	No
Waterbuck	594±403	1,554±876	No
Eland	3,029±2,582	1,526±983	?
STRATUM 2			
Impala	39,255±6,512	33,738±9,456	No
Topi	6,946±1,208	9,651±3,213	No
Kongoni	6,865±1,237	6,591±2,188	No
Giraffe	2,780±634	4,970±916	Yes
Buffalo males	1,423±354	1,646±569	No
Waterbuck	365±382	352±370	No
Eland	1,869±533	1,468±947	?

the central ranges of hills. Therefore, this count cannot be directly compared with the 1971 and 1976 results, but it does indicate that impala numbers were much lower in 1966 than in 1976. A further count was carried out by Duncan (pers. comm.) in 1973, returning an estimate of 20,422. In this case, the census area was identical to the 1971 and 1976 counts and the method was similar, that is, random aerial sampling using north-south transects (as in 1976).

The available counts in stratum 1 may be summarized as follows: more than 5181 in 1966, at least 6759 in 1971, 20,422 in 1973, and 30,440 in 1976. Because there is some doubt about the earlier counts, it may be unwise to draw conclusions concerning the nature of population growth, except to note that the population does seem to have increased considerably between 1966 and 1976.

As to reasons for the increase, the most important may have been a reduction in human hunting during the ten-year period, for other factors such as habitat change and disease do not seem to be involved in this case. Changes in the boundaries of the western Serengeti (see below) and effective law enforcement by the park authorities between 1966 and 1976 would have reduced the level of hunting.

Topi

In stratum 2, there has been no significant change in topi numbers. The picture is quite different for stratum 1, where there appears to have been a substantial increase in topi numbers. Censuses in the western Serengeti recorded 11,909 in 1966 (Bell 1969), 17,072 in 1971 (Sinclair 1972), 26,384 (Duncan, pers. comm.), and 51,633 in 1976. The census zone of the 66 count was smaller than stratum 1, as described above for impala. However, this would not have greatly affected the estimate for stratum 1, because few topi occur in the area not covered in the 1966 count. These counts suggest a fairly steady rate of population growth of around 15 percent a year between 1966 and 1976.

The increase is probably related to recent shifts in human settlement, together with domestic stock, and to a relaxation of hunting pressure. In regard to recent shifts in human settlement, the first change took place in 1959, when the southern boundary of the western Serengeti was expanded in part exchange for land lost when the eastern plains and the Ngorongoro Conservation Area were exercised from the park. The second change occurred in 1967 when the northern borders of the western Serengeti were widened; and the third change occurred in 1973–74 with the Tanzanian government's policy of settlement relocation and consolidation. All these events released more land for wildlife populations and at the same time competition for food between topi and cattle was reduced. These factors, together with the control of illegal hunting by park authorities, have probably largely contributed to the topi increase. Eradication of rinderpest, of such importance to wildebeest and buffalo, would not have affected topi as they are not susceptible to the disease.

Kongoni

Little change in Kongoni numbers has taken place in stratum 2. Although not significant statistically, there is a suggestion of increase in stratum 1, a suggestion which is supported by a count of 197 kongoni in 1966 (Bell 1969). As already mentioned, the 1966 census zone was smaller than stratum 1; but taking the figures as they stand, kongoni would appear to have increased at about the same rate as the topi population of the western Serengeti, possibly for the same reasons.

Giraffe

Giraffe appear to have remained relatively stable in stratum 1, but have shown a significant increase in stratum 2. This may be explained by more food becoming available to them in this area, through improved regeneration of young trees as a result of the general reduction in burning in recent years. For the Seronera area of stratum 2, Pellew (1977) has evidence of an increasing giraffe population. Some of the increase may be due to immigration, but the results from this census also suggest a general population increase through improved recruitment and survival. The age distribution of the Seronera population is suggestive of an increasing population (Pellew, 1977).

Buffalo Bachelor Males

No significant changes have occurred in either stratum, but there is a suggestion of an increase, particularly in stratum 1. This would be expected, as other census data show a continued increase of the buffalo population of the park (Grimsdell 1977). The increase has been most rapid in the western Serengeti, and this is also indicated by the present data.

Waterbuck

Waterbuck numbers seem to have remained constant in stratum 2, and although there is a suggestion of increase in stratum 1, this is not statistically significant. The relatively low estimate in 1971, as compared to 1976, could be partly due to the different sampling methods, as discussed for the impala results. Waterbuck are restricted to riverine habitats and the north-south tarnsects used in 1976 should have provided a more effective sampling of these habitats. Once again, if there has been a real increase, then the chief reason could be the relaxation of hunting pressure.

Eland

No significant changes have taken place in eland numbers. Unlike the other species in this census, eland are migratory, and, consequently, it is difficult to deduce trends from counts of only part of their range, a fact also stressed by Sinclair (1972).

Conclusion

Resident ungulates of the central and western Serengeti have either increased or remained stable between 1971 and 1976. No species has
decreased in number. The census should be repeated in a few years time to
monitor future changes. For stratum 1, it is recommended that the number
of transects be increased (say, to forty) so that the variances of the population estimates may be reduced. North-south transects should be flown for
this stratum.

The author is grateful to the following: M. Norton-Griffiths and H.
Croze for providing their respective aircraft and piloting the census;
G. Frame, F. Kurji, H. Hoeck, R. Pellew, and A. Diamond, who took part
as observers; J. Malcolm, D. Bygott, R. Pellew, E. Diamond, F. Kessy, and
R. Malleko, who took part as recorders; Ecosystems Ltd., Nairobi, for
helping with the data analysis; R. Bell and P. Duncan for helpful comments on an earlier draft; and the African Wildlife Leadership Foundation for funding.

References

Bell, R. H. V. 1969. The use of the herb layer by grazing ungulates in the
Serengeti National Park, Tanzania. Ph.D. dissertation, Manchester
University.

Grimsdell, J. J. R. 1977. Serengeti Research Institute, Annual Report,
1975–76, pp. 3–10. Arusha: Tanzania National Parks.

Jolly, G. M. 1969. Sampling methods for aerial census of wildlife populations. *E. Afr. Agric. For. J.* 34:46–49.

Pellew, R. A. 1977. Serengeti Research Institute, Annual Report, 1975–
76, pp. 49–67. Arusha: Tanzania National Parks.

Pennycuick, C. J. 1973. The shadowmeter: a simple device for controlling
an aircraft's height above the ground. *E. Afr. Wildl. J.* 11:109–12.

Sinclair, A. R. E. 1972. Long-term monitoring of mammal populations in
the Serengeti: census of nonmigratory ungulates, 1971. *E. Afr. Wildl. J.*
10:287–97.

Scientific and common names of mammal and bird species mentioned in the text.

MAMMALS

Order Carnivora

Lycaon pictus	Wild dog
Canis mesomelas	Black-backed jackal
Canis aureus	Golden jackal
Canis adustus	Side-striped jackal
Otocyon megalotis	Bat-eared fox
Crocuta crocuta	Spotted hyena
Panthera pardus	Leopard
Panthera leo	Lion
Acinonyx jubatus	Cheetah

Order Proboscidia

Loxodonta africana	African elephant

Order Perissodactyla

Equus burchelli	Zebra
Diceros bicornis	Black rhinoceros
Ceratotherium simum	White rhinoceros

Order Artiodactyla

Phacochoerus aethiopicus	Warthog
Hippopotamus amphibius	Hippopotamus
Giraffa camelopardalis	Giraffe
Tragelaphus scriptus	Bushbuck
Taurotragus oryx	Eland

Syncerus caffer	African buffalo
Sylvicapra grimmia	Gray duiker
Kobus defassa	Waterbuck
Redunca redunca	Bohor reedbuck
Aepyceros melampus	Impala
Hippotragus equinus	Roan antelope
Oryx beisa	Oryx
Damaliscus korrigum	Topi
Alcelaphus buselaphus	Kongoni, Coke's hartebeest
Connochaetes taurinus	Wildebeest
Oreotragus oreotragus	Klipspringer
Madoqua kirki	Dikdik
Raphicerus campestris	Steinbuck
Ourebia ourebi	Oribi
Gazella thomsoni	Thomson's gazelle
Gazella granti	Grant's gazelle

BIRDS

Gyps rueppellii	Ruppell's griffon vulture
Gyps africanus	White-backed vulture
Torgos tracheliotus	Lappet-faced vulture
Trigonoceps occipitalis	White-headed vulture
Neophron percnopterus	Egyptian vulture
Necrosyrtes monachus	Hooded vulture
Gyptaetus barbatus	Lammergeier
Leptoptilos crumeniferus	Maribou stork

Appendix C Bibliography of Serengeti Scientific Publications

This bibliography, compiled with the assistance of D. A. Kreulen, reflects the recent nature of biological research in East Africa. The first studies began in the 1950s, but the main work occurred during 1966–76, following the formation of the Serengeti Research Institute. The main objective was an understanding of ecosystem function, but, for conservation reasons, studies concentrated on the large-mammal fauna, their habitats, and the soils. There have been a few studies on other animal groups, particularly by the Max-Planck Institute scientists. The following list of 318 references includes the scientific publications that we know of relating to ecology and behavior up to mid 1978. For lack of space, we have omitted the more popular natural history articles, films, and papers on other subjects such as paleontology and archeology; and most of those relating to areas outside the Serengeti-Mara ecosystem.

Adamson, G. A. G. 1964. Observations of lions in Serengeti National Park, Tanganyika. *E. Afr. Wildl. J.* 2:160–61.

Albrecht, H. 1967. Freiwasserbeobachtungen an Tilapien (Pisces, Cichlidae) in Ostafrika, *Z. Tierpsychol.* 25:375–94.

Albrecht, H., and Wickler, W. 1968. Freilandbeobachtungen zur 'Begrussungszeremonie' des Schmuckbartvogels *Trachyphonus d'arnaudii* (Prévost u. Des Murs). *J. Ornithol.* 109:225–63.

Anderson, G. D., and Talbot, L. M. 1965. Soil factors affecting the distribution of the grassland types and their utilization by wild animals on the Serengeti plains, Tanganyika. *J. Ecol.* 53:33–56.

Atang, P. G., and Plowright, W. 1969. Extension of the JP-15 Rinderpest control campaign to Eastern Africa: the epizootiological background. *Bull. Epizoot. Dis. Afr.* 17:161–70.

Baerends, G. P. 1974. Het onderzoek programma van het "Serengeti

Research Institute" in Tanzania. Verslag gewone vergadering Afd. *Natuurkunde Kon. Ned. Akad. Wetensch.* 83:58–65.

Baker, J. R. 1969. Trypanosomes of wild mammals in the neighbourhood of the Serengeti National Park. In *Diseases in free-living wild animals,* ed. A. McDiarmid. *Symp. Zool. Soc. Lond.,* no. 24, pp. 147–58. London: Academic Press.

Baker, J. R.; Sachs, R.; and Laufer, I. 1967. Trypanosomes of wild mammals in an area northwest of the Serengeti National Park, Tanzania. *Z. Tropenmed. Parasit.* 18:280–84.

Beglinger, R.; Kauffmann, M.; and Müller, R. 1976. Culverts and trypanosome transmission in the Serengeti National Park (Tanzania). Part 2. Immobilization of animals and isolation of trypanosomes. *Acta Tropica* 33:68–73.

Bell, R. H. V. 1969. The use of the herb layer by grazing ungulates in the Serengeti National Park, Tanzania. Ph.D. dissertation, Manchester University.

———. 1970. The use of the herb layer by grazing ungulates in the Serengeti. In *Animal populations in relation to their food resources,* ed. A. Watson, pp. 111–23. Oxford: Blackwell.

———. 1971. A grazing ecosystem in the Serengeti. *Sci. Am.* 224(1): 86–93.

Bentley, P., and Huxley, J. 1961. The drama of Serengeti. *Unesco Courier* 14:18–23, 34.

Bertram, B. C. R. 1973a. Sleeping sickness survey in the Serengeti area (Tanzania) 1971. Part 3. Discussion of the relevance of the trypanosome survey to the biology of large mammals in the Serengeti. *Acta Tropica* 30:36–47.

———. 1973b. Lion population regulation. *E. Afr. Wildl. J.* 11:215–25.

———. 1975a. Social factors influencing reproduction in wild lions. *J. Zool.* (Lond.) 177:463–82.

———. 1975b. Weights and measures of lions. *E. Afr. Wildl. J.* 13: 141–43.

———. 1975c. The social system of lions. *Sci. Am.* 232:54–65.

———. 1976a. Lion immobilization using phencyclidine (Sernylan). *E. Afr. Wildl. J.* 14:233–35.

———. 1976b. Kin selection in lions and in evolution. In *Growing points in ethology,* ed. P. P. G. Batecon and R. A. Hinde, pp. 281–301. Cambridge: Cambridge Univ. Press.

———. 1976c. *Studying predators,* handbk. 3. Nairobi: American Wildlife Leadership Foundation.

―――. 1977. Variation in the wing-song of the flappet lark. *Anim. Behav.* 25:165–70.

―――. 1978. *Pride of Lions.* London: Dent.

Bertram, B. C. R., and King, J. M. 1976. Lion and leopard immobilization using CI-744. *E. Afr. Wildl. J.* 14:237–39.

Bindernagel, J. A. 1975. *Wildlife utilization in Tanzania: the ecology of three wildlife areas in Tanzania with special reference to wildlife utilization.* (project working doc. no. 1, URT: 72/011). Rome: F.A.O.

Boer, Th. A. de. 1975. Mosaikvegetation im National Park Serengeti. In *Vegetation und Substrat,* ed. H. Dierschke, pp. 503–10. Ber. Int. Symp. Int. Ver. Vegetationskunde, Rinteln 1969. Cramer.

Boreham, P. F. L., and Geigy, R. 1976. Culverts and trypanosome transmission in the Serengeti National Park (Tanzania). Part 3. Studies on the genus *Auchmeromyia* Brauer and Bergenstamm (Diptera: Calliphoridae). *Acta Tropica* 33:74–87.

Bower, J. R. F. 1973. Seronera: Excavations at a stone bowl site in the Serengeti National Park, Tanzania. *Azania* 8:71–101.

Bradley, R. M. 1977. Aspects of the ecology of the Thomson's gazelle in the Serengeti National Park, Tanzania. Ph.D. dissertation, Texas A & M University.

Branagan, D., and Hammond, J. A. 1965. Rinderpest in Tanganyika: a review. *Bull. Epizoot. Dis. Afr.* 13:225–46.

Braun, H. M. H. 1973. Primary production in the Serengeti: purpose, methods and some results of research. *Ann. Univ. d'Abidjan* (E) 6: 171–88.

Brooks, A. C. 1961. *A study of the Thomson's gazelle (Gazella thomsonii* Gunther) *in Tanganyika.* Colonial Res. Pub. no. 25. London: H.M.S.O.

Cooper, J. E., and Houston, D. C. 1972. Lesions in the crop of vultures associated with Bot fly larvae. *Trans. Roy. Soc. Trop. Med.* 66:515–16.

Croze, H. 1972. A modified photogrammetric technique for assessing age-structure of elephant populations and its use in Kidepo National Park. *E. Afr. Wildl. J.* 10:91–115.

―――. 1974a. The Seronera bull problem. Part 1. The elephants. *E. Afr. Wildl. J.* 12:1–27.

―――. 1974b. The Seronera bull problem. Part 2. The trees. *E. Afr. Wildl. J.* 12:29–47.

Darling, F. F. 1960. An ecological reconnaisance of the Mara Plains in Kenya Colony. Wildl. Monogr. no. 5. The Wildlife Society.

Dinnik, J. A., and Sachs, R. 1968. A gigantic Protostrongylys, *P. africanus* sp. nov., and other lung nematodes of antelopes in the Serengeti, Tanzania. *Parasitology* 58:819–29.

―――. 1969a. Zystizerkose der Kreuzbeinwirbel bei Antilopen und *Taenia olngojinei* sp. nov. der Tüpfelhyäne. *Z. Parasitenk.* 31:326–39.

―――. 1969b. Cysticercosis, Echinococcosis and Sparganosis in wild herbivores in East Africa. *Vet. Med. Rev.,* pp. 113–22.

Douglas-Hamilton, I. 1972. On the ecology and behaviour of the African elephant: the elephants of Lake Manyara. D. Phil. thesis, Oxford Univ.

Duncan, P. 1971. Immobilisation of topi. *E. Afr. Wildl. J.* 9:152.

―――. 1975. Topi and their food supply. Ph.D. dissertation, Univ of Nairobi.

Estes, R. D. 1966. Behaviour and life history of the wildebeest (*Connochaetes taurinus* Burchell). *Nature* (Lond.) 212:999–1000.

―――. 1969. Territorial behaviour of the wildebeest (*Connochaetes taurinus* Burchell, 1823). *Z. Tierpsychol.* 26:284–370.

―――. 1976. The significance of breeding synchrony in the Wildebeest. *E. Afr. Wildl. J.* 14:135–52.

Frame, L. H., and Frame, G. W. 1976. Female African wild dogs emigrate. *Nature* (Lond.) 263:227–29.

Frame, L. H.; Frame, G. W.; Malcolm, J. R.; and van Lawick, H. 1979. Park dynamics of African wild dogs (*Lycaon pictus*) on the Serengeti plains. *Z. Tierpsychol.,* in press.

Fosbrooke, H. A. 1968. Elephants in the Serengeti National Park: an early record. *E. Afr. Wildl. J.* 6:150–52.

Geertsema, A. A. 1976. Impressions and observations on serval behaviour in Tanzania. *Mammalia* 40:13–19.

Geigy, R. 1976. Culverts and trypanosome transmission in the Serengeti National Park (Tanzania). General introduction. *Acta Tropica* 33:54–56.

Geigy, R., and Boreham, P. F. L. 1976. Culverts and trypanosome transmission in the Serengeti National Park (Tanzania). Part 1. Survey of the culverts. *Acta Tropica* 33:57–67.

Geigy, R., and Kauffmann, M. 1973. Sleeping sickness survey in the Serengeti area (Tanzania) 1971. Part 1. Examination of large mammals for trypanosomes. *Acta Tropica* 30:12–23.

Geigy, R.; Kauffmann, M.; Mayende, J. S. P.; Mwambu, P. M.; and Onyango, R. J. 1973. Isolation of *Trypanosoma* (*Trypanozoon*) *rhodesiense* from game and domestic animals in Musoma district, Tanzania. *Acta Tropica* 30:49–56.

Geigy, R.; Mwambu, P. M.; and Kauffmann, M. 1971. Sleeping sickness survey in Musoma District, Tanzania. Part 4. Examination of wild mammals as a potential reservoir for *T. rhodesiense* infections. *Acta Tropica* 28:211–20.

Geigy, R.; Mwambu, P. M.; and Onyango, R. J. 1972. Additional animal reservoirs of *T. rhodesiense* sleeping sickness. *Acta Tropica* 29:199.

Gentry, A. W., and Gentry, A. 1977. Fossil Bovidae (Mammalia) of Olduvai Gorge, Tanzania, part 1. *Bull. Brit. Mus.* (geol. ser.), vol. 29, no. 4.

———. 1978. Fossil Bovidae (mammalia) of Olduvai Gorge, Tanzania, part 2. *Bull. Brit. Mus.* (geol. ser.), vol. 30, no. 1.

Gerresheim, K. 1972. Die Landschaftsgliederung als ökologischer Datenspeicher—angewandte Landschaftsökologie im Serengeti National Park, Tanzania. *S. Natur. Landschaft* 47(2):35–45.

Gerrescheim, K. 1974. *The Serengeti landscape classification—map and manuscript*. Nairobi: African Wildlife Leadership Foundation.

Gogan, P. J. P. 1973. Some aspects of nutrient utilization by Burchell's zebra (*Equus burchelli bohmi* Matschie) in the Serengeti-Mara Region, East Africa. Master's thesis, Texas A & M Univ.

Grzimek, B. 1960. Attrappenversucke mit Zebras und Löwen in der Serengeti. *Z. Tierpsychol.* 17:351–57.

———. 1963. Notizen über afrikanische Säugetiere. *Z. Säugetierk.* 28:13–15.

Grizimek, B., and Grzimek, M. 1960. *Serengeti shall not die*. London: Hamish Hamilton.

Grzimek, M., and Grzimek, B. 1960a. A study of the game of the Serengeti plains. *Z. Säugetierk.* 25:1–61.

———. 1960b. Census of plains animals in the Serengeti National Park, Tanganyika. *J. Wildl. Manage.* 24:27–37.

Gwynne, M. D., and Bell, R. H. V. 1968. Selection of vegetation components by grazing ungulates in the Serengeti National Park. *Nature* (Lond.) 220:390–93.

Haffner, K. von; Rack, G.; and Sachs, R. 1969. Verschiedene Vertreter der Familie Linguatulidae (Pentastomida) als Parasiten von Säugetieren der Serengeti. *Mitt. Hamb. Zool. Mus.* 66:93–144.

Haffner, K. von; Sachs, R.; and Rack, G. 1967. Das Vorkommen von Stachellarven aus der Familie Linguatulidae (Pentastomida) in afrikanischen Huftieren und ihr Parasitismus. *Z. Parasitenk.* 29:329–55.

Hagen, H. 1975. Flavinismus bei einer Massai giraffe, *Giraffa camelopar-*

dalis tippelskirchi Matschie 1898, im Serengeti National Park, Tanzania. *Säugetierk, Mitt.* 23(3):238–40.

———. 1976. Ungewöhnlicher Kampf bei Massai Giraffen Bullen, *Giraffa camelopardalis tippelskirchi* Matschie 1898. *Säugetierk. Mitt.* 24(1):43–45.

Hay, R. L. 1970. Pedogenic calcretes of the Serengeti Plain, Tanzania. Abstr. with Program, *Geol. Soc. Am.* 2:572.

———. 1976. *Geology of the Olduvai Gorge.* Los Angeles: Univ. of Calif. Press.

Helversen, D. von, and Wickler, W. 1971. Uber den Duettgesang des afrikanischen Drongo (*Dicrurus adsimilis*) Bechstein. *Z. Tierpsychol.* 29:301–21.

Hendrichs, H. 1970. Schätzungen der Huftierbiomasse in der Dornbusch-savanne nördlich und westlich der Serengetisteppe in Ostafrika nach einem neuen Verfahren und Bemerkungen zur Biomasse der anderen pflanzenfressenden Tierarten. *Säugetierk. Mitt.* 18:237–55.

Hendrichs, H. 1972. Beobachtungen und Untersuchungen zur Ökologie und Ethologie, insbesondere zur sozialen Organisation, ostafrikanischer Säugetiere. *Z. Tierpsychol.* 30:146–89.

Hendrichs, H., and Hendrichs, U. 1971. *Dikdik und Elefanten: Okologie und Soziologie zweier afrikanischer Huftiere.* Munich: Piper and Co.

———. 1975. Observations on a population of Bohor Reedbuck, *Redunca redunca* (Pallas 1767. *Z. Tierpsychol.* 38:44–54.

———. 1975. Changes in a population of dikdik, *Madoqua* (*Rhyncho-tragus*) *kirki* (Günther 1880). *Z. Tierpsychol.* 38:55–69.

Herlocker, D. J. 1976. Structure, composition, and environment of some woodland vegetation types of the Serengeti National Park, Tanzania. Ph.D. dissertation, Texas A & M Univ.

———. 1976. *Woody vegetation of the Serengeti National Park.* College Station, Texas: Texas A & M Univ.

Hoeck, H. N. 1975. Differential feeding behaviour of the sympatric Hyrax *Procavia johnstoni* and *Heterohyrax brucei. Oecologia* (Berlin) 22:15–47.

———. 1977. "Teat order" in Hyrax (*Procavia johnstoni* and *Hetero-hyrax brucei*). *Z. Säugetierk.* 42:112–15.

Houston, D. C. 1972. The ecology of Serengeti vultures. D. Phil. thesis, Oxford Univ.

———. 1974a. The role of griffon vultures as scavengers. *J. Zool.* (Lond.) 172:35–46.

———. 1974b. Food searching behaviour in griffon vultures. *E. Afr. Wildl. J.* 12:63–77.

———. 1974c. Mortality in *Gyps coprotheres* in southern Africa. *Ostrich* 45:57–62.

———. 1975. The moult of the white-backed and Ruppell's griffon vultures, *Gyps africanus* and *G. rueppellii. Ibis* 117:474–88.

———. 1976a. Breeding of the white-backed and Ruppell's griffon vultures. *Ibis* 118:14–39.

———. 1976b. Ecological isolation of African scavenging birds. *Ardea* 63:56–64.

———. 1979. Interrelations of African Scavenging animals. *Ostrich,* in press.

———. 1978. The effect of food quality on breeding strategy in griffon vultures. *J. Zool.* (Lond.) 186:175–84.

Houston, D. C., and Cooper, J. E. 1973. Use of the drug Metomidate to facilitate the handling of vultures. *Int. Zoo. Yearbook* 13:269–71.

———. 1975. The digestive tract of the white-backed vulture and its role in disease transmission among wild ungulates. *J. Wildl. Dis.* 11:306–13.

Hummel, P. H., and Staak, C. 1974. *Brucella abortus* biotype 3 in Tanzania. *Vet. Rec.* 90:579.

Huxley, J. 1965. Serengeti: a living laboratory. *New Scientist* 27:504–8.

Hvidberg-Hansen, H. 1970. Contribution to the knowledge of the reproductive physiology of the Thomson's gazelle (*Gazella thomsonii* Günther). *Mammalia* 34:551–63.

Hvidberg-Hansen, H., and de Vos, A. 1971. Reproduction, population and herd structure of two Thomson's gazelle (*Gazella thomsonii* Günther) populations. *Mammalia* 35:1–16.

Inglis, J. M. 1976. Wet season movements of individual wildebeests of the Serengeti migratory herd. *E. Afr. Wildl. J.* 14:17–34.

Irvin, A. D.; Omwoyo, P.; Purnell, R. E.; Pierce, M. A.; and Schiemann, B. 1973. Blood parasites of the impala (*Aepyceros melampus*) in the Serengeti National Park. *Vet. Rec.* 93:200–203.

Jager, Tj. 1979. The soils of the Serengeti woodlands. Thesis (in prep.), Wageningen.

Jarman, M. V. 1970. Attachment to home area in impala. *E. Afr. Wildl. J.* 8:198–200.

———. 1976. Impala social behaviour: birth behaviour. *E. Afr. Wildl. J.* 14:153–67.

Jarman, M. V., and Jarman, P. J. 1973. Daily activity of impala. *E. Afr. Wildl. J.* 11:75–92.

Jarman, P. J. 1972. The development of a dermal shield in impala. *J. Zool.* (Lond.) 166:349–56.

———. 1973. The free water intake of impala in relation to the water content of their food. *E. Afr. Agric. For. J.* 38:343–51.

———. 1974. The social organization of antelope in relation to their ecology. *Behaviour* 48:215–66.

———. 1976. Damage to *Acacia tortilis* seeds eaten by impala. *E. Afr. Wildl. J.* 14:223–25.

———. 1977. Behaviour of topi in a shadeless environment. *Zool. Afr.* 12:101–12.

Jarman, P. J., and Jarman, M. V. 1973. Social behaviour, population structure, and reproductive potential in impala. *E. Afr. Wildl. J.* 11:329–38.

———. 1974. A review of impala behaviour, and its relevance to management. In *The behaviour of ungulates and its relation to management,* ed. V. Geist and F. Walther, n.s. no. 24, pp. 871–81. Morges, Switzerland: I.U.C.N.

Jarman, P. J., and Mmari, P. E. 1971. Selection of drinking places by large mammals in the Serengeti woodlands. *E. Afr. Wildl. J.* 9:158–61.

Jolly, G. M. 1969. The treatment of errors in aerial counts of wildlife populations. *E. Afr. Agric. For. J.* 34:50–55.

Kaliner, G.; Sachs, R.; Fay, L. D.; and Schiemann, B. 1971. Untersuchungen über das Vorkommen von Sarcosporidien bei Ostafrikanischen wildtieren. *Z. Tropenmed. Parasitol.* 22:156–64.

Kaliner, G., and Staak, C. 1973. A case of orchitis caused by *Brucella abortus* in the African Buffalo. *J. Wildl. Dis.* 9:251–53.

Kalunda, M., and Plowright, W. 1972. Pathogenicity for cattle of Allerton-type Herpesvirus isolated from a Tanzanian buffalo (*Syncerus caffer*). *J. Comp. Path.* 82:65–72.

King, J. M.; Bertram, B. C. R.; and Hamilton, P. H. 1977. Tiletamine and Zalozepam for immobilization of wild lions and leopards. *J. Am. Vet. Med. Assoc.* 171:894–98.

Klingel, H. 1964a. Zur Sozialstruktur des Steppenzebras (*Equus quagga böhmi* Matschie). *Naturwiss.* 51:347.

———. 1965. Notes on the biology of the plains zebra *Equus quagga boehmi* Matschie. *E. Afr. Wildl. J.* 3:86–88.

———. 1965. Notes on tooth development and aging criteria in the plains zebra *Equus quagga boehmi* Matschie. *E. Afr. Wildl. J.* 3:127–29.

————. 1967. Soziale Organisation und Verhalten freilebender Steppen-zebras. *Z. Tierpsychol.* 24:580–624.

————. 1968a. Soziale Organisation under Verhaltensweisen von Hartmann-und Bergzebras (*Equus zebra hartmannae* und *Equus zebra zebra*). *Z. Tierpsychol.* 25:76–88.

————. 1968b. Die immobilisation von Steppenzebras (*Equus quagga böhmi*). *Zool. Gart.* 35:54–66.

————. 1968c. Das Sozialleben der Steppenzebras. *Naturwiss. und Med.* 24:10–21.

————. 1969a. The social organisation and population ecology of the plains zebra (*Equus quagga*). *Zool. Afr.* 4:249–64.

————. 1969b. Reproduction in the plains zebra *Equus burchelli boehmi:* behaviour and ecological factors. *J. Reprod. Fert.*, supp. 6, pp. 339–45.

————. 1972. Social behaviour of African equidae. *Zool. Afr.* 7:175–85.

————. 1974. A comparison of the social behaviour of the equidae. In *The behaviour of ungulates and its relation to management,* ed. V. Geist and F. Walther, n. s. no. 24, pp. 124–32. Morges, Switzerland: I.U.C.N.

————. 1975 Social organisation of the equidae. *Verh. Deutsch. Zool. Ges.* 68:71–80.

Klingel, H., and Klingel, U. 1966a. Die Geburt eines Zebras (*Equus quagga böhmi* Matschie). *Z. Tierpsychol.* 23:72–76.

————. 1966b. Tooth development and age determination in the plains zebra (*Equus quagga boehmi* Matschie). *Zool. Gart.* (Frankfurt) 33:34–54.

Krampitz, H. E.; Sachs, R.; Schaller, G. B.; and Schindler, R. 1968. Zur Verbreitung von Parasiten der Gattung Hepatozoon Miller, 1908 (Protozoa, Adeleidae) in ostakrikanischen Wildsäugetieren. *Z. Parasitenk.* 31:203–10.

Kreulen, D. A. 1975. Wildebeest habitat selection on the Serengeti plains, Tanzania, in relation to calcium and lactation: a preliminary report. *E. Afr. Wildl. J.* 13:297–304.

————. 1977. Taming of wild-captured wildebeests for food habit studies. *J. Wildl. Manage.* 41:793–95.

————. 1979. Factors affecting reptile biomass in African grasslands. *Amer. Nat.,* in press.

Kreulen, D. A., and Hoppe, P. P. 1979. Diurnal trends and relationships to forage quality of ruminal VFA concentrations, pH, and osmolarity in wildebeest on dry range in Tanzania. *E. Afr. Wildl. J.,* in press.

Kruuk, H. 1966. Clan system and feeding habits of spotted hyaenas
(*Crocuta crocuta erxleben*). *Nature* 209:1257–58.

————. 1966. A new view of the hyaena. *New Scientist* 30:849–51.

————. 1967. Competition for food between vultures in East Africa.
Ardea 55:171–93.

————. 1970. Interactions between populations of spotted hyaenas
(*Crocuta crocuta erxleben*) and their prey species. In *Animal popula-
tions in relation to their food resources,* ed. A. Watson, pp. 359–74.
Oxford: Blackwell.

————. 1972a. Surplus killing in carnivores. *J. Zool.* (London.) 166:
233–44.

————. 1972b. The urge to kill. *New Scientist* 215:735–37.

————. 1972c. *The spotted hyena.* Chicago: Univ. of Chicago Press.

————. 1975a. *Hyaenas.* Oxford: Oxford Univ. Press.

————. 1975b. Functional aspects of social hunting by carnivores. In
Function and evolution in behaviour, ed. G. Baerends, C. Beer, and
A. Manning, pp. 119–41. Oxford: Clarendon Press.

————. 1976a. Feeding and social behaviour of the striped hyaena
(*Hyaena vulgaris*). *E. Afr. Wildl. J.* 14:91–111.

————. 1976b. Carnivores and conservation. In *Proc. Symp. Endangered
Wildl. in Southern Africa, Univ. Pretoria,* pp. 1–8.

Kruuk, H., and Sands, W. A. 1972. The aardwolf (*Proteles cristatus*
Sparrmann) as predator of termites. *E. Afr. Wildl. J.* 10:211–27.

Kruuk, H., and Turner, M. I. M. 1967. Comparative notes on predation
by lion, leopard, cheetah, and wild dog in the Serengeti area, East
Africa. *Mammalia* 31:1–27.

Kühme, W. 1964a. Uber die soziale Bindung innerhalb eines
Hyänenhund-Rudels. *Naturwiss.* 51:567–68.

————. 1964b. Die Ernährungsgemeinschaft der Hyänenhunde (*Lycaon
pictus lupinus* Thomas 1902). *Naturwiss.* 52:495.

————. 1965. Communal food distribution and division of labour in
African hunting dogs. *Nature* (Lond.) 205:443–44.

————. 1965. Freilandstudien zur Soziologie des Hyänenhundes (*Lycaon
pictus lupinus* Thomas 1902). *Z. Tierpsychol.* 22:495–541.

————. 1966. Beobachtungen zur Soziologie des Löwen in der Serengeti-
Steppe Ostafrikas. *Z. Säugetierk.* 31:205–13.

Kuttler, K. 1965. Serological survey of anaplasmosis incidence in East
Africa, using the compliment-fixation test. *Bull. Epizoot. Dis. Afr.* 13:
257–62.

Lamprey, H. F. 1964. Estimation of the large mammal densities, biomass, and energy exchange in the Tarangire Game Reserve and the Masai steppe in Tanganyika. *E. Afr. Wildl. J.* 2:1–46.

———. 1967. Notes on the dispersal and germination of some tree seeds through the agency of mammals and birds. *E. Afr. Wildl. J.* 5:179–80.

———. 1969a. Ecological research in the Serengeti National Park. *J. Reprod. Fert.,* supp. 6, pp. 487–93.

———. 1969b. The range of possible observations. *E. Afr. Agric. For. J.* 34:64–69.

———. 1972. On the management of flora and fauna in National Parks. Background paper for session 8 (part 1) of the Second World Conference on National Parks, September 1972, Fort Collins, Colo.

———. 1978. The Serengeti region: a semi-arid grassland ecosystem in Africa. In *State of knowledge report on tropical grazing land ecosystems.* Geneva: UNESCO.

Lamprey, H. F.; Glover, P. E.; Turner, M.; and Bell, R. H. V. 1967. Invasion of the Serengeti National Park by elephants. *E. Afr. Wildl. J.* 5:151–66.

Lamprey, H. F.; Halevy, G.; and Makacha, S. 1974. Interactions between *Acacia,* bruchid seed beetles, and large herbivores. *E. Afr. Wildl. J.* 12:81–85.

Lamprey, H. F.; Kruuk, H.; and Norton-Griffiths, M. 1971. Research in the Serengeti. *Nature* (Lond.) 230:497–99.

Laurie, W. A. 1971. The food of the barn owl in the Serengeti National Park, Tanzania. *J. E. Afr. Nat. Hist. Soc.* 28:1–4.

Ledger, H. P., and Sachs, R. 1965. Wildnutzung in trockenen Gebieten Ostafrikas. *Die Fleischwirtschaft* 45:1421–24.

Losos, G. J., and Gwamaka, B. 1973. Histological examination of wild animals naturally infected with pathogenic African trypanosomes. *Acta Tropica.* 30:57–63.

McCulloch, B.; Suda, B'O. J.; Tungaraza, R.; and Kalaye, W. J. 1968. A study of East Coast Fever, drought, and social obligations in relation to the need for the economic development of the livestock industry in Sukumaland, Tanzania. *Bull. Epizoot. Dis. Afr.* 16:303–26.

McCulloch, J. S. G., and Talbot, L. M. 1965. Comparison of weight estimation methods for wild animals and domestic livestock. *J. Appl. Ecol.* 2:59–69.

Macfarlane, A. 1969. Preliminary report on the geology of the Central Serengeti, northwest Tanzania. 13th Ann. Rep. Inst. Afr. Geol., Univ. Leeds, pp. 14–16.

McNaughton, S. J. 1976. Serengeti migratory wildebeest: facilitation of energy flow by grazing. *Science* 191:92–94.

———. 1977. Diversity and stability of ecological communities: a comment on the role of empiricism in ecology. *Amer. Nat.* 111:515–25.

———. 1978. Serengeti ungulates: feeding selectivity influences the effectiveness of plant defense guilds. *Science* 199:806–7.

———. 1979. Grazing as an optimization process: grass-ungulate relationships in the Serengeti. *Amer. Nat.* 113:691–703.

Makacha, S., and Schaller, G. B. 1969. Observations on lions in the Lake Manyara National Park, Tanzania. *E. Afr. Wildl. J.* 7:99–104.

Malcolm, J. R., and Lawick, H. van. 1975. Notes on wild dogs hunting zebras. *Mammalia* 39:231–40.

Martin, P. S. 1967. Overkill at Olduvai Gorge. *Nature* 215:212–13.

Misonne, X., and Verschuren, J. 1966. Les rongeurs et lagomorphes de la région du Parc National de Serengeti. *Mammalia* 30:517–37.

Mollel, C. L. 1977. A possible case of anthrax in a Serengeti lioness. *E. Afr. Wildl. J.* 15:331.

Mukinya, J. G. 1977. Feeding and drinking habits of the black rhinoceros in Masai Mara Game Reserve. *E. Afr. Wildl. J.* 15:125–38.

Norton-Griffiths, M. 1973. Counting the Serengeti migratory wildebeest using two-stage sampling. *E. Afr. Wildl. J.* 11:135–49.

———. 1974. Reducing counting bias in aerial censuses by photography. *E. Afr. Wildl. J.* 12:245–48.

———. 1976. Further aspects of bias in aerial census of large mammals. *J. Wildl. Manage.* 40:368–71.

———. 1978. *Counting animals,* 2d ed. Nairobi: African Wildlife Leadership Foundation.

Norton-Griffiths, M.; Herlocker, D.; and Pennycuick, L. 1975. The patterns of rainfall in the Serengeti ecosystem, Tanzania. *E. Afr. Wildl. J.* 13:347–74.

Olivier, R. C. D., and Laurie, W. A. 1974. Habitat utilization by hippopotamus on the Mara River. *E. Afr. Wildl. J.* 12:249–71.

Owen, J. S. 1971. Fire management in the Tanzania national parks. *Proc. Ann. Tall Timbers Fire Ecol. Conf.* 11:233–41.

———. 1972. Some thoughts on management of national parks. *Biol. Conserv.* 4:241–46.

Pearsall, W. H. 1957. Report on an ecological survey of the Serengeti National Park. *Oryx* 4:71–136.

Pennycuick, C. J. 1969. The mechanics of bird migration. *Ibis* 111:525–56.

————. 1971a. Gliding flight of the white-backed vulture *Gyps africanus*. *J. Exp. Biol.* 55:13–38.

————. 1971b. Control of gliding angle in Ruppell's griffon vulture, *Gyps rueppellii*. *J. Exp. Biol.* 55:39–46.

————. 1972. Soaring behaviour and performance of some East African birds, observed from a motor-glider. *Ibis* 114:178–218.

————. 1973. The shadowmeter: a simple device for controlling an aircraft's height above the ground. *E. Afr. Wildl. J.* 11:109–12.

————. 1975. On the running of the gnu (*Connochaetes taurinus*) and other animals. *J. Exp. Biol.* 63:775–99.

————. 1976. Breeding of the lappet-faced and white-headed vultures (*Torgos tracheliotus* Forster and *Trigonoceps occipitalis* Burchell) on the Serengeti plains, Tanzania. *E. Afr. Wildl. J.* 14:67–84.

Pennycuick, L. 1975. Movements of the migratory wildebeest population in the Serengeti area between 1960 and 1973. *E. Afr. Wildl. J.* 13:65–87.

Pennycuick, L., and Norton-Griffiths, M. 1976. Fluctuations in the rainfall of the Serengeti ecosystem, Tanzania. *J. Biogeogr.* 3:125–40.

Phillipson, J. 1973. The biological efficiency of protein production by grazing and other land based systems. In *The biological efficiency of protein production*, ed. J. G. W. Jones, pp. 217–35. London: Cambridge Univ. Press.

Pitman, C. R. S. 1965. A communal nest of lappet-faced vulture, *Torgos tracheliotus* (Forster) and East African greater kestrel, *Falco rupicoloides arthuri* (Gurney). *Bull. Brit. Ornith. Club* 85:93–95.

Plowright, W. 1963. The role of game animals in the epizootiology of rinderpest and malignant catarrhal fever in East Africa. *Bull. Epizoot. Dis. Afr.* 11:149–62.

————. 1965. Malignant catarrhal fever in East Africa. Part 1. Behaviour of the virus in free-living populations of blue wildebeest (*Gorgon taurinus taurinus*, Burchell). *Res. Vet. Sci.* 6:56–68.

Plowright, W., and Ferris, R. D. 1961. Serological evidence for enzootic rinderpest in wildebeest of the Serengeti-Mara-Loita plains area. *E. Afr. Agric. For. J.* 27:100.

Plowright, W., and Jessett, D. M. 1971. Investigations of Allerton-type herpes virus infection in East African game animals and cattle. *J. Hyg.* (Camb.) 69:209–22.

Plowright, W., and McCulloch, B. 1967. Investigations on the incidence of rinderpest virus infection in game animals of N. Tanganyika and S. Kenya, 1960–63. *J. Hyg.* (Camb.) 65:343–58.

Plowright, W.; Parker, J.; and Pierce, M. A. 1969a. African swine fever virus in ticks (*Ornithodoros moubata,* Murray) collected from animal burrows in Tanzania. *Nature* (Lond.) 221:1071–73.

———. 1969b. The epizootiology of African swine fever in Africa. *Vet. Rec.* 85:668–74.

Plowright, W., and Scott, G. R. 1960. Blue wildebeest and the aetiological agent of bovine malignant catarrhal fever. *Nature* (Lond.) 188: 1167–69.

Podoler, H., and Rogers, D. 1975. A new method for the identification of key factors from life-table data. *J. Anim. Ecol.* 44:85–114.

Purnell, R. E.; Schiemann, B.; Brown, C. G. D.; Irvin, A. D.; Ledger, M. A.; Payne, R. C.; Radley, D. E.; and Young, A. S. 1973. Attempted transmission of *Theileria gorgonis,* Brocklesby and Vidler 1961, from blue wildebeest (*Connochaetes taurinus*) to cattle. *Z. Tropenmed. Parasit.* 24:181–85.

Reid, H. W.; Plowright, W.; and Rowe, L. W. 1975. Neutralizing antibody to herpesviruses derived from wildebeest and hartebeest in wild animals in East Africa. *Res. Vet. Sci.* 18:269–73.

Rogers, D., and Boreham, P. F. L. 1973. Sleeping sickness survey in the Serengeti area (Tanzania) 1971. Part 2. The vector role of *Glossina swynertoni* Austen. *Acta Tropica* 30:24–35.

Rweyemamu, M. M. 1974. The incidence of infectious bovine rhinotracheitis antibody in Tanzanian game animals and cattle. *Bull. Epizoot. Dis. Afr.* 22:19–22.

Sachs, R. 1966. Note on cysticercosis in game animals of the Serengeti. *E. Afr. Wildl. J.* 4:152–53.

———. 1967. Liveweights and body measurements of Serengeti game animals. *E. Afr. Wildl. J.* 5:24–36.

———. 1969a. Serosal cysticercosis in East African game animals. *Bull. Epizoot. Dis. Afr.* 17:337–39.

———. 1969b. Finnenbefall bei ostafrikanischen Antilopen und die Verteilung der Muskelfinnen im Wildtierkörper. *Fleichwirtschaft* 49: 1331–35.

———. 1969c. Untersuchungen zur Artbestimmung und Differenzierung der Muskelfinned ostafrikanischer Wildtiere. *Z. Tropenmed. Parasit.* 20:39–50.

———. 1969d. Über den Muskelfinnen-befall wildebender Herbivoren des Serengetigebietes im Norden von Tanzania. *Z. Jagdwissenschaft* 15:151–57.

Sachs, R., and Debbie, J. G. 1969. A field guide to the recording of parasitic infestation of game animals. *E. Afr. Wildl. J.* 7:27–37.

Sachs, R.; Frank, H.; and Bindernagel, J. A. 1969. New host records for *Mammomonogamus* in African game animals through application of a simple method of collection. *Vet. Rec.* 84:562–63.

Sachs, R., and Gibbons, L. M. 1973. Species of *Haemonchus* from domestic and wild ruminants in Tanzania, East Africa, including a description of H. *dinniki* n.sp. *Z. Tropenmed. Parasit.* 24:467–75.

Sachs, R.; Hoffmann, R. M.; and Sorheim, A. O. 1969. Stilesia infestation in East African antelopes. *Vet. Rec.* 82:233–34.

Sachs, R.; Rack, G.; and Woodford, M. H. 1973. Observations on pentastomid infestation of East African game animals. *Bull. Epizoot. Dis. Afr.* 21:401–10.

Sachs, R., and Sachs, C. 1968. A survey of parasitic infestation of wild herbivores in the Serengeti region of N. Tanzania and the Lake Rukwa region in S. Tanzania. *Bull. Epizoot. Dis. Afr.* 16:455–72.

Sachs, R.; Schaller, G. B.; and Baker, J. R. 1967. Isolation of typanosomes of the *T. brucei* group from lion. *Acta Tropica* 24:109–12.

Sachs, R.; Schaller, G. B.; and Schindler, R. 1971. Untersuchungen über das vorkommen von typanosomen bei wildkarnivoren des Serengeti National Parks in Tanzania. *Acta Tropica* 28:323–28.

Sachs, R., and Staak, C. 1966. Evidence of brucellosis in antelopes of the Serengeti. *Vet. Rec.* 79:857–58.

Sachs, R.; Staak, C.; and Groocock, C. M. 1968. Serological investigation of brucellosis in game animals in Tanzania. *Bull. Epizoot. Dis. Afr.* 16:93–100.

Sachs, R., and Taylor, A. S. 1966. Trichinosis in a spotted hyaena (*Crocuta crocuta*) of the Serengeti. *Vet. Rec.* vol. 78, no. 20.

Schaller, G. B. 1968. Hunting behaviour of the cheetah in the Serengeti National Park, Tanzania. *E. Afr. Wildl. J.* 6:95–100.

——. 1972. *The Serengeti lion.* Chicago: Univ. of Chicago Press.

Schaller, G. B., and Lowther, G. R. 1969. The relevance of carnivore behavior to the study of early hominids. *Southwest. J. Anthropol.* 25: 307–41.

Schiemann, B.; Plowright, W.; and Jessett, D. M. 1971. Allerton-type herpes virus as a cause of lesions of the alimentary tract in a severe disease of Tanzanian buffaloes (*Syncerus caffer*). *Vet. Rec.* 89:17–22.

Schiemann, B., and Staak, C. 1971. *Brucella melitensis* in Impala. *Vet. Rec.* 88:344.

Schindler, R.; Sachs, R.; Hilton, P. J.; and Watson, R. M. 1969. Some veterinary aspects of the utilization of African game animals. *Bull. Epizoot. Dis. Afr.* 17:215–21.

Schmidt, H. 1966. *Agama atricollis* subsp. inc. aus der Serengeti. *Salamandra* 2:57–68.

Schmidt, W. 1975a. The vegetation of the northeastern Serengeti National Park, Tanzania. *Phytocoenologia* 3:30–82.

———. 1975b. Plant communities on permanent plots of the Serengeti Plains. *Vegetatio* 30:133–45.

Seibt, U. 1975. Instrumental Dialekte der Klapperlerche, *Mirafra rufocinnamomea* (Salvadori). *J. Ornith.* 116:103–7.

Shifrine, M.; Stone, S. S.; and Staak, C. 1970. Contagious bovine pleuropneumonia in African buffalo (*Syncerus caffer*). *Bull. Epizoot. Dis. Afr.* 18:201–5.

Sinclair, A. R. E. 1969. Serial photographic methods for population, age and sex structure. *E. Afr. Agric. For. J.* 34:87–93.

———. 1970. Studies of the ecology of the East African buffalo. D. Phil. thesis, Oxford Univ.

———. 1971. Wildlife as a resource. *Outlook on Agriculture* 6(6): 261–66.

———. 1972. Long-term monitoring of mammal populations in the Serengeti: census of non-migratory ungulates, 1971. *E. Afr. Wildl. J.* 10:287–97.

———. 1973a. Population increases of buffalo and wildebeest in the Serengeti. *E. Afr. Wildl. J.* 11:93–107.

———. 1973b. Regulation, and population models for a tropical ruminant. *E. Afr. Wildl. J.* 11:307–16.

———. 1974a. The natural regulation of buffalo populations in East Africa. Part 1. Introduction and resource requirements. *E. Afr. Wildl. J.* 12:135–54.

———. 1974b. The natural regulation of buffalo populations in East Africa. Part 2. Reproduction, recruitment and growth. *E. Afr. Wildl. J.* 12:169–83.

———. 1974c. The natural regulation of buffalo populations in East Africa. Part 3. Population trends and mortality. *E. Afr. Wildl. J.* 12: 185–200.

———. 1974d. The natural regulation of buffalo populations in East Africa. Part 4. The food supply as a regulating factor, and competition. *E. Afr. Wildl. J.* 12:291–311.

———. 1974e. The social organization of the East African buffalo (*Syncerus caffer* Sparrman). In *The behaviour of ungulates and its relation to management,* ed. V. Geist and F. Walther, n.s. no. 24, pp. 676–689. Morges, Switzerland: I.U.C.N.

———. 1975a. The resource limitation of trophic levels in tropical grassland ecosystems. *J. Anim. Ecol.* 44:497–520.

———. 1975b. The effect of food supply on the breeding of resident bird species and the movement of palaearctic migrants in a tropical African savannah. 16th Ornith. Congr., Canberra, Aust., 1974. *Emu* 74:318.

———. 1977a. *The African buffalo.* Chicago: Univ. of Chicago Press.

———. 1977b. Lunar cycle and timing of mating season in Serengeti wildebeest. *Nature* (Lond.) 267:832–33.

———. 1978. Factors affecting the food supply and breeding season of resident birds and movements of Palaearctic migrants in a tropical African savannah. *Ibis* 120:480–97.

Sinclair, A. R. E., and Duncan, P. 1972. Indices of condition in tropical ruminants. *E. Afr. Wildl. J.* 10:143–49.

Sinclair, A. R. E., and Gwynne, M. D. 1972. Food selection and competition in the East African buffalo (*Syncerus caffer* Sparrman). *E. Afr. Wildl. J.* 10:77–89.

Staak, C.; Sachs, R.; and Groocock, C. M. 1968. Bruzellose beim afrikanischen Buffel (*Syncerus caffer*) in Tanzania. *Vet. med. Nachr.* 68:245–49.

Stewart, D. R. M., and Talbot, L. M. 1962. Census of wildlife on the Serengeti, Mara, and Loita plains. *E. Afr. Agric. For. J.* 28:58–60.

Swynnerton, G. H. 1958. Fauna of the Serengeti National Park. *Mammalia* 22:435–50.

Talbot, L. M. 1956. Report on the Serengeti National Park, Tanganyika. Mimeographed. P. 9. Brussels: I.U.C.N.

———. 1960. Field immobilization of some East African wild animals and cattle. *E. Afr. Agric. For. J.* 26:92–102.

———. 1961. Preliminary observations on the population dynamics of the wildebeest in Narok district, Kenya. *E. Afr. Agric. For. J.* 27:108–16.

———. 1962a. Resultats préliminaires d'une étude de la dynamique des populations de gnous du district Narok, Kenya. *Terre et la vie* (1962), pp. 161–74.

———. 1962b. Food preferences of some East African wild ungulates. *E. Afr. Agric. For. J.* 27:131–38.

Talbot, L. M., and Lamprey, H. F. 1961. Immobilization of free-ranging

East African ungulates with succinylcholine chloride. *J. Wildl. Manage.*
25:303–10.

Talbot, L. M., and McCulloch, J. S. G. 1965. Weight estimations for
East African wild animals from body measurements. *J. Wildl. Manage.*
29:84–89.

Talbot, L. M., and Stewart, D. R. M. 1964. First wildlife census of the
entire Serengeti-Mara Region, East Africa. *J. Wildl. Manage.* 28:
815–27.

Talbot, L. M., and Talbot, M. H. 1962a. Flaxadil and other drugs in
field immobilization and translocation of large mammals in East Africa.
J. Mammal. 43:76–88.

———. 1962b. A hoisting apparatus for weighing and loading large
animals in the field. *J. Wildl. Manage.* 26:217–18.

———. 1963b. The high biomass of wild ungulates on East African
savanna. *Trans. N. Am. Wildl. Conf.* 28:465–76.

———. 1963b. The wildebeest in western Masailand. *Wildl. Monogr.*
no. 12. The Wildlife Society.

Taylor, W. P., and Watson, R. M. 1967. Studies on the epizootiology of
rinderpest in blue wildebeest and other game species of Northern Tanzania and Southern Kenya. *J. Hyg.* (Camb.) 65:537–45.

Turner, M. I. M. 1964. Some observations on bird behaviour made from
an aircraft in the Serengeti National Park, Tanzania. *Bull. Brit. Ornith.
Cl.* 84:65–67.

Turner, M. I. M., and Watson, R. M. 1964. A census of game in Ngorongoro Crater. *E. Afr. Wildl. J.* 2:165–68.

———. 1965a. An introductory study on the ecology of hyrax (*Dendro-
hyrax brucei* and *Procavia johnstonii*) in the Serengeti National Park.
E. Afr. Wildl. J. 3:49–60.

———. 1965b. Game management and research by aeroplane. *Oryx*
8:13–22.

Usenik, E. A.; Kreulen, D. A.; and Duncan, P. 1977. Oesophageal fistula-
tion of topi and wildebeest. *E. Afr. Wildl. J.* 15:207–12.

Verschuren, J. 1965. Contribution à l'étude des chiroptères du Parc National de Serengeti (Tanzanie). *Rev. Zool. Bot. Afr.* 71:371–75.

———. 1967. Note sur les oiseaux de la région du Parc National de
Serengeti (Tanzanie). *Le Gerfaut* 57:77–80.

Vesey-Fitzgerald, D. 1971. Fire and animal impact on vegetation in Tanzania National Parks. *Proc. Ann. Tall Timbers Fire Ecol. Conf.* 11:
297–317.

Walther, F. R. 1966. *Mit Horn und Huf*. Hamburg: Paul Parey.

———. 1968. *Verhalten der Gazellen*. Neue Brehm-Bücherei, Wittenberg: Ziemsen Verlag.

———. 1969. Flight behaviour and avoidance of predators in Thomson's gazelle (*Gazella thomsoni* Guenther 1884). *Behaviour* 34:184–221.

———. 1972a. Social grouping in Grant's gazelle (*Gazella granti* Brooke 1827) in Serengeti National Park. *Z. Tierpsychol*. 31:348–403.

———. 1972b. On age class recognition and individual identification of Thomson's gazelle in the field. *J. S. Afr. Wildl. Manage. Assoc*. 2: 9–15.

———. 1972c. Territorial behaviour in certain horned ungulates, with special reference to the examples of Thomson's and Grant's gazelles. *Zool. Afr*. 7:303–7.

———. 1973. Round-the-clock activity of Thomson's gazelle (*Gazella thomsoni* Günther 1884). *Z. Tierpsychol*. 32:75–105.

———. 1977. Sex and activity dependency of distances between Thomson's gazelles (*Gazella thomsoni* Günther 1884). *Anim. Behav*. 25: 713–19.

Watson, R. M. 1965a. Game utilization in the Serengeti: Preliminary investigation, part 1. *Brit. Vet. J*. 121:540–46.

———. 1965b. Observations on the behaviour of young spotted hyaena (*Crocuta crocuta*) in the burrow. *E. Afr. Wildl. J*. 3:122–23.

———. 1965/66. Population ecology in the Serengeti. *CIBA Journal* 36:46–51.

———. 1966a. Game utilization in the Serengeti: Preliminary investigations. Part 2. Wildebeest. *Brit. Vet. J*. 122:18–27.

———. 1966b. Air photography in East African game management and research. In *The use of air photography*, ed. J. K. St. Joseph. London: John Baker Publishers.

———. 1967. The population ecology of the wildebeest (*Connochaetes taurinus albojubatus* Thomas) in the Serengeti. Ph.D. dissertation, Cambridge Univ.

———. 1969a. Aerial photographic methods in censuses of animals. *E. Afr. Agric. For. J*. 34:32–37.

———. 1969b. Reproduction of wildebeest, *Connochaetes taurinus albojubatus* Thomas, in the Serengeti and its significance to conservation. *J. Reprod. Fert.*, supp. 6, pp. 287–310.

———. 1970. Generation time and intrinsic rates of natural increase in wildebeeste (*Connochaetes taurinus albojubatus* Thomas). *J. Reprod. Fert*. 22:557–61.

Watson, R. M., and Bell, R. H. V. 1969. The distribution, abundance, and status of elephant in the Serengeti region of Northern Tanzania. *J. Appl. Ecol.* 6:115–32.

Watson, R. M.; Graham, A. D.; Bell, R. H. V.; and Parker, I. S. C. 1971. A comparison of four East African crocodile (*Crocodylus niloticus* Laurenti) populations. *E. Afr. Wildl. J.* 9:25–34.

Watson, R. M.; Graham, A. D.; and Parker, I. S. C. 1969. A census of the large mammals of Loliondo Controlled Area, Northern Tanzania. *E. Afr. Wildl. J.* 7:43–59.

Watson, R. M., and Kerfoot, O. 1964. A short note on the intensity of grazing of the Serengeti Plains by plains game. *Z. Säugetierk.* 29: 317–20.

Watson, R. M., and Turner, M. I. M. 1965. A count of the large mammals of the Lake Manyara National Park—results and discussion. *E. Afr. Wildl. J.* 3:95–98.

Wickler, W. 1966. Freilandbeobachtungen an der Uferschrecke *Tridactylus madecassus* in Ostafrika. *Z. Tierpsychol.* 23:845–52.

———. 1967. Der "Fluegelgesang" der ostafrikanischen Klapperlerche *Mirafra rufocinnamomea* (Salvadori). *Vogelwelt* 88:161–65.

———. 1972a. Aufbau und Paarspeziffität des Gesangsduettes von *Laniarius funebris* (Aves, Passeriformes, Laniidae). *Z. Tierpsychol.* 30:464–76.

———. 1972b. Duettieren zwischen artverschiedenen Vögeln im Freiland. *Z. Tierpsychol.* 31:98–103.

———. 1973. Artunterschiede im Duettgesang zwischen *Trachyphonus d'arnaudii usambiro* und den anderen Unterarten von *T.d'arnaudii*. *J. Ornith.* 114:123–28.

Wickler, W., and Seibt, U. 1972. Zur Ethologie afrikanischer Stielaugenfliegen (Diptera, Diopsidae). *Z. Tierpsychol.* 31:113–30.

———. 1974. Rufen und Antworten bei *Kassina senegalensis, Bufo regularis* und anderen Anuran. *Z. Tierpsychol.* 34:524–37.

Wickler, W., and Uhrig, D. 1969a. Betteln, Antwortszeit und Rassenunterschiede im Begrussungsduett des Schmuckbartvogels *Trachyphonus d'arnaudii*. *Z. Tierpsychol.* 26:651–61.

———. 1969b. Verhalten und ökologische Nische der Gelbfluegelfledermaus, *Lavia frons* (Geoffrey) (Chiroptera, Megadermatidae). *Z. Tierpsychol.* 26: 726–36.

Wit, H. A. de. 1977. Soil map of The Serengeti Plain. Appendix to: Soils and grassland types of the Serengeti Plain (Tanzania). Thesis, Mededelingen Landbouwhogeschool, Wageningen (1978).

Wright, B. S. 1960. Predation on big game in East Africa. *J. Wildl. Manage.* 24:1–15.

Young, A. S.; Branagan, D.; Brown, C. G. D.; Burridge, M. J.; Cunningham, M. P.; and Purnell, R. E. 1973. Preliminary observations on a theilerial species pathogenic to cattle isolated from buffalo (*Syncerus caffer*) in Tanzania. *Brit. Vet. J.* 129:382–89.

Young, A. S.; Brown, C. G. D.; Burridge, M. J.; Cunningham, M. P.; Kirimi, I. M.; and Irvin, A. D. 1973. Observations on the cross-immunity between *Theileria lawrencei* (Serengeti) and *Theileria parva* (Muguga) in cattle. *Int. J. Parasitol.* 3:723–28.

Young, A. S., and Purnell, R. E. 1973. Observations on *Babesia equi* in the salivary glands of *Rhipicephalus evertsi*. *Bull. Epizoot. Dis. Afr.* 21:377–83.

Index